D0989821

BIOLOGICAL AND MEDICAL PHYSICS SERIES

Springer
Berlin
Heidelberg
New York
Barcelona
Hong Kong
London
Milan
Paris
Tokyo

BIOLOGICAL AND MEDICAL PHYSICS SERIES

The field of biological and medical physics is a broad, multidisciplinary, and dynamic one, touching on many areas of research in physics, biology, chemistry, and medicine. The Biological and Medical Physics Series is intended to be comprehensive, covering a broad range of topics important to the study of biological and medical physics. Its goal is to provide scientists, medical doctors and engineers with text books, monographs and reference books to address the growing need for information.

Y. Taniguchi   H.E. Stanley   H. Ludwig
(Eds.)

Biological Systems Under Extreme Conditions

Structure and Function

With 123 Figures and 18 Tables

Professor Dr. Yoshihiro Taniguchi
Ritsumeikan University
Department of Applied Chemistry
Faculty of Science and Engineering
1-1-1, Nojihigshi, Kusatsu
Shiga, 525-8577
Japan

Professor Harry E. Stanley
Center for Polymer Studies
and Department of Physics
Boston University
Boston, MA 02215
USA

Professor Dr. Horst Ludwig
University of Heidelberg
Institute of Pharmaceutical Technology
and Biopharmacy
Im Neuenheimer Feld 346
69120 Heidelberg
Germany

Library of Congress Cataloging in Publication Data applied for.

Die Deutsche Bibliothek - CIP-Einheitsaufnahme
Biological systems under extreme conditions : structure and function ; with 18 tables / Y. Taniguchi ... (ed.) - Berlin ; Heidelberg ; New York ; Barcelona ; Hong Kong ; London ; Milan ; Paris ; Tokyo : Springer, 2002
(Biological and medical physics series)
(Physics and astronomy online library)
ISBN 3-540-65992-7

ISSN 1618-7210

ISBN 3-540-65992-7 Springer-Verlag Berlin Heidelberg New York

Springer-Verlag Berlin Heidelberg New York
a member of BertelsmannSpringer Science+Business Media GmbH

http://www.springer.de

Typesetting by the authors/editors
Cover concept by eStudio Calamar Steinen using a background picture from The Protein Databank (1 Kzu). Courtesy of Dr. Antoine M. van Oijen, Department of Molecular Physics, Huygens Laboratory, Leiden University, The Netherlands. Reprinted with permission from Science 285 (1999) 400–402 ("Unraveling the Electronic Structure of Individual Photosynthetic Pigment-Protein Complexes", by A. M. van Oijen et al.) Copyright 1999, American Association for the Advancement of Science.
Cover production: *design & production* GmbH, Heidelberg

Printed on acid-free paper SPIN 10731051 57/3141/di - 5 4 3 2 1 0

Preface

Biological systems are regulated by the basic thermodynamic parameters of pressure and temperature. Life exists only within a closed range of pressure and temperature parameters: a pressure range from 0.1 to more than 1000 MPa, and a temperature range from –22 to more than 100°C, depending on the phase behavior of liquid water as the pressure and temperature change. Pressure has an effect on the environment of biological systems in a continuous, controlled way by changing only intermolecular distances. Temperature produces simultaneous changes in both volume and thermal energy.

In recent years, there has been increasing interest in exploiting the effects of pressure itself and pressure combined with temperature on biological systems. One of the reasons is the possibility of applying pressure in food processing. Also, pressure can be used to provide more detailed and unique information on biological systems. Modern technical progress with various kinds of spectroscopes, FTIR, NMR, solution X-ray analysis, molecular dynamics simulation, flash photolysis, and stopped-flow methods, has made it possible to study the structure and function of biological systems under extreme conditions of pressure and temperature.

The deep sea was perceived as a cold (around 0°C), dark, and high-pressure (around 100 MPa) environment inhabited by barophilic microbial communities. In contrast, deep-sea hydrothermal vent areas are warm to hot reaching around 350°C at a depth of 2000 m (20 MPa), and are inhabited by thermophilic communities. The survival and adaptive mechanisms of microorganisms at extreme conditions are important in the understanding of the function of deep-sea microorganisms. The explanations of these functions are anticipated to explicate the origins and evolution of life.

There has been a long history of heating and freezing in food processing. During the past several years, pressure and temperature effects on biological systems have been used in research into high-pressure processing in the food industries, especially since the Japanese Research Association for High Pressure Technology in Food Industry was established in 1989.

In this book, we have collected together chapters covering in a wide range of subjects from the basics to applications of the structure and function of biological systems under extreme pressure and temperature conditions. This book is composed of four main research fields. The first field is on the basic theory of biological systems: water under extreme conditions. The second field is the recent advanced results and methods in the study of the structure and function of proteins with changes in pressure and temperature. The third field deals with the recent aspects of molecular biology in the deep sea and in deep-sea hydrothermal vents, and with the origins and evolution of life. The final field highlights one of the many trends in the use of pressure and/or temperature for microorganisms, a trend that appears to have many important future applications of the yeast *Saccharomyses cerevisiae*.

This book is intended to stimulate research in this field and to demonstrate the basics and applications of pressure and temperature effects. It is intended for students and researchers studying and exploring the new world of biological systems in extreme environments.

Kusatsu,
Boston,
Heidelberg,
October 2001

Yoshihiro Taniguchi
H. Eugene Stanley
Horst Ludwig

Contents

1 Liquid Water at Low Temperature: Clues for Biology?

H. Eugene Stanley

Center for Polymer Studies and Department of Physics
Boston University, Boston, MA 02215 USA
e-mail: hes@bu.edu

Abstract. Although H_2O has been the topic of considerable research since the beginning of the century, the peculiar physical properties are still not well understood. First we discuss some of the anomalies of this 'complex fluid'. Then we describe a qualitative interpretation in terms of percolation concepts. Finally, we discuss recent experiments and simulations relating to the hypothesis that, in addition to the known critical point in water, there exists a 'second' critical point at low temperatures. In particular, we discuss very recent measurements of the compression-induced melting and decompression-induced melting lines of high-pressure forms of ice. We show how knowledge of these lines enables one to obtain an approximation for the Gibbs potential $G(P,T)$ and the equation of state $V(P,T)$ for water, both of which are consistent with the possible continuity of liquid water and the amorphous forms of solid water.

1.1 Introduction

Liquid water at low temperature is at first sight not relevant to life, and hence has no place in this book. However, understanding of the highly anomalous equilibrium and dynamical properties are generally connected to the view that water, even above its melting temperature, is a transient gel with structural heterogeneities of very short length scales. Moreover, understanding the properties of water is important for understanding phenomena in 'aqueous solutions', such as the structure of micelles and microemulsions. For these and other reasons, water is generally included under the rubric of complex fluids.

I have organized the chapter around three questions:

Question 1: *'What is the puzzle of liquid water?'*

Question 2: *'Why do we care about this puzzle?'*

Question 3: *'What do we actually do?'*

The 'we' in this case is a rather large number of individuals. I'll focus mostly on very recent work that's been published, or at least submitted, this year. This work was done in collaboration with M.-C. Bellissent-Funel, S. V. Buldyrev, M. Canpolat, M. Meyer, O. Mishima, R. Sadr-Lahijany, A. Scala, and F. W. Starr. It's also based on earlier research a few years back with C. A. Angell, P. Debenedetti, A. Geiger, P. H. Poole, S. Sastry, F. Sciortino, and J. Teixeira. Any of these 15 valued collaborators could probably give this lecture as well as, if not better than, I will.

We will organize the answer to the third and principal question in the following way: First, we will attempt to identify, in the spirit of Sherlock Holmes, some interesting clues. Then, we'll formulate a working hypothesis, and then we'll test the hypothesis, both by experiments on real water, and by simulations of what sometimes is called 'computer water'. These simulations are useful because some of the problems associated with doing experiments on real water can be avoided using computer simulations.

Before beginning, let's note that there are some very good general references on liquid and glassy water. A recent book entitled *Metastable Liquids* [1. 1] is a kind of sequel to Skripov's classic on this subject. There's an even more recent book called *Supercooled Liquids* published by the American Chemical Society [1. 2]. And there are many recent reviews [1. 3–5], including a very recent seven-page mini-review [1. 6]. Finally, and perhaps most interestingly, the accomplished science writer Philip Ball has just prepared a book that surveys the entire story of water, from the earliest times to the present [1. 7] – responding perhaps to the fact that water is one of the most appealing of the open puzzles in all of science.

1.2 What is the Puzzle of Liquid Water?

We start with the first question, 'What is the puzzle of liquid water?' There are many puzzles associated with water. Firstly, water exists in many forms. All of us are familiar with one important control parameter that distinguishes these forms, namely temperature. Stable water is the water that one is familiar with when one discusses biology (in reality, much of biology takes place when water is in a supercooled regime). Why is that? Because in Boston nothing would survive the winter were it not for the fact that living systems are able to survive even when the water inside them is metastable, namely supercooled. But below about $-38°$C, we hit a kind of Berlin Wall, beyond which there is a 'no-man's land' where water exists only in the solid, crystalline phase. That temperature is called the temperature of homogeneous nucleation and denoted T_H. At very low temperature, there's a region in which water exists in the glassy phase. If one heats glassy water gently above its glass transition temperature, $T_g \approx 135$ K, one finds a narrow window of 10-15 degrees in which water is liquid, although very, very viscous [1. 8]. If one heats it further, one reaches a temperature, $T_X \approx 150$ where the mobility of the ultraviscous water reaches a sufficiently large value that the molecules can readily find out that the crystalline phase indeed has the lowest free energy – and it freezes. Between T_X and T_H, an interval of roughly 80 degrees, there's a kind of no-man's land in the sense that water is stable only in its solid crystalline phases. In short, only ice cubes exist in no-man's land!

1.2.1 Volume Fluctuations

So let's begin by asking about the nature of the puzzle of liquid water. What are the features on which we would like to focus? In any short presentation, our choices are necessarily subjective, so we start by considering a simple response function, the isothermal compressibility K_T, which is a thermodynamic response function – the response $\delta\bar{V}$ of the volume per particle to an infinitesimal pressure change δP. As one lowers the temperature of a typical liquid, K_T decreases. We understand this feature because in statistical mechanics courses we learn that K_T is also related to something statistical, namely the expectation value of the fluctuations in $\bar{V}$. As the temperature is lowered, we expect the fluctuations to decrease. Water is an anomaly in the following:.

- First, the compressibility, when normalized by mass and other material parameters, is rather larger than one would expect.
- Second, and very dramatically, although at high temperatures the compressibility decreases on lowering the temperature, below about 46°C, it actually starts to increase, and this is not a small effect. By the time one reaches the Berlin Wall temperature of −38°C, the compressibility has actually increased by a factor of 2. It's a hundred percent larger than its value at 46°C.
- The third remarkable feature is that if one plots one's data, not linearly, but rather on double-logarithmic paper – where the x-axis is the logarithm of $|T - 228|$ K and the y-axis the logarithm of the compressibility – one finds a region of straight-line behavior, hinting at some sort of critical behavior. In mentioning these experiments, I should emphasize the name of Austen Angell who, together with numerous students, pioneered in the identification of this apparent power-law singularity [1. 9].

1.2.2 Entropy Fluctuations

These anomalies associated with the compressibility response function is also shared by other response functions. Consider, for example, the specific heat at constant pressure, C_P, which is the thermodynamic response, $\delta\bar{S}$, of the entropy per particle to an infinitesimal temperature change, δT (apart from proportionality factors). Once again, for a typical liquid, this quantity gently decreases as one lowers the temperature, which we understand because this quantity is also related to a fluctuation – the entropy fluctuations. And one imagines that entropy fluctuations also decrease when one cools. Water is again anomalous in the same three respects. The specific heat for water is, as every engineer knows, rather large. Secondly, below about 35°C the specific heat actually starts to increase. Lastly, that increase can be approximated by a power law.

1.2.3 Volume-Entropy Cross-Correlations

The last of the three response functions is the one that's exemplified by the experiment measuring the temperature at the bottom of a glass of ice water. That's the coefficient of thermal expansion, α_P, the response, $\delta\bar{V}$, of the volume to an infinitesimal temperature change, δT. For a typical liquid α_P is positive. And again, one can understand that positivity, in a statistical mechanics sense, because this coefficient of thermal expansion is proportional to the cross-fluctuations of entropy and specific volume, $\langle \delta\bar{V}\delta\bar{S}\rangle$. Normally, if one has a region of the system in which the specific volume is a little larger than the average, then the local entropy in that same region of the system is also larger. For example, I often complain that the reason my office is so messy is that it's too small, and if I could only get a bigger one it would be less messy. But in reality, I think we all know what would happen if I had a larger office – it would be even messier because there would be even more ways to arrange my huge quantity of material.

So $\langle \delta\bar{V}\delta\bar{S}\rangle$ is positive because in local regions of the liquid in which the volume is larger than the average ($\delta\bar{V}$ is positive), then the entropy is also larger (so $\delta\bar{S}$ is positive). The product of two positive numbers is positive. Water is unusual in the same three respects. First of all, its coefficient of thermal expansion is about 3 times smaller than that of the typical liquid. Secondly, as one lowers the temperature, the anomaly gets stronger and stronger, and at the magic number of 4°C (which has been known for over three centuries), α_P actually becomes negative; once again an apparent divergence exists in the sense that an imminent apparent singularity appears to occur. From a simple kitchen experiment, we can understand that the 0°C water in a glass of ice water (or a pond in a Boston winter) 'floats' on top of heavier layers of 4°C water – allowing the fish to survive the long winter.

1.3 Why Do We Care About Liquid Water?

Let's turn now to Question 2, *'Why do we care about this puzzle?'* There are always lots of reasons to care. Almost all of them are, in some sense, personal. For me, certainly, scientific curiosity is always one reason to care. Why on earth could water have such a difference compared to that of a typical liquid no matter what kind of fluctuating quantity we look at? It doesn't make sense.

A second reason, which I alluded to in the beginning, is that water is a kind of model tetrahedral liquid. It's a model in the sense that it's relatively easy to study. Most of the temperature regions in which one studies water are reasonably accessible. There are other tetrahedral liquids. In fact, the second most common material on our planet is SiO_2, and SiO_2 has many features that are remarkably parallel to those of water. Water and SiO_2 both share the fact that they are tetrahedral liquids; that is to say, they are liquids whose local structure approximates that of a four-coordinated, open, loose-packed lattice.

There are also numerous practical reasons to care about water. I already alluded to the fact that engineering depends very much on the remarkable properties of water, and so also does life itself. Glassy water is relevant to drug preservation: if we make a glass, the dynamics are slowed down by a factor of $\approx 10^{13}$. Hence the time scales of biological processes are similarly slowed down.

1.4 Clues for Understanding Water

With this brief introduction behind us, let's consider what we actually do. First we seek for clues. The fundamental clue, perhaps, for the mystery of liquid water dates back more than 50 years to Linus Pauling, who recognized that the distinguishing feature of H_2O, when compared to other materials that otherwise look chemically similar, is the preponderance of hydrogen bonds. A useful cartoon illustrates the implications of a high degree of hydrogen bonding. Each molecule has two arms, corresponding to the two protons, and two attractive spots, corresponding to the lone pairs. The arms reach out to the attractive spots. Since the fraction of intact bonds exceeds the percolation threshold, the basic structure of water is a hydrogen-bonded gel. Because these hydrogen bonds have very, very short lifetimes – typically a picosecond – water doesn't behave like Jell-O We can pour water and it flows just like any other liquid.

Computer simulations also support the idea that water is a network. In traditional simulations, we have 216 water molecules in a little box, 18Å on each edge, and we can visualize the hydrogen bonds that are intact at room temperature. For any reasonable definition of bonds, we still come back to the fact that it is made up of a strongly hydrogen-bonded network, not isolated molecules.

There is one sense in which water is, locally, very much like ice. The distance between nearest-neighbor molecules in liquid water is very similar to what we find in ice, and if we calculate the positions of second-neighbors in liquid water, assuming the orientation of these two molecules and the distance between them to be exactly the same as in ice, we get 4.5 Å. If we measure a histogram experimentally, counting the number of molecules a distance R from a central molecule, we find peaks of both the nearest- and second-neighbors, and, if we push our luck and the temperature is not too high, we even see a peak at the third neighbors. Thus the hydrogen-bonded network has a remarkable feature: it constrains the positions of the water molecules sufficiently that within "small neighborhoods" (seeing a locale in terms of only a few neighbors) the situation is not that different from what we find in ice.

1.5 Qualitative Picture: Locally Structured Transient Gel

Although there have been many different detailed models used by researchers to describe the structure of liquid water, many would agree that, on a qualitative level, water is a *locally structured transient gel* [1. 11].

1.5.1 'Locally Structured'

We describe liquid water as *locally structured* because its structure – such as it is – extends out only to about 4-8 Å. Within this local structure, the entropy, S, is less than the global entropy, $\langle S \rangle$, of the entire network, and the specific volume, V, is larger than the specific volume, $\langle V \rangle$, of the entire network. Thus $\delta S \equiv S - \langle S \rangle$ is negative, while $\delta V \equiv V - \langle V \rangle$ is positive.

1.5.2 'Transient Gel'

We describe liquid water as a *transient gel* because, although water is a connective network, it does not behave like your average bowl of 'Jell-O.' I can tip a glass of water and it will start to flow. If I tip a container of Jell-O, it will not. Unlike the bonds in Jell-O, the bonds in liquid water have a characteristic life that is remarkably short, on the order of a picosecond. Nevertheless, we still call water a gel because its connected network is a random system far above its percolation threshold.

1.6 Microscopic Structure: Local Heterogeneities

To represent this structure dynamically and in three dimensions we need computer-simulation resources and a good ability to visualize three dimensions. To represent it as an instantaneous picture in two dimensions, we can use a square lattice, even a chess board – with the edges of the chess board's squares representing 'bonds', and the corners of the squares representing 'sites' (molecules).

If we take our chessboard and randomly break 20% of its bonds, leaving the other 80% intact, we get a snapshot of the structure of liquid water. The sites with all four bonds still intact we designate 'black' sites. These have nearest-neighbor distances almost identical to those of solid water. The areas on the grid exhibiting contiguous black sites we call 'patches'. The patches have properties that differ from the global properties of the heterogeneous gel: a lower entropy and a larger specific volume. To calculate the number of black molecules – if we neglect the possibility of correlations in this bond-breaking – is fairly simple. The probability that the four bonds of a given site are intact is simply the bond probability to the power four ($f_4 = p_B^4$). If the bond probability is ≈ 0.8, as we saw earlier, the probability that four bonds of a given site are intact is $\approx 0.8^4$ or ≈ 0.4. So about 40% of the black sites are a part of the patches.

This picture of water is sufficient to *qualitatively* rationalize experimental data. For example, this picture can explain the anomalous compressibility behavior of water – water's compressibility minimum at 46°C (typical liquids have no minimum) and the unusually large size of water's compressibility (twice that of typical liquids). The patches in water's hydrogen-bonded network influence the behavior of its compressibility – the fluctuations in its specific volume; these black sites are present with a probability of p_B^4. As temperature decreases, the bond probability increases. For each 1% rise in bond probability there is a 4% rise in f_4 and thus an amplification in the number of black sites. As we know from percolation theory, when the number of black sites increases, the characteristic size of a cluster or patch increases still more dramatically, providing a second amplification mechanism. The greater increase in compressibility is a result of these two amplification mechanisms. Were there no patches, the increase would not be as great. The amount of this increase goes up as we lower the temperature. If I take a function with constant slope and add this increase I end up with a function that exhibits a minimum.

We can use this picture of water to rationalize other anomalies, e.g., an anomaly in water's specific heat (an anomaly we did not mention earlier). The specific heat is proportional to the fluctuations in entropy. Entropy is often imagined to be a constant quantity, one without fluctuation. But if we treat entropy as a fluctuating quantity, we have a measure of the specific heat at constant pressure – and that also increases dramatically as one lowers the temperature below 35°C, for approximately the same reasons. As we lower the temperature, the build-up of the little patches is more and more dramatic, causing an additional contribution to the entropy fluctuation due their increasing presence. The anomalous behavior of specific heat is not just the increase at low temperatures, but also the fact that the average value of its specific heat is larger than one would expect if these patches were not present (water is industrially important because of its high specific heat).

But perhaps the most dramatic of these anomalies – the one students first learn about – is the 4°C anomaly in the coefficient of thermal expansion, the response of the volume to infinitesimal changes in temperature. This coefficient of thermal expansion is anomalous not only because it becomes negative below 4°C, but because it is 2–3 times smaller – even at high temperatures – than it would be in a typical fluid. How can this be understood in terms of these patches?

When we lower the temperature, the number and size of the structured patches increase. These structured little regions have the following properties: that their entropy is smaller than the average (a negative quantity) and their specific volume is larger than the average (a positive quantity). Multiplying the negative quantity by the positive one, we get a negative contribution from this macroscopic thermodynamic response function. As we lower the temperature, and the patches become more numerous, the magnitude of this negative response function gets bigger and bigger, and it just happens to pass through zero at a temperature of 4°C.

This is qualitative, but we can test for the existence of these patches in various ways. Simulations provide unambiguous evidence [1. 12], but in simulations the results always could be wrong if the potential is wrong. Experimentally, we can't actually see them, but by beaming X-rays into supercooled water (−25°C), Bosio and Teixeira have observed a characteristic Ornstein-Zernike behavior, indicating a build-up in correlated regions with a characteristic size proportional to the inverse of the width of that Lorentzian. The characteristic size measured using this experimental technique turns out to be a diameter of ≈ 8 Å, i.e., 2–3 atomic spacings; recent data suggests this number may be even smaller (at least at atmospheric pressure).

Another approach to testing this picture is to dilute our water sample with some other liquid. To do that, we go into that wonderfully ordered hydrogen-bonded network with its little patches and replace 10% of the water molecules with something that does *not* form four hydrogen bonds at tetrahedral angles. If that 10% replacement is hydrogen peroxide, the anomalies almost disappear [1. 15]! Evidently the patches are broken up by the impurity.

Another simple but striking example is the dependence of the adiabatic sound velocity, $v_s \propto K_S^{-1/2}$, upon the mole fraction x of ethanol. Although v_s for pure ethanol ($x = 1$) is much *smaller* than v_s for pure water ($x = 0$), one finds a substantial *increase* in v_s as ethanol is added to pure water. We would interpret this finding as follows: ethanol is 'breaking up' the patches, and thereby reducing the compressibility. When v_s is plotted against x for a range of temperature from 5°C to 45°C, one finds that all the isotherms intersect at a single 'isosbestic point' with $x = 0.17$ and $v_s = 1.6\,\mathrm{km/s}$. Thus, at 17% ethanol, increasing T from 5°C to 45°C serves to decrease the fluctuations due to the patches, but this is precisely compensated for by an increase in the fluctuations of the normal regions of the network. It would be desirable to extend these observations to a wider range of temperature and pressure, to other impurities, and to properties other than v_s. Careful study of judiciously-chosen two-component systems may serve to provide useful clues relevant to ' the puzzle of liquid water'.

To summarize thus far: When looking at the bond connectivity problem, water appears as a large macroscopic space-filling hydrogen bond network, as expected from continuum models of water. However when we focus on the four-bonded molecules ('sites'), we find that water can be regarded as having certain clustering features – the clusters being not isolated 'icebergs' in a sea of dissociated liquid (as postulated in mixture models dating back to Röntgen) but rather patches of four-bonded molecules embedded in a highly connected network or 'gel'. Similar physical reasoning applies if we generalize the concept of four-bonded molecules to molecules with a smaller than average energy [1. 18] or to molecules with a larger than average 'local structure' [1. 19].

1.7 Liquid-Liquid Phase Transition Hypothesis

This qualitative picture is sufficient to explain that there is a minimum, but not sufficient to relate at all to the presence of the apparent power-law singularity. This power-law singularity is reminiscent of critical phenomena. It suggests there might be some sort of critical point – but how could there be a critical point? Water does have a critical point, but water's critical point is a very ordinary critical point. The critical point of water is up around 374°C at a pressure of approximately 227 atm. Above the critical point, water is a homogeneous fluid, while below the critical point, water can exist in either of two distinct phases. That critical point is known and well-studied by individuals such as Anneke Levelt-Sengers at the National Institutes of Standards and Technology in Gaithersburg, Maryland. Water is not unusual in connection with this critical point.

What could this apparent power-law singularity have to do with a critical point? Notice that the temperature here is about −50°C, nowhere near 374°C. So there's a kind of puzzle here. Why would one find critical point phenomena in experimental data?

A possible resolution to this paradox was proposed a couple of years ago. A gifted Boston University graduate student, Peter Poole, working with post-docs Francesco Sciortino and Uli Essmann, made computer simulations in this low-temperature region with the goal of exploring in detail with a computer what might happen [1. 20]. Why a computer? Because 'computer water' does not suffer from this Berlin Wall catastrophe at −38°C. Computer water has a wonderful virtue: in simulations of sufficiently small systems with sufficiently small time scales, the nucleation phenomenon that plagues real experiments simply doesn't occur. One can go as low in temperature as one has computer time to simulate. What Poole and collaborators discovered in computer water was the apparent existence of a second critical point. That second critical point was indeed in the range of −50°C. Exactly where is difficult to say because in computer simulations you get out what you put in to some degree. It's very difficult to put in an accurate temperature scale, because the actual model that's used to simulate water is itself not terribly accurate. So when I say $\approx -50°$C, I mean −50°C ±10°C, or even −50°C ±20°C. Nevertheless, what was discovered is that, below this second critical point, the liquid phase separates into two distinct phases – a low-density liquid (LDL) and a high-density liquid (HDL).

1.8 Plausibility Arguments

A non-interacting gas has no critical point, but a gas with arbitrarily weak attractive interactions does since at sufficiently small temperature, the ratio of the interaction to kT will become sufficiently significant to condense the liquid out of the gas. That all interacting gases display a critical point below which a distinct liquid phase appears was not always appreciated. Indeed, in the early years of the 20th century one spoke of 'permanent gases' – to describe gases that had never been liquefied. Helium is an example of what was once thought to be a permanent gas [1. 21].

Nowadays, we understand that permanent gases cannot exist since all molecules exert some attractive interaction, and at sufficiently low temperature this attractive interaction will make a significant contribution. To make the argument more concrete, one can picture droplets of lower specific volume, $\bar{V}$, forming in a single-component fluid. Once the interaction between molecules is fixed (and P is fixed at some value above P_C), then the only remaining control parameter is T; as T decreases the high-density droplets increase in number and size and eventually below T_C they coalesce as a distinct liquid phase.

Water differs from most liquids due to the presence of a line of maximum density (TMD line) in the PT phase diagram. This TMD is physically very significant, as it divides the entire PT phase diagram into two regions with remarkably different properties: the coefficient of thermal expansion – which is proportional to the thermal average ('correlation function') $\langle \delta \bar{V} \delta \bar{S} \rangle$ – is *negative* on the low-temperature side of the TMD line, while it is *positive* on the high-temperature side. Here $\bar{V}$ is the volume per molecule, $\bar{S}$ the entropy per molecule, and the δX notation indicates the departure of a quantity X from its mean value.

That $\langle \delta \bar{V} \delta \bar{S} \rangle$ is negative is a thermodynamic necessity given the presence of a TMD line. What microscopic phenomenon causes it? One not implausible explanation is related to the presence of local regions of the hydrogen-bonded network that are characterized by four 'good' hydrogen bonds – and these local regions can be considered as droplets, just as the high-density droplets in a gas above C. Stated more formally: the sensitivity of hydrogen bonds to the orientation of the molecules forming them encourages local regions to form that are partially ordered in the sense that, if there is a region of the water network where each molecule has four 'good' (strong) hydrogen bonds, then the local entropy is lower (so $\delta \bar{S} < 0$) and the local specific volume is larger (so $\delta \bar{V} > 0$), so the contribution to $\langle \delta \bar{V} \delta \bar{S} \rangle$ is negative for such regions.

As the temperature is lowered, there is no *a priori* reason why the 'droplets' characterized by negative values of $\delta \bar{V} \delta \bar{S}$ should not increase in number and size, just as the droplets associated with a normal phase transition increase in number, since all water molecules exert mutual interactions on one another. These interactions – because of their sensitivity to orientation and well as distance – favor the open clusters characterized by $\delta \bar{S} \delta \bar{V} < 0$. It is thus plausible that at sufficiently low temperature these orientation-sensitive interactions will make a larger and larger contribution, and at sufficiently low temperature (and for sufficiently low pressure) a new phase – having roughly the density of the fully hydrogen-bonded network – will "condense" out of the one-fluid region.

This intuitive picture has received striking support from a recent generalization of the van der Waals theory. Specifically, Poole et al. [1. 22] allowed each water molecule to be in many bonding states, only one of which corresponds to a 'good' quality hydrogen bond (with a larger number of states corresponding to 'poor' quality bonds). To build in this feature, Poole et al. adopted the approach of Sastry and co-workers [1. 23, 24] and assumed that there are $\Omega \gg 1$ configurations of a weak bond, all having $\epsilon = 0$, and only *a single configuration* in which the HB is strong, with $\epsilon = \epsilon_{\mathrm{HB}}$. Thus the thermal behavior of the HBs is represented by independent $(\Omega + 1)$–state systems, each described by a partition function, $Z = \Omega + \exp(-\epsilon_{\mathrm{HB}}/kT)$. Poole et al. found that for small values of the parameter ϵ_{HB} there is no critical point (but rather a re-entrant spinodal of the form first conjectured by Speedy [1. 25]). However for ϵ_{HB} above a threshold (about 16 kJ/mol), a critical point appears.

The possibility of a second critical point has received recent support by phenomenological analysis of Ponyatovskii and colleagues[1. 26] and by lattice gas models[1. 24, 27]. Also, Roberts and co-workers [1. 28] have shown that simulation results for a microscopic 'water-like' Hamiltonian confirms the presence of a second phase transition, previously deduced from approximate calculations[1. 27].

1.9 Tests of the Hypothesis: Computer Water

Simulation studies of liquid water have a rich history and have contributed greatly to our understanding of the subject. In fact, over a quarter century ago, the ST2 ('Stillinger-2') potential was introduced. Water is represented by a central point from which emanate 4 arms – two carrying positive charges to represent the two protons associated with each water molecule, and two carrying negative charges to represent the two lone electron pairs [1. 29]. The central points interact via a Lennard-Jones potential, while the point charges and the arms interact via a Coulomb potential. Thus every pair of water-like particles has $4^2 + 1 = 17$ interaction terms. Corresponding to the rather 'cumbersome' nature of such a potential is the fact that most studies are limited to extremely small systems – a typical number being $N = 6^3 = 216$ water-like particles. Recently some studies have considered larger systems, but the typical size rarely exceeds $N = 12^3 = 1728$. It is hoped that by using fast multipole methods one can begin to simulate much larger systems[1. 30].

One way to obtain less cumbersome simulations is to simplify the intermolecular potential. To this end, the simpler TIP4P potential [1. 31] and the much simpler SPC/E potential [1. 32] have enjoyed considerable popularity. However the opposite direction is also under active investigation: simulating more realistic potentials, such as polarizable potentials[1. 33]. The researcher is left with the perplexing problem of which model potential to adopt!

With these caveats, let us very briefly summarize some recent work that might be interpreted as being consistent with (or at least not contradicting) the hypothesis that a HDL-LDL critical point, C', exists. We emphasize that most of this work has not reached the stage that it can be interpreted as 'evidence' favoring the hypothesis, so we also outline appropriate avenues where future work may strengthen the argumentation.

1.9.1 Does $1/K_T^{\max}$ Extrapolate to Zero at $(T_{C'}, P_{C'})$?

The compressibility K_T diverges at a second-order critical point. Thus, we expect $1/K_T^{\max}$ to extrapolate to zero at the 'new' HDL-LDL critical point, C', exactly as it does for the 'old' liquid-gas critical point, C. Recent ST2 calculations [1. 34] are consistent with a plausible extrapolation to a single point in the phase diagram at which $K_T^{\max} = \infty$. The caveat is that one can never know that a given quantity is approaching infinity – it could as well just be approaching a very large number. Indeed, the possibility has been raised, and seriously discussed, that there is no genuine singularity [1. 35]; this possibility will be discussed briefly at the end of this chapter.

1.9.2 Is There a 'Kink' in the $P\rho$ Isotherms?

If there is a critical point, then we expect to find a kink in the $P\rho$ isotherms when T is below $T_{C'}$. Indeed, such a kink appears to exist for the ST2 potential, at a temperature of 235 K but not at a temperature of 280 K, consistent with $T_{C'}$ somewhere between 235 K and 280 K. This finding, originally made for simulations of 216 ST2 particles [1. 36, 37], has very recently been strikingly confirmed for a system 8 times larger [1. 38]. An analogous kink has not been found for the TIP4P potential, but a prominent inflection occurs at the lowest temperature studied – suggesting that such a kink may be developing. Work is underway testing for inflections and possible kinks for other water potentials in three, and also in two, dimensions.

1.9.3 Is There a Unique Structure of the Liquid near the Kink?

If there exists a critical point C', then we would expect a two-phase coexistence region below C'. To investigate the possible structural difference between these two phases, Sciortino et al. [1. 34] studied the structure of the liquid at a temperature just below the estimated value of $T_{C'}$ at two values of ρ on the two sides of $\rho_{C'}$. They found that the structure of the liquid state of ST2 at $\rho = 1.05$ g/cm^3 is similar to the experimental data on high-density amorphous (HDA) solid water, while the structure of the structure at $\rho = 0.92$ g/cm^3 resembles the data on low-density amorphous (LDA) solid water. The correspondence between the HDA ice phase and ST2 water just above $\rho_{C'}$ and between the LDA phase and ST2 water just below $\rho_{C'}$ suggests that the two phases that become critical at C' in ST2 water are related to the known HDA and LDA phases of amorphous ice[1. 39, 40].

1.9.4 Does the Coordination Number Approach Four as C' is Approached?

Sciortino et al [1. 34] have studied the coordination number N_{nn} of the ST2 liquid as a function of T and V, where N_{nn} is the average number of nearest-neighbors found in the first coordination shell of an O atom. For the high-T isotherms, their results show that a four-coordinated 'LDL'-like configuration is approached at negative P, in agreement with previous simulations of Geiger and co-workers [1. 41]. For $T \leq 273$ K, N_{nn} also approaches four at positive P. That is, if T is low enough, it appears that a four-coordinated network can form in liquid water even for $P > 0$. This result is consistent with an experimental study of the evolution of the structure function, $S(Q)$, as water is supercooled at atmospheric pressure, in which it was found that the structure tends toward that of the LDA ice [1. 39].

1.9.5 Is It Possible that Two Apparent 'Phases' Coexist Below C'?

Convincing evidence for a HDL-LDL critical point, C', would be the presence of two coexisting phases below C'. This search is the focus of ongoing work. One can, e.g., partition the water molecules into two groups ('red' and 'blue' molecules), those with fewer than the average number of nearest neighbors and those with more than the average. We find that the red molecules and the blue molecules segregate to different regions of the 18 Å box in which they are residing. The preliminary investigations at a temperature somewhat below $T_{C'}$ do not prove phase coexistence [1. 38,42], but work is underway to establish this possibility. In particular, one must first rule out the likelihood that the two 'phases' are merely large fluctuations due to a large correlation length (because near a critical point there should be fluctuations of all sizes and shapes, while the sample separating into two distinct regions is rather different). Also, one must seek to find the phase separation occurring in much larger systems. To be conclusive, firstly one must demonstrate that phase separation occurs in a much larger system, and secondly one must study systematically the time dependence of $S(Q)$ as one quenches into the two-phase region from a large temperature value.

Separate calculations of the weighted correlation function $h(r)$ for the two tentatively identified HDL and LDL phases suggest similarities with experimental results on the two amorphous solid phases, HDA and LDA[1. 42]. Additional work remains to be done to establish this point.

1.9.6 Do Fluctuations Appear on All Time Scales?

For the ST2 potential, a histogram of hydrogen-bond lifetimes reveals power-law behavior over as much as two decades, with the region of 'scale-free behavior' extending over a larger time domain as T is decreased [1. 43]. For the TIP4P potential, no calculations have yet been carried out, but for the SPC/E potential, non-Arrhenius behavior has also been found at high temperatures [1. 44]. At low temperatures, it is possible that power-law behavior is found [1. 30,38]. An important caveat in interpreting these results is that this scale-free behavior is exactly what one would expect if the hydrogen-bonded network were regarded as possessing defects (corresponding to molecules with fewer than four good bonds), and these defects were allowed to diffuse randomly [1. 45,46]. Possibly some of these ambiguities will be resolved by applying to this problem Sasai's 'local structure index', which permits one to study in some detail the local dynamics [1. 47].

1.9.7 Is There 'Critical Slowing Down' of a Characteristic Time Scale?

For the ST2 potential, the characteristic value of hydrogen-bond lifetime, defined as the value of time at which the power-law distribution of bond lifetimes is cut off by an exponential, depends sensitively on temperature and in fact is consistent with a power-law divergence as T approaches $T_{C'}$ [1. 43]. The temperature dependence of the cutoff has not been studied for other potentials.

Appearing to diverge at roughly the same temperature is a less ambiguous measure of the characteristic time – the inverse of the self-diffusion coefficient

D [1. 48]. This slowing down of the dynamics is consistent with what one expects near a critical point. Specifically, $1/D$ strongly increases as $N_{\mathrm{nn}} \to 4$. Consistent with this picture, it was found [1. 49, 50] that additional nearest neighbors beyond four have a 'catalytic' effect on the mobility of the central molecule, in that they lower the local energy barrier of the molecular exchanges that are the microscopic basis of diffusion, demonstrating the importance for molecular mobility of molecular environments having more than four nearest neighbors.

Because of the relation between $1/D$ and $(N_{\mathrm{nn}} - 4)$, the manner in which $N_{\mathrm{nn}} \to 4$ is also significant. At high T the decrease of N_{nn} with P is relatively uniform. However, as T decreases, N_{nn} is observed to vary more and more abruptly from a high-coordinated structure ($N_{\mathrm{nn}} > 6$) to $N_{\mathrm{nn}} \simeq 4$. It should be possible to collapse this family of curves onto a single 'scaling function' if the two axes are divided by appropriate powers of $T - T_{C'}$; these tests are underway.

1.9.8 Is the Characteristic Dynamics of Each 'Phase' Different?

We can identify molecules as 'red'/'blue' if they are in a region of locally high/low density for a specified amount of time (say 100 ps). Looking at the mean square displacement of the red and blue 'phases', we see that the red molecules (corresponding to high densities) move much further than blue molecules (corresponding to low densities)[1. 38,42]. The nature of transport in each phase is under active investigation, particularly in light of recent proposals for the nature of the anomalous dynamics taking place in low-temperature water [1. 51].

1.9.9 Is There Evidence for a HDL-LDL Critical Point from Independent Simulations?

Recently, Tanaka independently found supporting evidence of a critical point by simulations for the TIP4P potential [1. 52,53]. Tanaka's value of the critical temperature $T_{C'}$ agrees with the earlier estimates, but his critical pressure $P_{C'}$ occurs at roughly atmospheric pressure, or perhaps at negative pressures [1. 52, 53]. The resolution between the two different values of $P_{C'}$ is an open question that will hopefully be resolved shortly.

It was actually possible to identify these two phases in the computer. I emphasize *in the computer* because what you find in the computer can be suggestive and frequently can confirm experiments. In this case it actually motivated some experiments – the search in real water for evidence for such a critical point deep in the no-man's land between T_H and T_X.

The pressure of the second critical point is around one kilobar. That's easy to remember because that happens to be the pressure at the bottom of the Marianas Trench! Because it's at a high pressure, I'm going to gradually ramp up the complexity of our phase diagram. I warned you that this was a slice at one atmosphere and that sooner or later we'd have to look at pressures other than one atmosphere, and that's what this diagram does. In fact, it shows pressures all the way up to about 0.3 GPa, or approximately 3000 atm.

The awkward location of the second critical point at approximately 1000 atm presents us with perplexities. First of all, on which side of the Berlin Wall does this mythical beast lie? It's lying on the side that we can't reach experimentally! I will spend the rest of this chapter discussing experiments, but there are experimental attempts to identify a mythical beast that we can't get close to at all, no closer than 5 or 10 or perhaps even 15 degrees from this actual critical point.

If we can't get close to it, why does it matter? It matters for the same reason that Mount Everest matters. Mount Everest, approximately 10,000 m high, is not a delta function somewhere in Nepal and Tibet. Rather, Mount Everest is a peak that exerts an influence on its neighboring altitudes, both high above the timberline and well below the timberline. In other words, if we have a singularity in our phase diagram at a well-defined critical point, it's going to have an effect on an entire region around the critical point – a so-called critical region. It is not required that the system is exactly *at* its critical point in order that the system exhibits remarkable behavior, such as the phenomenon of critical opalescence discovered and correctly explained in 1869 by Andrews [1. 54] in terms of increased fluctuations away from (but close to) the critical point. It is for this reason that critical phenomena are particularly interesting.

1.10 Tests of the Hypothesis: Real Water

1.10.1 A Cautionary Remark

The first statement we must make concerns the presence of the impenetrable 'Berlin wall': the line, $T_H(P)$, of homogeneous nucleation temperatures [1. 55]. By careful analysis of experimental data above $T_H(P)$, Speedy and Angell[1. 25, 56, 57] pioneered the view that some sort of singular behavior is occurring in water at a temperature $T_s(P)$ some 5-10 degrees below $T_H(P)$. Our belief is that, even though the region below T_H is experimentally inaccessible, we want to learn about the liquid equation of state in this region, since anything that might occur in this region (such as a line of spinodal singularities [1. 57, 58] or a critical point) will influence the equation of state in a large neighborhood.

1.10.2 Previous Work

Bellissent-Funel and Bosio have recently undertaken a detailed structural study of D_2O using neutron scattering to study the effect of decreasing the temperature on the pair correlation function and structure factor $S(Q)$ [1. 39, 40]. For experimental paths, they chose a family of isobars ranging from 0.1 MPa up to 600 MPa (well above the critical point pressure of about 100 MPa). They plotted the temperature dependence of the first peak position Q_o of $S(Q)$ for each isobar. They found that for the 0.1 MPa isobar, Q_o approaches 1.7 Å^{-1} – the value for LDA, low-density amorphous ice. In contrast, for the 465 MPa and 600 MPa isobars, Q_o approaches a 30% larger value, 2.2 A^{-1} – the value for HDA, high-density amorphous ice. For the 260 MPa isobar, $Q_o \rightarrow 2.0A^{-1}$, as if the sample were a two-phase mixture of HDA and LDA.

Since the critical point occurs below $T_H(P)$, it is not possible to probe the two phases experimentally. However two analogous solid amorphous phases of H_2O have been studied extensively by Mishima and co-workers [1. 59]. In particular, Mishima has recently succeeded in converting the LDA phase to the HDA phase by increasing the pressure, and then reversing this conversion by lowering the pressure. The jump in density was measured for a range of temperatures from 77 K to 140 K. Moreover, the magnitude of the density jump decreases as the temperature is raised, just as would occur if, instead of making measurements on the HDA and LDA amorphous solid phases, one were instead considering the HDL and LDL liquid phases. These results have been independently corroborated by computer simulations performed using both the ST2 and TIP4P intermolecular potentials [1. 36].

If we assume that HDA and LDA ice are the glasses formed from the two liquid phases discussed above, then the HDA–LDA transition can be interpreted in terms of an abrupt change from one microstate in the phase space of the high–density liquid, to a microstate in the phase space of the low–density liquid. The experimentally detected HDA–LDA transition line would then be the extension into the glassy regime of the line of first-order phase transitions separating the HDL and LDL phases.

1.10.3 Recent Work

We discuss now the very recent measurements of the compression-induced melting and decompression-induced melting lines of high-pressure forms of ice. We show how knowledge of these lines enables one to obtain an approximation for the Gibbs potential, $G(P,T)$, and the equation of state, $V(P,T)$, for water, both of which are consistent with the possible continuity of liquid water and the amorphous forms of solid water.

When liquid water is supercooled below the homogeneous nucleation temperature, T_H, crystal phases nucleate homogeneously, and the liquid freezes spontaneously to the crystalline phase. When amorphous solid ice is heated, it crystallizes above the crystallization temperature, T_X. Therefore, amorphous forms of H_2O do not exist in the 'no-man's land' between T_H and T_X.

When we compress the crystalline ice I_h at low temperatures, it transforms to a supercooled liquid on its metastable melting line above T_H. Between T_H and T_X, it transforms to a high-pressure crystalline ice at the smoothly extrapolated melting line[1. 60]. Below T_X, ice I_h amorphizes to HDA at a pressure higher than the smoothly extrapolated melting line [1]. To avoid the complication of the usual crystal-crystal transformations interrupting the melting process, we use an ice emulsion (1-10 μm ice particles in oil[1. 61]).

Mishima created 1 cm^3 emulsified high-pressure ices in a piston-cylinder apparatus, decompressed the sample at a constant rate of 0.2 GPa/min, and – because melting is endothermic – observed their transitions by detecting a change in the sample temperature using an attached clomel-alumel thermocouple during the decompression. Then, he determined melting pressures at different temperatures. The melting curves he obtained agree with previously reported data [1. 62,63], which confirms the accuracy of this method. Moreover, he can determine the location of metastable melting lines to much lower temperatures. Unexpectedly, he found what appear to be two possible new phases (PNP) of solid H_2O, denoted PNP-XIII and PNP-XIV.

Using the measured melting lines of ice phases at low temperatures, we calculate the Gibbs energy and the equation of state [1. 64]. The P-V-T relation is consistent with (but of course does not prove) the existence of a line of first-order liquid-liquid transitions which continues from the line of LDA-HDA transitions and terminates at an apparent critical point, C'. The P-V-T relation is also consistent with other known experimental data and also with simulation results [1. 20,22,24,28,35,47,52,53,59,64–66]

One somewhat speculative argument arises from a recent interpretation of the work of Lang and Lüdemann, who made a series of careful NMR measurements of the spin-lattice relaxation time, T_1 [1. 67–70]. This argument emphasizes the pressure dependence of T_1 for a family of isotherms. The Lang-Lüdemann data are *extrapolated* into the experimentally inaccessible region (below the homogeneous nucleation boundary), which is consistent with, but of course by no means proves, the hypothesized second critical point. This extrapolation is made by eye, not by formula. The extrapolated inflection corresponds to the occurrence of a singularity or critical point. This occurs at roughly the same coordinates as found in the experiments reported in [1. 64] – possibly a coincidence, but the estimated coordinates of C', 220 K and 100 MPa, are the same as those obtained by analysis of the metastable melting lines.

In summary, we know the free energy surface to some level of approximation. Since we know the Gibbs potential as a function of pressure and temperature, by differentiation, we know the volume as a function of pressure and temperature. Having the volume as a function of temperature is just where we want to end this chapter, because volume as a function of pressure and temperature is the equation of state of the liquid. So what ultimately comes out from these experiments is the complete equation of state. This may look a little bit complicated, so let's take it in stages.

At high temperature, we have a familiar ideal gas behavior – the volume is approximately inversely proportional to the pressure. If we look at very low temperatures, we see the two known glassy phases of water: low-density amorphous and high density amorphous. We have two phases here, and even the color-coding is the same. The low density is gray; the high density is pink. They're separated by a first-order transition. 'First order' means the discontinuity in the first derivative. Volume is the first derivative, so there's a discontinuity

– we've already mentioned about 25 %. So this part is known experimentally without a shadow of a doubt. The other part has been known experimentally for perhaps one hundred years. There's only one little thing missing, which is the connection between them. The connection between them is what's provided by the experiments that traverse the no-man's land, and some of those experiments are indicated by these little lines that shoot across from one phase, the known part of the phase diagram, to the other part of the phase diagram.

To summarize, we started with a one-dimensional phase diagram, went to a two-dimensional one, and then to a three-dimensional one. Qualitative features of liquid water appear to be connected to the behavior of a locally structured transient gel. This locally structured transient gel has the feature that little clusters are formed inside the hydrogen-bonded network. Locally, these clusters have distances and bond angles similar to those of ice and therefore have smaller density, larger specific volume, and lower entropy, and contribute to explaining qualitatively the anomalies of water.

1.11 Discussion

The most natural response to the concept of a second critical point in a liquid is bafflement – such a thing just does not make sense. To make the concept more plausible, we offer the following remarks. Consider a typical member of the class of inter-molecular potentials that go by the name of core-softened potentials [1. 71]. These are potentials with two wells, an outer well that is deep and an inner well that is more shallow. Recently, Sadr-Lahijany and collaborators have re-visited such potentials with a view toward applications to water [1. 72]. These simple potentials might capture the essential physics of water-water interactions because, in the case of water, a hydrogen-bonded interaction leads to a larger inter-molecular spacing (say 2.8Å) compared to a 'non-hydrogen-bonding' interaction. Since at low temperatures, hydrogen bonds predominate – increasing the volume – it follows that the outer well of a core-softened potential must be deeper. Then, as temperature is lowered, the system finds itself more likely in the outer 'deep' well than in the inner 'shallow' well. Further, pressure has the same effect as raising the temperature, since, for a fixed temperature, applying pressure favors the inner shallow well.

An advantage of such double-well potentials is that they can be solved analytically in one dimension and are tractable to study using approximation procedures (and simulations) in higher dimensions [1. 72].

To complete the intuitive picture, let us imagine two (or more) local structures, one favored at low pressure (the outer deeper well) and the other favored at high pressure (the inner well). If a system is cooled at a fixed *low* value of pressure, then it will settle into a phase whose properties are related to the parameters of the outer well. If, on the other hand, the system is cooled at a fixed *high* value of pressure, it will settle into a phase whose properties are related to the parameters of the inner well. Thus, it becomes plausible that, depending on the pressure, the liquid could approach different phases as the temperature is lowered. Moreover, if the outer well is deep and narrow, then we anticipate that when $\delta\bar{V} > 0$, $\delta S < 0$ – i.e., volume and entropy fluctuations will be anticorrelated, leading to $\alpha_P < 0$.

A clear physical picture has by no means emerged. However, recent work has addressed the question of whether we can characterize (or at least "caricature") the local structural heterogeneities that appear in liquid water. Specifically, Canpolat and collaborators [1. 73] considered state points of liquid water at different pressures – especially near its phase boundaries with ice I_h and with ice VI (a high-pressure polymorph of solid H_2O). To this end, in the spirit of the 'Walrafen pentamer', they developed a model of interacting water pentamers, and found two distinct local energy minima, which they identified with two well-defined configurations of neighboring pentamers. The 'Walrafen pentamer' is defined by four water molecules located at the corners of a tetrahedron that are hydrogen-bonded to a central molecule – see, e.g., [1. 74]; the corner molecules are separated from the central molecule by 2.8 Å, corresponding to the first peak in the oxygen-oxygen radial distribution function. They advance the hypothesis that these configurations may be related to the local 'high-density' and 'low-density' structural heterogeneities occurring in liquid water. These results are consistent with recent experimental data on the effect of high pressure on the radial distribution function, and are further tested by molecular-dynamics simulations.

Although such a picture may seem to be oversimplified, very recent Work of Bellissent-Funel [1. 75] successfully shows that detailed neutron structure data agree with it. Also, the simulation results are in good accord with neutron results (see, e.g., [1. 40]), so one can be optimistic that some day a unified coherent picture may emerge.

1.12 Outlook

Many open questions remain, and many experimental results are of potential relevance to the task of answering these questions. For example, the dynamics of water is only beginning to receive some possible clarification – see, e.g., [1. 30, 51, 76] and references therein). In particular, the effect on the hypothesized second critical point of modifying water in ways that are interesting is unclear. For example, trees survive arctic temperatures because the water in the cell does not freeze, even though the temperature is below the homogeneous nucleation temperature of −38°C. The effect of confinement on the second critical point is just now beginning to be studied [1. 77]. Similarly, the effect on the second critical point of adding a second component is of interest, especially because the interior of the cell is anything but pure water!

Before concluding, we ask 'What is the requirement for a liquid to have such a second critical point?' In fact, by the arguments above, some other liquids should display second critical points, namely, systems which at low temperature and low pressure have anticorrelated entropy and specific volume fluctuations.

Thus, a natural extension to our work is to consider other tetrahedrally coordinated liquids. Examples of such systems are SiO_2 and GeO_2, known for their geological and technological importance. Both of these systems display features in their equations of state similar to those found in simulations of water and that can be traced to their tetrahedral configurations. This tetrahedrality of local structure has the implication that locally ordered regions of the liquid will have a specific volume *larger* rather than *smaller* than the global specific volume (as in most liquids, for which the local structure, also resembling the global structure of the solid, has a specific volume smaller than the global specific volume). Whenever we are at a state point in the P-T phase diagram to the left of the locus of points where the coefficient of thermal expansion is zero (the 'TMD line'), then of necessity the volume fluctuations are most unusual in that they are anticorrelated with the entropy fluctuations. These unusual fluctuations grow as one moves further into the 'anomalous' region to the left of the TMD line, and ultimately a new phase condenses out of the fluid which has the property that, although the entropy of the new phase is low, the specific volume is large – this is what is called the 'low-density liquid'. Since other tetrahedral liquids have similar features, we might anticipate similar critical points occur on the liquid free energy surface of these liquids. Simulation evidence in favor of this possibility has been reported recently for SiO_2 [1. 78] and a two-level model has been developed for amorphous GaSb [1. 79]. Understanding one such material, water, may help in understanding others – whether they be other materials with tetrahedral local structures (and corresponding TMD lines) such as SiO_2 or whether they be more complex local structures like amorphous GaSb, which appears to display strikingly ordered local heterogeneities.

Acknowledgements

In addition to my 15 valued collaborators named at the outset, I thank S. H. Chen, T. Grande, H.-D. Lüdemann, P. F. McMillan, J. K. Nielsen, C. J. Roberts, R. L. B. Selinger, F. H. Stillinger, Y. Suzuki, H. Tanaka, P. Tartaglia, G. E. Walrafen, B. Widom, and R. Zhang for guidance throughout the course of this work. This work was supported by the National Science Foundation grant CH9728854.

References

1. 1 P. G. Debenedetti: *Metastable Liquids* (Princeton University Press, Princeton 1996)
1. 2 J. T. Fourkas, D. Kivelson, U. Mohanty, K. A. Nelson, (eds.) *Supercooled Liquids: Advances and Novel Applications* (ACS Books, Washington DC 1997)
1. 3 J. C. Dore, J. Teixeira: *Hydrogen-Bonded Liquids* (Kluwer Academic Publishers, Dordrecht 1991)
1. 4 S.-H. Chen, J. Teixeira, *Adv. Chem. Phys.* **64**, 1 (1985)
1. 5 M.-C. Bellissent-Funel, J. C. Dore, *Hydrogen Bond Networks* (Kluwer Academic Publishers, Dordrecht 1994)

1. 6 O. Mishima, H.E. Stanley, Nature **396**, 329 (1998)
1. 7 P. Ball: *Water: A Biography* (Farrah Strauss, New York 1999)
1. 8 R. S. Smith, B. D. Kay, Nature, **398**, 788 (1999); K.P. Stevenson, G.A. Kimmel, Z. Dohnalek, R.S. Smith, B.D. Kay, Science **283**, 1505 (1999)
1. 9 C. A. Angell, in *Water: A Comprehensive Treatise* Vol. 7, ed. by F. Franks (Plenum Press, New York 1980) pp.1-81
1. 10 R. Waller, trans., *Essayes of Natural Experiments* [original in Italian by the Secretary of the Academie del Cimento]. Facsimile of 1684 English translation (Johnson Reprint Corporation, New York 1964)
1. 11 H.E. Stanley, J. Phys. A **12**, L329 (1979); H.E. Stanley, J. Teixeira, J. Chem. Phys. **73**, 3404 (1980); H.E. Stanley, J. Teixeira, A. Geiger, R.L. Blumberg, Physica A **106**, 260 (1981)
1. 12 A. Geiger, H.E. Stanley, Phys. Rev. Lett. **49**, 1749 (1982); H.E. Stanley, R. L. Blumberg, A. Geiger, Phys. Rev. B **28**, 1626 (1983); H.E. Stanley, R. L. Blumberg, A. Geiger, P.Mausbach, J. Teixeira, J. de Physique **45**, C7[3] (1984)
1. 13 L. Bosio, J. Teixeira, H.E. Stanley, Phys. Rev. Lett. **46**, 597 (1981)
1. 14 Y. Xie, K. F. Ludwig, Jr., G. Morales, D. E. Hare, C. M. Sorensen, Phys. Rev. Lett. **71**, 2050 (1993)
1. 15 C. A. Angell, M. Oguni, W. J. Sichina, J. Phys. Chem. **86**, 998 (1982)
1. 16 A. Giacomini, J. Acoustical Soc. Am. **19**, 701 (1947)
1. 17 G.W. Willard, J. Acoustical Soc. Am. **19**, 235 (1947)
1. 18 R.L. Blumberg, H.E. Stanley, A. Geiger, P. Mausbach, J. Chem. Phys. **80**, 5230 (1984)
1. 19 E. Shiratani, M. Sasai, J. Chem. Phys. **104**, 7671 (1996)
1. 20 P.H. Poole, F. Sciortino, U. Essmann, H.E. Stanley, Nature **360**, 324 (1992)
1. 21 K. Mendelssohn: *The Quest for Absolute Zero: The Meaning of Low-Temperature Physics* (McGraw, New York 1966)
1. 22 P.H. Poole, F. Sciortino, T. Grande, H.E. Stanley, C.A. Angell, Phys. Rev. Lett. **73**, 1632 (1994); C. F. Tejero, M.Baus, Phys. Rev. **57**, 4821 (1998)
1. 23 S. Sastry, F. Sciortino, H.E. Stanley, J. Chem. Phys. **98**, 9863 (1993)
1. 24 S. S. Borick, P. G. Debenedetti, S. Sastry, J. Phys. Chem. **99**, 3781 (1995); S. S. Borick, P. G. Debenedetti, J. Phys. Chem. **97**, 6292 (1993)
1. 25 R.J. Speedy, J. Phys. Chem. **86**, 3002 (1989)
1. 26 E.G. Ponyatovskii, V.V. Sinitsyn, T.A. Pozdnyakova, JETP Lett. **60**, 360 (1994)
1. 27 C.J. Roberts, P.G. Debenedetti, J. Chem. Phys. **105**, 658 (1996)
1. 28 C.J. Roberts, A.Z. Panagiotopoulos, P.G. Debenedetti, Phys. Rev. Lett. **77**, 4386 (1996)
1. 29 F.H. Stillinger, A. Rahman, J. Chem. Phys. **57**, 1281 (1972)
1. 30 F.W. Starr, J.K. Nielsen, H.E. Stanley, Phys. Rev. Lett. **82**, 2294 (1999)
1. 31 W.L. Jorgensen, J. Chandrasekhar, J. Madura, R. W. Impey, M. Klein, J. Chem. Phys. **79**, 926 (1983)
1. 32 H.J. C. Berendsen, J.R. Grigera, T.P. Straatsma, J. Phys. Chem. **91**, 6269 (1987)
1. 33 U. Niesar, G. Corongiu, E. Clementi, G.R. Kneller, D. Bhattacharya, J. Phys. Chem. **94**, 7949 (1990)
1. 34 F. Sciortino, P.H. Poole, U. Essmann, H.E. Stanley, Phys. Rev. E **55**, 727 (1997)
1. 35 S. Sastry, P.G. Debenedetti, F. Sciortino, H.E. Stanley, Phys. Rev. E, **53**, 6144 (1996); L. P. N. Rebelo, P. G. Debenedetti, S. Sastry, J. Chem. Phys. **109**, 626, (1998)

1. 36 P. H. Poole, U. Essmann, F. Sciortino, H. E. Stanley, Phys. Rev. E **48**, 4605 (1993)
1. 37 P. H. Poole, F. Sciortino, U. Essmann, H.E. Stanley, Phys. Rev. E **48**, 3799 (1993)
1. 38 S. T. Harrington, R. Zhang, P. H. Poole, F. Sciortino, H.E. Stanley, Phys. Rev. Lett. **78**, 2409 (1997)
1. 39 M. C. Bellissent-Funel, L. Bosio, J. Chem. Phys. **102**, 3727 (1995)
1. 40 F. W. Starr, M.-C. Bellissent-Funel, H.E. Stanley, Phys. Rev. E **60**, 1084 (1999)
1. 41 A. Geiger, P. Mausbach , J. Schnitker, in *Water and Aqueous Solutions*, ed. by G. W. Neilson, J. E. Enderby (Adam Hilger, Bristol 1986) pp. 15
1. 42 S.T. Harrington, P.H. Poole, F. Sciortino, H.E. Stanley, J. Chem. Phys. **107**, 7443 (1997)
1. 43 F. Sciortino, P. H. Poole, H. E. Stanley, S. Havlin, Phys. Rev. Lett. **64**, 1686 (1990)
1. 44 A. Luzar, D. Chandler, Phys. Rev. Lett. **76**, 928 (1996); B. A. Luzar, D. Chandler, Nature **379**, 55 (1996)
1. 45 H. Larralde, F. Sciortino, H.E. Stanley, "Restructuring the Hydrogen Bond Network of Water" (preprint); H. Larralde, Ph.D. Thesis (Boston University, 1993)
1. 46 F. Sciortino, A. Geiger, H.E. Stanley, Phys. Rev. Lett. **65**, 3452 (1990)
1. 47 E. Shiratani, M. Sasai, J. Chem. Phys. **108**, 3264 (1998)
1. 48 F.W. Starr, C.A. Angell, R.J. Speedy, H.E. Stanley, "Entropy, Specific Heat, and Relaxation of Water at 1 atm between 136K and 236K" (preprint).
1. 49 F. Sciortino, A. Geiger, H.E. Stanley, Nature **354**, 218 (1991)
1. 50 F. Sciortino, A. Geiger, H.E. Stanley, J. Chem. Phys. **96**, 3857 (1992)
1. 51 P. Gallo, F. Sciortino, P. Tartaglia, S.-H. Chen, Phys. Rev. Lett. **76**, 2730 (1996); F. Sciortino, P. Gallo, P. Tartaglia, S.-H. Chen, Phys. Rev. E, **54**, 6331 (1996)
1. 52 H. Tanaka, Nature **380**, 328 (1996)
1. 53 H. Tanaka, J. Chem. Phys. **105**, 5099 (1996)
1. 54 T. Andrews, Philos. Trans. **159**, 575 (1869)
1. 55 *Water: A Comprehensive Treatise*, Vol. 1–7, ed. by F. Franks (Plenum Press, New York 1972); *Water Science Reviews*, Vol. 1–4, ed. by F. Franks (Cambridge University Press, Cambridge 1985)
1. 56 R. J. Speedy, J. Phys. Chem. **91**, 3354 (1987)
1. 57 C. A. Angell, Supercooled water, in: Water: A Comprehensive Treatise. ed. by F. Franks (Plenum Press, New York 1982)
1. 58 P.H. Poole, T. Grande, F. Sciortino, H.E. Stanley, C.A. Angell, J. Comp. Mat. Sci. **4**, 373 (1995)
1. 59 O. Mishima, J. Chem. Phys. **100**, 5910 (1994)
1. 60 O. Mishima, Nature **384**, 546 (1996)
1. 61 H. Kanno, R. Speedy, C. A. Angell, Science **189**, 880 (1975)
1. 62 P. W. Bridgman, Proc. Amer. Acad. Arts Sci. **47**, 441 (1912)
1. 63 L. F. Evans, J. Appl. Phys. **38**, 4930 (1967)
1. 64 O. Mishima, H.E. Stanley, Nature **392**, 164 (1998)
1. 65 O. Mishima , H.E. Stanley, "Discontinuity in Decompression-Induced Melting of Ice IV" [Proc. Symposium on Water and Ice, Int'l Conf. on High Pressure Science and Technology AIRAPT-16 & HPCJ-38] J. Rev. High Press. Sci. Tech. **6**, 1103 (1998)

1. 66 H.E. Stanley, S.T. Harrington, O. Mishima, P.H. Poole, F. Sciortino, "Cooperative Molecular Motions in Water: The Second Critical Point Hypothesis" [Proc. 1997 Symposium on Water and Ice, Int'l Conf. on High Pressure Science and Technology AIRAPT-16 & HPCJ-38], J. Rev. High Press. Sci. Tech. **7**, 1090 (1998)
1. 67 E. Lang, H.-D. Lüdemann, Ber. Bunsenges. Phys. Chem. **84**, 462 (1980)
1. 68 E. Lang, H.-D. Lüdemann, Ber. Bunsenges. Phys. Chem. **85**, 1016 (1981)
1. 69 E. Lang, H.-D. Lüdemann, J. Chem. Phys. **67**, 718 (1977)
1. 70 E. Lang, H.-D. Lüdemann, in *NMR Basic Principles and Progress* Vol.24 (Springer-Verlag, Berlin 1990) pp. 131-187
1. 71 P. C. Hemmer, G. Stell, Phys. Rev. Lett. **24**, 1284 (1970); G. Stell, P. C. Hemmer, J. Chem. Phys. **56**, 4274 (1972); C. K.Hall, G. Stell, Phys. Rev. A **7**, 1679 (1973).
1. 72 M.R. Sadr-Lahijany, A. Scala, S.V. Buldyrev, H.E. Stanley, Phys. Rev. Lett. **81**, 4895 (1998); M. R. Sadr-Lahijany, A. Scala, S.V. Buldyrev, H.E. Stanley, "Water-Like Anomalies for Core-Softened Models of Fluids: One Dimension" (preprint); A. Scala, M.R. Sadr-Lahijany, S.V. Buldyrev, H.E. Stanley, "Water-Like Anomalies for Core-Softened Models of Fluids: Two Dimensions" (preprint).
1. 73 M. Canpolat, F.W. Starr, M.R. Sadr-Lahijany, A. Scala, O. Mishima, S. Havlin, H.E. Stanley, Chem. Phys. Lett. **294**, 9 (1998); M. Canpolat, F.W. Starr, M.R. Sadr-Lahijany, A. Scala, O. Mishima, S. Havlin, H.E. Stanley, "Structural Heterogeneities and Density Maximum of Liquid Water" (preprint).
1. 74 G. E. Walrafen, J. Chem. Phys. **40**, 3249 (1964); G. E. Walrafen, J. Chem. Phys. **47**, 114, 1967; W. B. Monosmith and G. E. Walrafen, J. Chem. Phys. **81**, 669 (1984)
1. 75 M.C. Bellissent-Funel, Europhys. Lett. **42**, 161 (1998)
1. 76 F.W. Starr, S. Harrington, F. Sciortino, H.E. Stanley, Phys. Rev. Lett. **82**, 3629 (1999); F. W. Starr, F. Sciortino, H.E. Stanley, preprint.
1. 77 M. Meyer, H.E. Stanley, J. Phys. Chem. in press , 1999.
1. 78 P. H. Poole, M. Hemmati, C. A. Angell, Phys. Rev. Lett. **79**, 2281 (1997); H. E. Stanley, C. A. Angell, U. Essmann, M. Hemmati, P. H. Poole, F. Sciortino, Physica A **205**, 122 (1994)
1. 79 E.G. Ponyatovskii, JETP Lett. **66**, 281 (1997)

2 *Ab Initio* Theoretical Study of Water: Extension to Extreme Conditions

Fumio Hirata and Hirofumi Sato

Department of Theoretical Study, Institute for Molecular Science, Okazaki National Research Institutes, Okazaki 444-8585, Japan
E-mail: hirata@ims.ac.jp, hirofumi@ims.ac.jp

Abstract. The electronic, liquid structure of water and its thermodynamic properties including the ionic product (pK_w) are studied over a wide range of temperatures and densities based on *ab initio* molecular orbital theory combined with the integral equation method for a molecular liquid. For the neat liquid system, it is found that the molecular dipole moments and electronic polarization energies of a water molecule decrease with increasing temperature and/or density, being in quantitative accord with the experimental data determined based on the NMR chemical shift coupled with molecular dynamics simulation. The temperature and density dependence of the number of hydrogen-bonds is discussed in terms of the liquid structure of water. The pK_w obtained from the theory shows a monotonical decrease with increasing density at all the temperatures investigated, in good accord with experimental observation.

2.1 Introduction

'Water' is undoubtedly the most unique substance, and has attracted attention from many different branches of science, including chemistry, physics, biology, and their interdisciplinary fields [1]. The uniqueness of liquid water, represented by the existence of a temperature of maximum density, comes from the hydrogen-bond network, which in turn has its origin in the molecular structure and electronic properties. According to the well-established models proposed based on quantum-chemistry and molecular-simulation studies, a water molecule can be described by a classical model consisting of point charges embedded in a van der Waals sphere; two positive (partial) charges represent hydrogen atoms, and one or two negatively charged sites, depending on the model, mimic the negative charge distribution. The electrostatic interaction between oxygen and hydrogen sites gives rise to the hydrogen bond between a pair of molecules. The directionality of hydrogen bonds and the steric constraints due to volume-exclusion effects impose the optimum condition for the number of hydrogen bond. It is the origin of the tetrahedral structure which enables water molecules to make a three-dimensional network.

The electronic structure of a water molecule leads to the other unique properties of the substance: the auto-ionization. Water molecules in a liquid state dissociate into ionic species (OH^- and H_3O^+) by themselves, something like partially dissociated molten salt. Although the dissociation constant is very small in the ambient condition, this property of water is directly responsible for the variety of chemistries observed in the laboratory as well as in living systems. Therefore, the auto-ionization of water represents unique chemical properties for this substance.

All the unique properties of water have an ultimate origin in the electronic structure of the constituent molecules. It is also common understanding that the electronic structure of water molecules undergoes a significant change when the thermodynamic state alters from a gas to a liquid phase: the dipole moment of a molecule increases by approximately 38 %. Therefore, the unique properties of water can be a result of cooperative interplay between the electronic properties of individual molecules and the liquid structure promoted by the hydrogen-bonding. The auto-ionization of water molecules mentioned above can be considered in a sense as an extreme manifestation of such cooperative coupling or interplay.

Since the electronic structure depends sensitively on the thermodynamic states of the liquid, theoretical treatment of the liquid requires special care with regard to the intermolecular interactions on which all molecular theories are based. For example, typical models used in molecular simulations, such as SPC, TIPS and ST2, include atomic (partial) charges which have been adjusted such that the model gives the best agreement with experiments regarding some physicochemical observation in a particular thermodynamic state. Therefore, it is not entirely clear if the parameters are valid or not when the thermodynamic state is changed. It is particularly true when an extreme thermodynamic condition, such as the super-critical state, is considered. In such a condition, the state dependence of the molecular property may jeopardize all the theoretical results obtained, assuming fixed molecular parameters. All such considerations require a theoretical treatment which does not depend on an empirical determination of interaction parameters.

During last few years, we have developed a method which meets this demand. The method hybridizes the *ab initio* electronic structure theory (SCF) with the statistical mechanics of molecular liquids, or more specifically, the RISM theory. The method referred to as the RISM-SCF determines simultaneously the electronic properties and the liquid structure of water in a self-consistent manner. The self-consistent determination of the intra- and intermolecular properties enables us to treat the situation described above.

In this chapter, we present our theoretical study of water over wide range of densities and temperatures, including part of the super-critical state, based on the RISM-SCF method.

The chapter consists of two sections, both of which clarify the properties of water from a microscopic viewpoint: the first section puts the stress on the physical aspects of water, such as liquid structure, while the second section focuses on the chemical equilibrium in the liquid, or the auto-ionization.

2.2 Liquid Structure, Electronic and Thermodynamic Properties of Water

The modern structural study of liquid water was initiated by Bernal and Fowler, with the X-ray diffraction method [2], who first clarified the existence of the ice-I-like tetrahedral coordination in liquid water. This monumental work motivated succeeding studies, experimental as well as theoretical, to elucidate the structure of water and its relation to thermodynamic and dynamic properties of the liquid. The combined X-ray and neutron diffraction study finally established a picture concerning water structure in terms of the atom-atom pair correlation functions, the essential features of which remain valid [3].

The pair correlation function (PCF) between oxygen and hydrogen has a distinct peak at ≈1.8 Å, which is assigned as the hydrogen bond. The oxygen-oxygen PCF makes a second peak at a separation of around $2\sqrt{2}\sigma/\sqrt{3} = 1.63\sigma$ (σ = diameter of a molecule), which is characteristic of the ice-I-like tetrahedral coordination, and is regarded as the fingerprint of water structure. The first peak of the O-O PCF is rather sharp, from it the coordination number has been determined to be around 4.4 in ambient conditions. This number is very close to that of ice, 4, but not those typical to normal liquids, 10 to 11. A variety of models and theories regarding the origin of the difference in the coordination number, 0.4, between ice and water, have been proposed. The earlier theories of water assumed the existence of the tetrahedral ice-I-like coordination in one way or the other, and attributed the deviation in the coordination number from that of ice to some kind of imperfection in tetrahedral structure due to breaking or distortion of hydrogen bonds. Roughly speaking, three models have been proposed to explain the deviation in the coordination number: the bent hydrogen-bond model [4], the interstitial model [5] and the mixture model [6]. Each of these models explain relations between the structural characteristics of water and its thermodynamic properties to some extent. However, these theories cannot solve a more challenging problem: How is the characteristic structure of water formed from intermolecular interactions?

The problem can be resolved only by the theories based on first principles. One of the theoretical approaches based on the Hamiltonian model is molecular simulation, or molecular dynamics and Monte-Carlo methods. The methods have made great contribution to the heuristic understanding of structural, dynamical as well as some thermodynamic properties of water. The other set of theories which tries to explain water structure from first principles is the integral equation theories for molecular liquids. After several attempts based on the Ornstein-Zenike type equations [7, 8] three different approaches have successfully achieved the goal to the same level, at least qualitatively: the extended reference interaction site model (XRISM) [9], RHNC [10] and HNC with the central-force model [11]. Although all three approaches are able to qualitatively reproduce the experimentally observed structural features of water, XRISM proves to have a clear advantage over the other methods in terms of applicability and extensibility to chemistry and biophysical chemistry [12].

In this section, we extend the XRISM study of the liquid to a wide range of thermodynamic conditions, including the super-critical region. As has been emphasized in the preceding section, it is crucial to take the polarizability of water molecules into account in order to treat such a wide range of thermodynamic conditions. Unlike the classical polarizable model commonly employed in molecular simulation studies, we rely on quantum chemistry to describe intramolecular events. The method employed is the combination of RISM theory and *ab initio* MO method described in the previous section. The polarization and other many-body effects are automatically taken into account through the self-consistent procedure for determining the liquid structure and electronic properties of the system.

2.2.1 *Ab Initio* Polarizable Model of Water

The RISM-SCF procedure consists of the following steps [13-15]:
1. The electronic structure of an isolated water molecule is calculated with the standard MO method.
2. Effective charges of the atoms in the water are determined by the least-squares fitting procedure from the electronic structure.
3. The RISM equation for the neat liquid water system is solved to determine the pair correlation functions using the effective charges.
4. The microscopic reaction field on the atoms is calculated from the pair correlation functions using the same definition for the Fock operator in the original RISM-SCF procedure.
5. The electronic structure of the molecule is calculated under the influence of the microscopic reaction field.

Steps 2 to 5 are iterated until mutual consistencies in both the electronic structure and the pair correlation functions are achieved. This procedure enables us to take into account polarization and other many-body effects automatically. Computations were repeated under several thermodynamic conditions. All the calculations in this study are carried out with the quantum description in the Hartree-Fock level.

2.2.2 Electronic Polarization and Energetics

Temperature and Density Dependence of Electronic Polarization

In Fig. 2.1 we show the difference in the dipole moment of a water molecule between liquid and gas phases. Since the electronic polarization depends on the density and/or temperature due to changes in thermal motions of water molecules and/or in the intermolecular interactions, the dipole moment of a water molecule in liquid directly reflects the electronic polarization due to surrounding molecules. At room temperature and standard density, the dipole moment is increased by about 0.7 Debye from that in the gas phase, which can be compared with the value widely known experimentally and theoretically: $2.5 - 1.8 = 0.7$ Debye [16]. With increasing temperature and decreasing density, the polarization effect becomes less, and the electronic structure of a water molecule approaches that in the gas phase. The decreased effect is attributed to the activation of molecular motion, especially to rotational motion. As the rotational motion of a water molecule increases, the average electric field produced by the molecule becomes weaker and less anisotropic. The reduced electric field polarizes surrounding media less, which in turn gives rise a reduced reaction field for the molecule concerned. Our theoretical results for average change in the dipole moment of a molecule due to solvent as well as its density and temperature dependence show quantitative accord with those determined by the new empirical method proposed by Matubayashi et al [17].

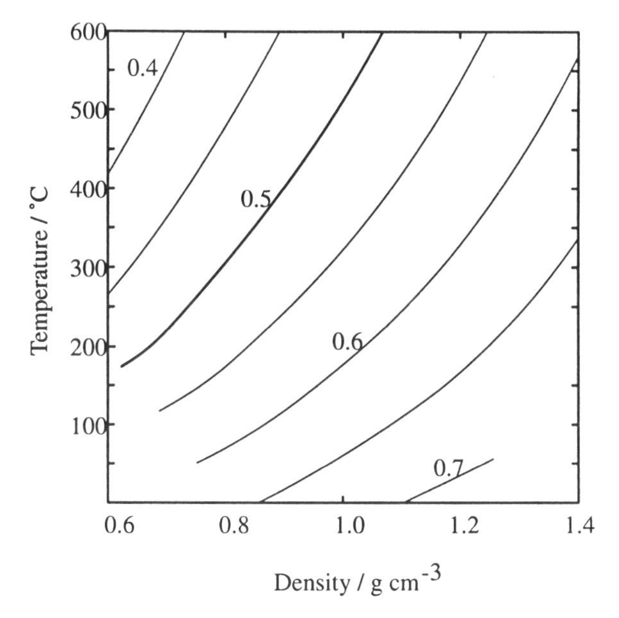

Fig. 2.1. Deviation of the dipole moment in a liquid relative to the value in the gas phase. Unit is Debye. It is worth noting the left and right corners in the bottom of the figure, where no values are plotted. These regions are outside of coexistence lines in the actual liquid system. Convergence of RISM calculations could not be attained in these regions

Energetics of the Liquid Water System

The total free energy of the system is defined as the sum of three terms: the electronic-energy change associated with the electronic polarization due to the solvation (ΔE^{elec}), the solvation free energy (excess chemical potential: $\Delta\mu$) and the ideal part of the chemical potential (μ^{id}). Among these three contributions, the profile of ΔE^{elec} is very similar to that of the dipole moment change, as is expected (not shown). In ambient conditions, the energy due to distortion of the electronic structure amounts to about 4.0 kcal/mol, but the value is reduced to less than 2.0 kcal/mol in low density and high temperature conditions.

The total free energy shows a marked dependency upon density but is rather insensitive to a temperature change (Fig. 2.2). Unfortunately, we could not find any experimental data for the free energy change in the extended range of temperature and density corresponding to the present study. However, we believe the results reported here predict experimental observations at least qualitatively in light of our previous study, in which the theory reproduced fairy well the experimental results for the dependence of the free energy on temperature in the range from 0°C to 100°C [15].

Solvation free energy, shown in Fig. 2.3, attributed to the intermolecular interaction increases monotonically as temperature increase. This can be explained in terms of the dominant contribution from the thermal disruption of the hydrogen bonding. In the high-density region, the oxygen core-core repulsion, or the packing effect, is dominant in the intermolcular interaction, and the hydrogen-bond interaction is no longer important. This point will be discussed later in some detail.

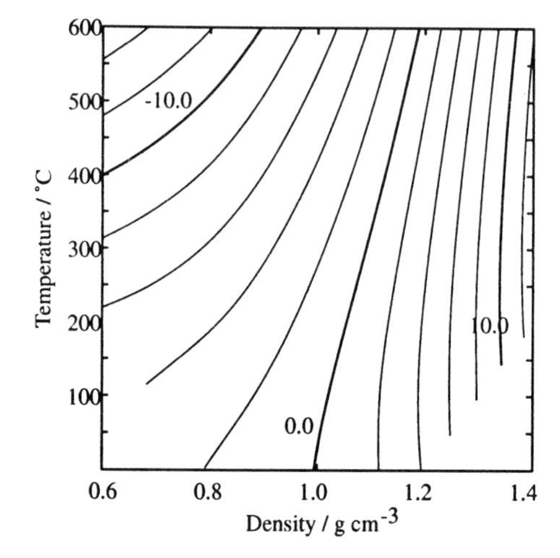

Fig. 2.2. Total free energy of liquid water. The contour spacing is 2.0 kcal/mol. See the comment in Fig. 2.1 regarding the lack of data in the bottom corners

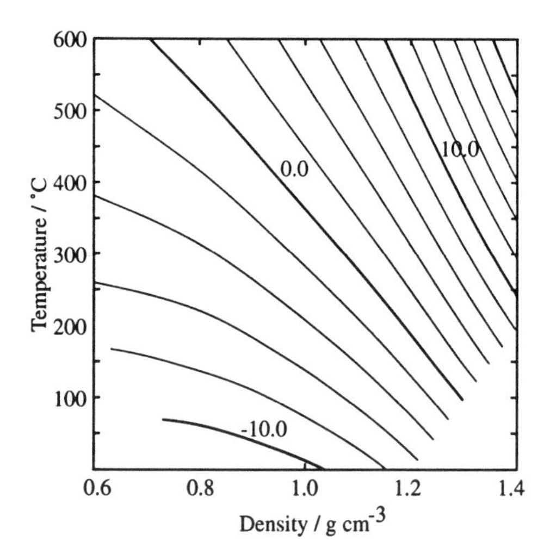

Fig. 2.3. Solvation free energy (excess chemical potential) of liquid water. Contour spacing is 2.0 kcal/mol. See the comment in Fig. 2.1 regarding the lack fo data in the bottom corners

2.2.3 Solvation Structure of Liquid Water

Temperature and Density Dependence of the Oxygen-Oxygen Pair Correlation Functions

The temperature dependence of the O–O PCFs at the three different densities is shown in Fig 2.4. The temperature ranges from 0°C to 600°C in the case of normal density (1.0 g/cm^3), and 200°C to 600°C both in the high (1.4 g/cm^3) and the low (0.6 g/cm^3) density conditions. Let us first focus on the case of density = 1.0 g/cm^3. The characteristics of water structure stated above is clearly seen at T = 0°C: the narrow first peak and a discernible shoulder at $R \approx 4.5$ ($\approx 1.63\sigma$) Å, which are the manifestation of the tetrahedral ice-like configuration. As temperature increases, the ice-like structure is taken over more and more by a configuration characteristic of simple liquids, in which the second peak appears at $R \approx 2\ \sigma$. Since the oscillation of the distribution functions corresponding to these two configurations are out of phase, isosbestic-like points are conspicuous.

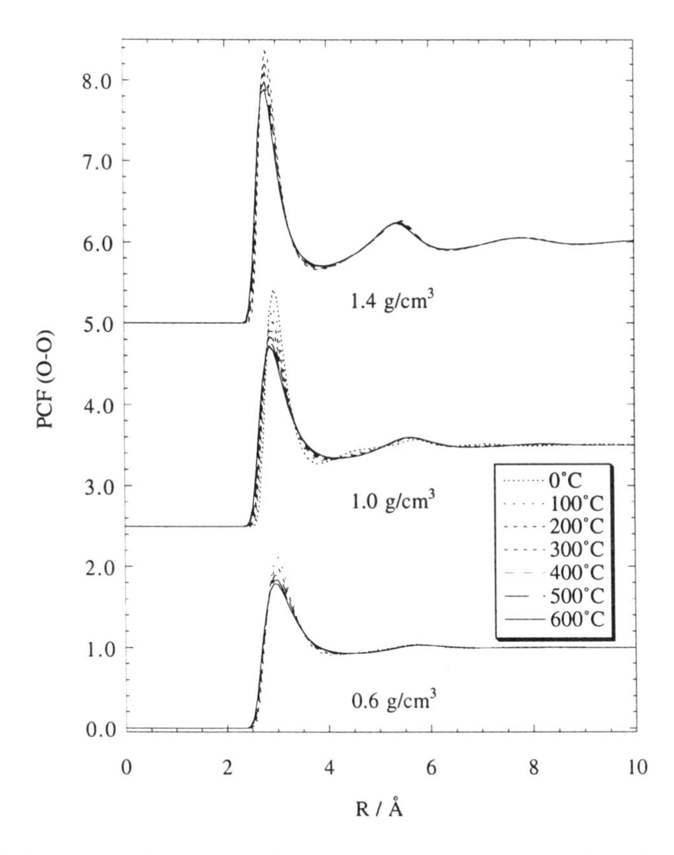

Fig. 2.4. Temperature dependence of O–O PCFs in liquid water at various densities. Note that temperature varies from 0 to 600°C at 1.0 g/cm^3, while it ranges from 200 to 600°C at 0.6 and 1.4 g/cm^3

It is instructive to compare the O–O PCFs with a Lennard-Jones (L–J) fluid, in which packing effect is dominant. The L–J fluid is a hypothetical liquid in which the L-J parameter for a pair of molecules is the same as that between an oxygen pair of water. Plotted in Fig. 2.5 are the first peak positions of the O–O PCFs (solid lines) and those corresponding to the L–J liquid (dotted lines) against the cubic root of the density. The x-axis measures the average distance, between molecules in the liquid, while the y-axis signifies the nearest neighbor distance reflecting all the interactions among molecules, attractive and repulsive, including many-body effects. The experimental results for water obtained by an X-ray diffraction study [18, 19] are also plotted crosses. At higher density (left hand side), the peak positions of water and the L-J liquid are close each other, indicating the importance of the packing effect. Both of their peak positions shift to a larger distance as the density decreases (right-hand side), but the positions in water are rather insensitive to the density change compared to those in the L–J fluid. This is because that the distance of closest approach between molecules in water is primarily determined by a balance of two forces, the core repulsion and the hydrogen bond; the core repulsion due to the many-body packing effect becomes important only at high density. In contrast, the distance in L–J fluids is determined essentially by the many-body packing effect, which of course is very sensitive to density: the distance increases rapidly as the packing effect disappears with decreasing density. The results are in qualitative accord with corresponding experimental data for water, indicating that this is a good quantity to characterize water structure in the wide range of density.

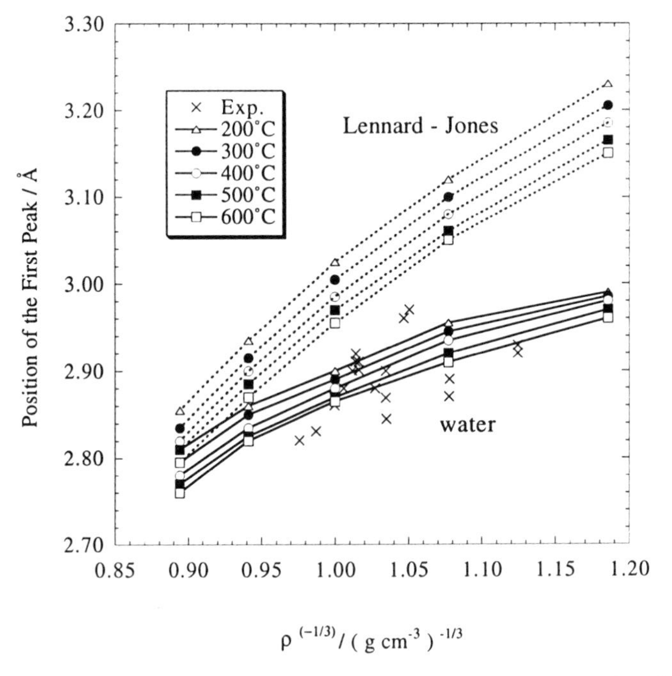

Fig. 2.5. The nearest O–O distance plotted against cubic root of the density. Water (*solid lines*), and a Lennard-Jones fluid (*dotted lines*) are shown with corresponding data obtained from an X-ray diffraction method for water (×)

Temperature and Density Dependence of O–H Pair Correlation Function

In Fig. 2.6 the temperature dependence of PCFs between oxygen and hydrogen at various densities is shown. Distinct peaks are found in the PCFs at around 1.8 Å at all densities, which are attributed to the hydrogen bond of an oxygen atom in a water molecule with a hydrogen atom in the nearest neighbor. Broad peaks positioned around ≈3.5 to ≈3.8 Å, depending on density and temperature, can be assigned to the other hydrogen atom of the nearest neighbor which is not involved in the hydrogen bond. The decrease in the height of the hydrogen-bond peak with increasing temperature, which is commonly observed at all densities, is due to disruption or weakening of the hydrogen bond, which in turn is caused by the concerted effect of enhanced thermal motion and reduced electronic polarization of water molecules. The decrease in the second peak with temperature is mainly due to increased thermal motion.

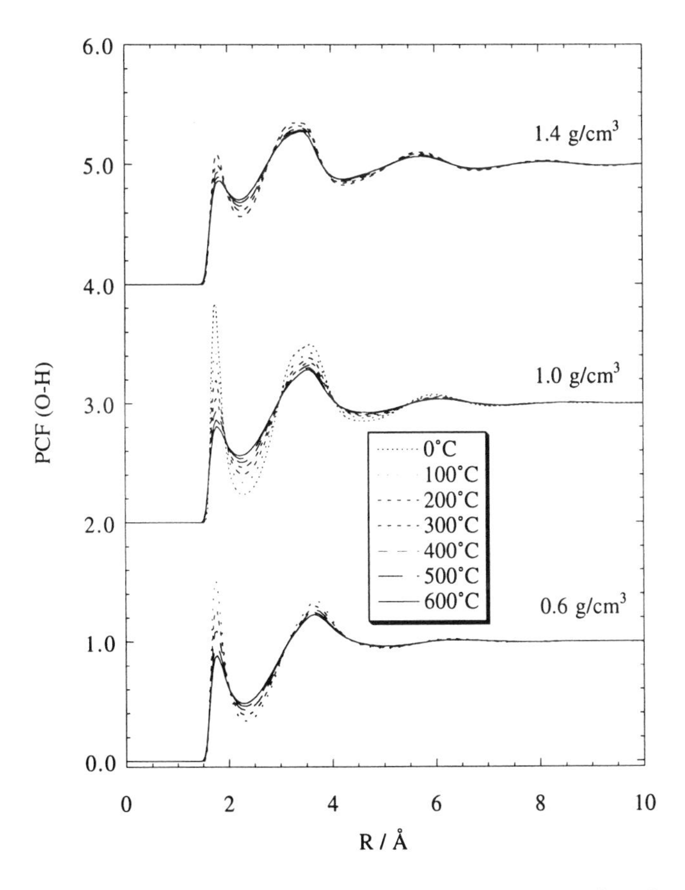

Fig. 2.6. Temperature dependence of O–H PCFs in liquid water at various density. Note that temperature varies from 0 to 600°C at 1.0 g/cm^3, while it ranges from 200 to 600°C at 0.6 and 1.4 g/cm^3

There is an interesting point to be noted concerning density dependence on the strength of the hydrogen bond between a pair of water molecules. The hydrogen bond at lower density (0.6 g/cm^3) is stronger than that at normal density (1.0 g/cm^3) at the same temperature. In contrast, the bond at higher density (1.4 g/cm^3) is weaker than that at normal density. This behavior cannot be explained from the change in the electronic polarization discussed earlier associated with Fig. 2.1, because a change in the electronic polarization predicts just the opposite trend: the higher the density, the greater the polarization. In general, the density increase will give rise to two effects on the hydrogen bond in water: an increase in the molecular polarization and an enhancement of 'packing' effect. The former promotes the hydrogen bond, while the latter disturbs the bond sterically. The strength of the hydrogen bond is determined by an interplay between those two effects with increasing density. At higher density, the packing effect dominates and weakens the hydrogen bond. In contrast, at lower density, the electronic polarization effect dominates over the packing effect to make the hydrogen bond stronger, although the polarization itself is less than that at higher density. The statement should not be confused with the discussion concerning the extension of the hydrogen-bond network. The above conclusion applies only to the hydrogen bond between a pair of neighboring water molecules. For the extension of hydrogen bond network, a better measure should be the correlation length in the O–O pair correlation functions.

2.2.4 Hydrogen Bonding in Liquid Water

Number of Hydrogen Bonds

Since the definition of the bond is not entirely free from ambiguity, determining the number of hydrogen bonds in water is always a tricky business. A definition which is most commonly employed is (definition I)

$$N_I = \rho \int_0^{R_{\min}} 4\pi r^2 g_{\mathrm{OH}}(r)dr \, , \tag{2.1}$$

where ρ is the number density of hydrogen atoms and R_{min} is the first minimum of the O–H PCF [$g_{\mathrm{OH}}(r)$]. Although the analysis appears to account well for the number of hydrogen bonds, there is some concern about the definition. The above definition does not separate well the number of hydrogen atoms which actually make hydrogen bonds from those which are just in close contact with an oxygen atom without making bonds. Here, we propose another definition of the number of hydrogen bonds, which gives reasonable results for the number of "real" hydrogen bonds.

In a recent study, we introduced a function concerning the response of the PCF to temperature, which clarifies the liquid structure of water. The function S_{OH} signifies a response of the PCFs to temperature change at constant density and is defined as follows:

$$S_{OH}{}^{i,j}(r) = \frac{1}{T_i - T_j}\left[g^{\,i}_{OH}(r) - g^{\,j}_{OH}(r)\right], \tag{2.2}$$

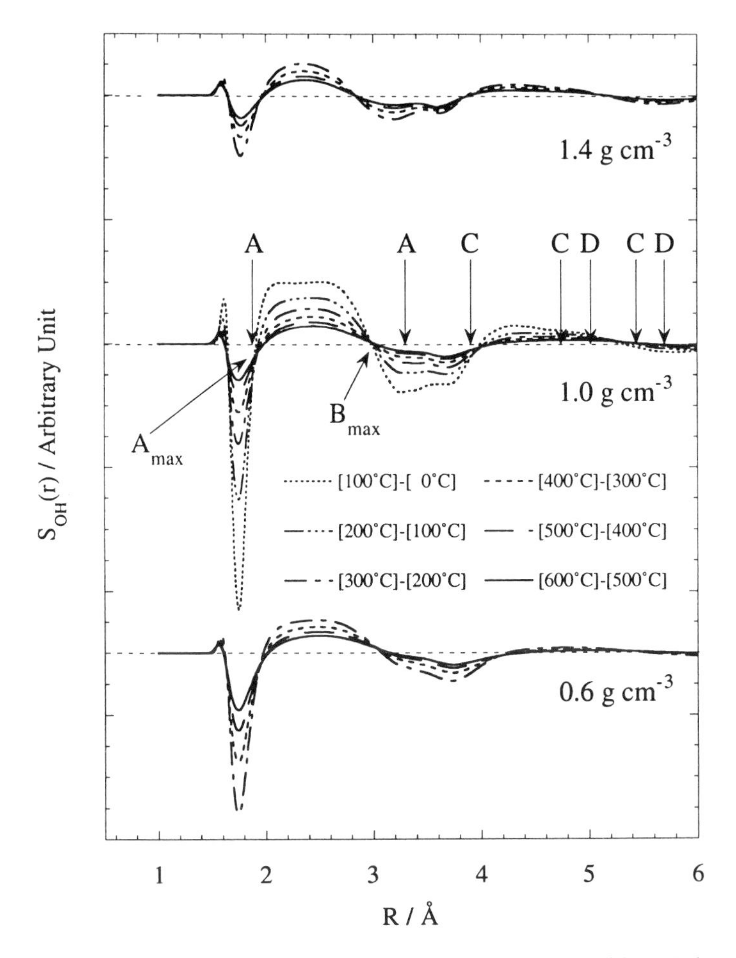

Fig. 2.7. Response of PCF to temperature change, $S_{OH}(r)$, at various densities. Hydrogen atom positions of the ice-I structure measured in the radial direction from a molecule at the origin are shown (arrows)

where *i* and *j* are the temperatures at which the RISM calculations were performed. S_{OH} at all the temperatures is plotted in Fig. 2.7. As we showed in our previous study, this function enables us to separate the actual number of hydrogen atoms involved in hydrogen bonds, N_{II}^{HB}, and that of interstitial molecules, N_{II}^{NB} (definition II).

$$N_{II}^{HB} = \rho \int_0^{A_{max}} 4\pi r^2 g_{OH}(r) dr \tag{2.3}$$

and

$$N_{II}^{NB} = \rho \int_{A_{max}}^{B_{max}} 4\pi r^2 g_{OH}(r) dr. \tag{2.4}$$

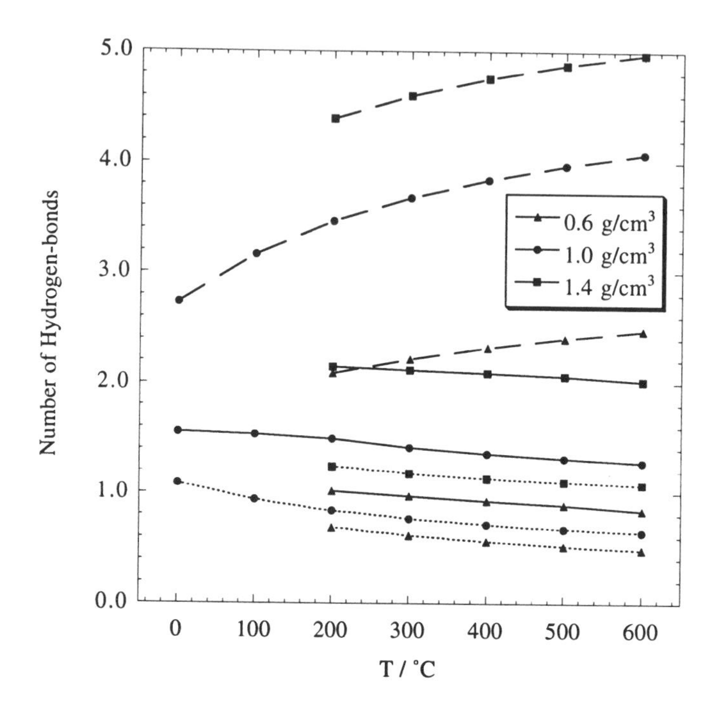

Fig. 2.8. Temperature and density denpendence of the number of hydrogen bonds. Solid lines show the numbers by "definition I". The dotted and dashed lines show the number by "definition II". Boundaries used in this computations are 1.995 and 3.110 Å (0.6 g/cm^3), 1.943 and 3.003 Å (1.0 g/cm^3), and 1.980 and 2.845 Å (1.4 g/cm^3)

The lower and upper limits of the integrals are defined by the points at which S_{OH} intersects the x-axis. The positions of hydrogens estimated from geometrical consideration of ice-structure are marked by arrows in Fig. 2.7. It is noteworthy that existence of hydrogen atoms is not expected between 2 and 3 Å in the ice-I structure. The region denoted as "A" can be assigned to the hydrogen atoms in the nearest water molecule which forms hydrogen bonds, while the region corresponding to "B" is due to hydrogen atoms in the "interstitial" water molecules, which are just close in contact and not making hydrogen bonds.

N_I , N_{II}^{HB} and N_{II}^{NB} are plotted against temperature in Fig. 2.8 for different densities. The values for an oxygen are less than two, which is in good accordance with estimation in other studies (1.6–1.9 [20]) at ambient conditions. The number of hydrogen atoms bonded to an oxygen atom decreases with increasing temperature and/or decreasing density. The behavior is consistent with the experimental results, and the temperature dependence is attributed to the thermal activation of intermolecular motions. As expected, N_{II}^{HB} is slightly smaller than N_I, but the difference of the numbers due to the two definitions are not so large. N_{II}^{NB} increases monotonically with temperature for all the densities, which is consistent with the intuitive picture of a hydrogen-bond network breaking with increased thermal motions.

2.3 Theoretical Prediction of pK_w

The auto-ionization of water in aqueous solution,

$$H_2O + H_2O \Leftrightarrow H_3O^+ + OH^-, \tag{2.5}$$

is a reaction which bears fundamental importance in various fields of chemistry, biochemistry and biology. The ionic product, $K_w = [H_3O^+][OH^-]$, and its logarithm, pK_w = –log K_w, are not only measures of the equilibrium, but the standard quantities for the protonation (or deprotonation) processes of other chemical species in aqueous solutions. This quantity can be related to the free energy change (ΔG) associated with (2.5) by the standard thermodynamic relations:

$$\Delta G = 2.303RT\,\mathrm{p}K_w. \tag{2.6}$$

pK_w measured in the extended thermodynamic conditions shows remarkable behavior: the value changes over a wide range, from 2 to as high as 15. Its temperature dependence along isobaric lines show a minimum, that is, pK_w first decreases and then increases with rising temperature [21, 22]. A microscopic interpretation of the behavior is by no means simple. Around room temperature, pK_w decreases with increasing temperature, or the equilibrium of (2.5) leans towards the right-hand side. This observation apparently contradicts our intuitive consideration of the reaction: the reaction field including the hydrogen-bond exerted on a water molecule should be reduced due to activated thermal motion with rising temperature. Why should pK_w decrease with increasing temperature and density in the first place? And then, why has it a minimum? It is the main purpose of the present study to answer these questions from a microscopic point of view.

In this section, we describe the *ab initio* investigation for pK_w over a wide range of thermodynamical conditions, temperature and density. Determination of pK_w as a function of thermodynamical conditions is deeply concerned with the electronic structure of the chemical species involved in the reaction. However, stability of the species in solution is also another important factor in determining the reaction, to which the solvation free energy makes an essential contribution. Therefore, the problem inevitably involves analyses concerning the combined effects of the electronic-structure change associated with the reaction and of the influence of solvent on the reacting species.

The theoretical treatment of liquid water including its electronic structure is one of important issues in modern physical chemistry, and numerous studies based on the dielectric continuum model, molecular simulations, and integral equation theories have been reported [23–25]. However, to our knowledge, non-empirical determination of pK_w and its thermodynamical variations has not been reported except for our study [26], because of the difficulties involved with accurate description of both electronic structure and solvation.

2.3.1 Description of the Auto-ionization Process in Water

We consider the thermodynamic cycle of the free energy changes in aqueous solution (ΔG^{aq}) associated with the auto-ionization reaction illustrated in Fig. 2.9.

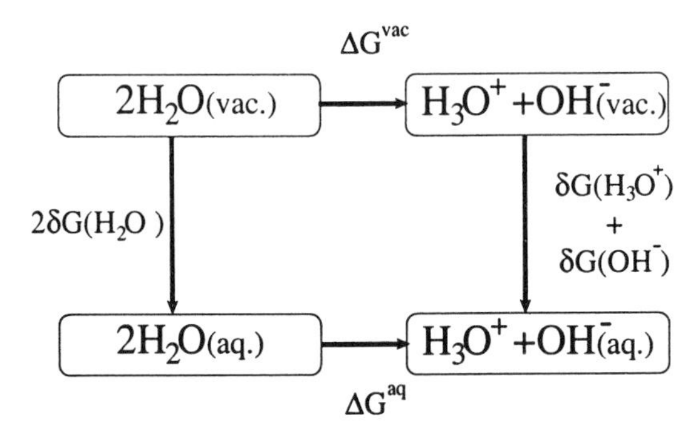

Fig. 2.9. Thermodynamic cycle of the free energy change in auto-ionization process of water

We regard the species involved in the reaction (i.e., H_3O^+, H_2O, OH^-) as "solutes" in aqueous solutions at infinite dilution. Hereafter, we use "Δ" for changes of quantities associated with the chemical reaction and "δ" for changes due to the solvation. Note that both these quantities depend upon temperature and density. Thus, ΔG^{aq} can be written in terms of the free energy changes associated with the reaction in vacuo (ΔG^{vac}) and those due to the solvation of reacting species [$\delta G(H_2O)$, $\delta G(OH^-)$, $\delta G(H_3O^+)$]:

$$\Delta G^{aq} = \Delta G^{vac} + \delta G(H_3O^+) + \delta G(OH^-) - 2\delta G(H_2O). \tag{2.7}$$

The energy change can be also rewritten as

$$\Delta G^{aq} = \Delta E^{vac}_{elec} + \Delta\delta E_{reorg} + \Delta\delta\mu + \Delta G^{vac}_{kinetic}, \tag{2.8}$$

where ΔE^{vac}_{elec} is the potential energy difference obtained by standard *ab initio* MO calculations and concerns the electronic structures of the isolated species in the gas phase. $\Delta\delta\mu$ corresponds to difference in the solvation free energies between the reactant and product systems. Electronic distortion or polarization upon solvation is described by the electronic reorganization energy, δE_{reorg}, and its change in the auto-ionization process is given by $\Delta\delta E_{reorg}$. These quantities are calculated in the course of a self-consistent determination of the electronic structure and solvent distribution by means of the RISM-SCF procedure. $\Delta\delta G^{vac}_{kinetic}$ is the kinetic contribution (translational, rotational and vibrational), which can be readily obtained from the elementary statistical mechanics.

Since our primary concern in the present study is the temperature and density dependence of pK_w, we concentrate our attention on the quantities relative to pK_w, at temperature $T = 273.15$K and density $\rho = 1.0$ g/cm^3, that is,

$$\Delta pK_w(\rho,T) = pK_w(\rho,T) - pK_w(\rho = 1.0, T = 273.15)\,. \tag{2.9}$$

The quantity is also decoupled into four components corresponding to the contributions from respective free energy changes.

$$\begin{aligned}\Delta pK_w(\rho,T) = {} & \Delta pK^{vac}_{w,elec}(\rho,T) + \Delta pK_{w,reorg}(\rho,T) \\ & + \Delta pK_{w,\mu}(\rho,T) + \Delta pK_{w,kin}(\rho,T)\end{aligned} \tag{2.10}$$

2.3.2 Solvation Structure of H_2O, H_3O^+, and OH^-

PCFs of Solute Oxygen and Solvent Hydrogen

The 'solute'-solvent PCFs between O (solute) and H (solvent) atoms, and their dependence on temperature and density, are of particular interest, because they are believed to be directly related to the behavior of pK_w. The O (solute) – H (solvent) PCFs are shown in Fig. 2.10 for T = 200°C. The extremely well-defined peak around R = 1.8 Å in the PCF of the O (OH^-)-H (H_2O) pair represents the hydrogen bond between those atoms, which is enhanced by molecular polarization of OH^-, as will be discussed later. The peak also shows remarkable density and temperature dependence, as one can easily see from the figure. Compared to the OH^- case, the PCFs for H_2O exhibit rather weak hydrogen-bonding, and only a weak dependence on density and temperature. The hydrogen-bond peak entirely disappears in the case of H_3O^+, because the positive charge in H_3O^+ repels the positive partial charges of water hydrogen electrostatically.

The density dependence of the height as well as the position of the first peaks can be explained in terms of an interplay of the two forces acting between the O (solute) and H (solvent) atoms: the hydrogen bond originating from the electrostatic interaction between the partial charges on the two atoms, and the many-body packing effects due essentially to the core repulsions. For OH^-, the hydrogen bond dominates over the packing effect across the entire density range, due to enhanced electronic polarization of the species, and the first-peak height decreases monotonically with decreasing density. These two effects are in balance for H_2O: the packing effect dominates at the higher density, while the hydrogen bond becomes superior in the lower density. Since there is no hydrogen bond between the two atoms for H_3O^+, the first-peak height and position are determined by the packing effect as well as the electrostatic repulsions between solute and solvent. The peak position shifts toward a greater distance as density decreases due to the electrostatic replusions. The essential nature of the density dependence remains the same at higher temperature, although the effect from hydrogen bonding becomes less important due to enhanced thermal motions (not shown).

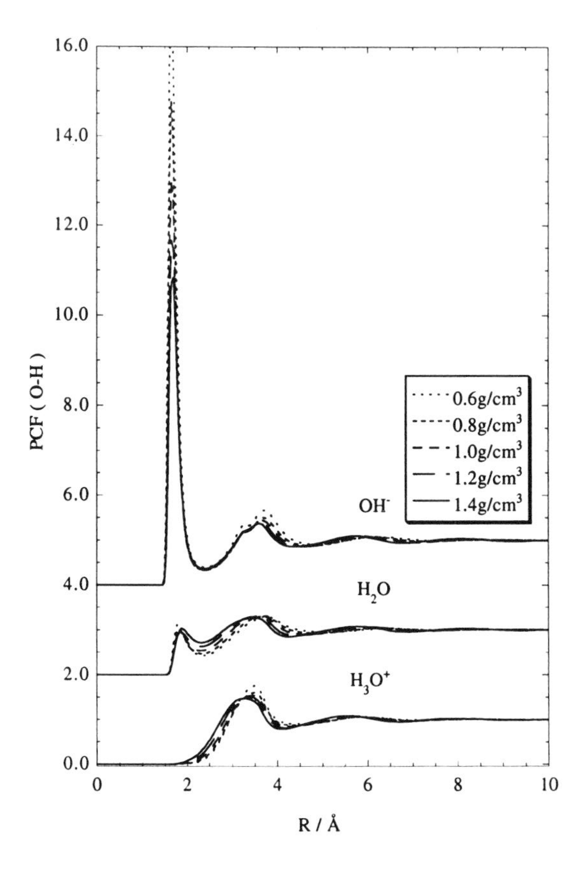

Fig. 2.10. Density dependence of O(solute) – H(solvent) PCFs at 200°C

PCFs of Solute Hydrogen and Solvent Oxygen

Dependence of the H (solute)-O (solvent) PCFs upon density can be explained in a similar manner with the case of O(solute)-H(solvent) in terms of the interplay between the two forces (Fig. 2.11).For OH^-, the two atoms, H (OH^-) and O (solvent), do not make hydrogen bonds due to the electrostatic repulsion between OH^- and O (solvent). Therefore, the peak height and position of the PCF is entirely determined by the many-body packing effect and the electrostatic repulsion. At higher density, the packing effect pushes the two atoms close together, while the two atoms are repelled farther apart at lower density. This is apparent in the PCF by the shift of the population toward a greater distance as density decreases. The two effects are in competition for H_2O as in the case of O(solute)-H(solvent) PCFs. In fact, behaviors in the two cases should be exactly the same if both the solute and solvent water molecules are polarizable. However, the water molecules as solvent and 'solute' are treated differently in the present model; therefore the behaviors are not necessarily the same. Nevertheless, essential features of the two cases are very similar, and it is not necessary to repeat the discussion. For H_3O^+, the first peak height and position

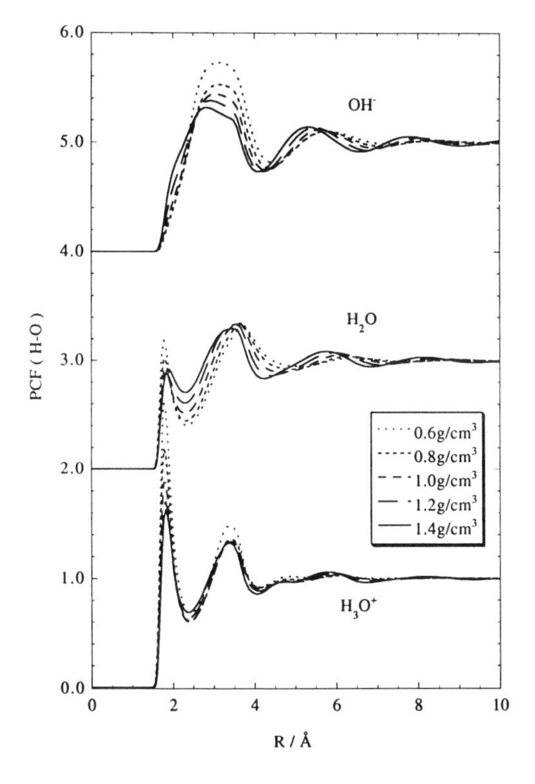

Fig. 2.11. Density dependence of H(solute) – O(solvent) PCFs at 200°C

are completely dominated by hydrogen bonding, and the peak height increases monotonically with decreasing density, although the behavior is not as prominent as is the case for O (OH^-) and H (solvent). The difference in the behavior between the two cases has its origin in the electronic structures of the two species, OH^- and H_3O^+, as has been discussed in the previous section.

PCFs of Solute Oxygen and Solvent Oxygen

The 'solute'-solvent PCFs between oxygen atoms are shown in Fig.2.12. Comparing the O–O PCFs for the three solute species, H_3O^+, H_2O and OH^-, at T = 200°C, one finds some variations in their density dependence. As density decreases, width of the first peak in the O–O PCFs broadness and its height lowers in a monotonic manner for the solute H_2O. The position of the first peak also shifts monotonically toward greater distance. In contrast, the PCFs do not show such a monotonic behavior for the other species, OH^- and H_3O^+. In both cases, the peak height first lowers and then it turns upward with decreasing density. The behavior can be explained in terms of an interplay between essentially two different 'solvation' structures, the contributions of which alter as density decreases. The first of those is the characteristic of a simple fluid in which the packing effect dominates the configuration, which is favored at higher density. The other structure is due to

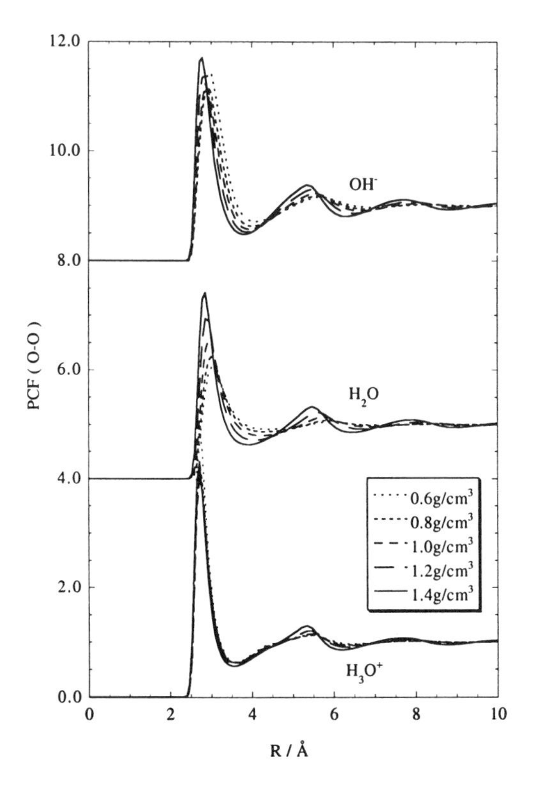

Fig. 2.12. Density dependence of O(solute) – O(solvent) PCFs at 200°C

hydrogen bonding between solute and solvent molecules, the contribution of which is greater at lower density. For H_2O, the packing effect dominates in this density range, and the height of first peak shows monotonic lowering with decreasing density. For the other species, OH^- and H_3O^+, the strength of the two configurations alters as density changes. As temperature rises, the contribution from hydrogen bonding becomes less compared to that from the packing effect.

It is worthwhile to mention an essential difference in the solvation structures among the three solute species, which can be deduced from the above discussion. Solvent water is characterized by a well-developed, hydrogen-bond network structure with a tetrahedral coordination. H_2O as a solute can be naturally a member of the three-dimensional network. However, as we have seen above, the solute species, OH^- and H_3O^+, can make hydrogen bonds only in one direction with solvent water, with either H or O.

2.3.3 Free Energy, Its Components and pK_w

Electronic Reorganization Energy

Temperature and density dependence of the electronic structure of the solute species, represented by δE_{reorg}, is shown for H_3O^+ (Fig. 2.13), H_2O (Fig. 2.14) and OH^- (Fig. 2.15).

In general, polarization or distortion of the electronic structure is reduced with rising temperature or decreasing density, and approaches the electronic structure in the isolated molecule. There are three factors which contribute to the temperature and density dependence of the electronic structure to be considered: the hydrogen bonding between solute and solvent, the many-body packing effect, and the solvent polarization around the solutes. By solvent polarization, we mean the sum of dipole moments of solvent per unit volume, which should not be confused with the electronic polarization of a molecule. Combination of the factors creates complicated behavior in the temperature and density dependence. For instance, the hydrogen bonding becomes weak with rising temperature, but strengthens with decreasing density. In contrast, the solvent polarization is enhanced with increasing density for an obvious reason. The partial charges on oxygen in all the species are reduced with rising temperature because the hydrogen bonds and the solvent polarization, and thereby the reaction field of the solvent, are reduced due to enhanced thermal motions. The charges also become less with decreasing density because the solvent polarization around the solute reduces with the density change.

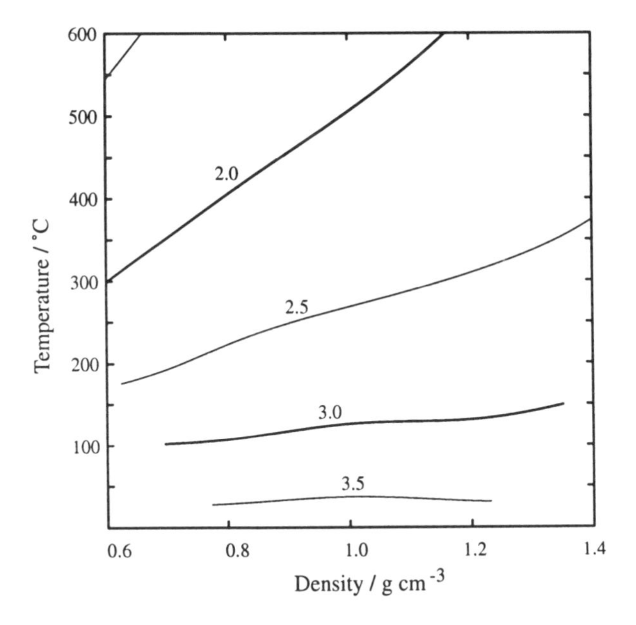

Fig. 2.13. Temperature and density dependence of the electronic reorganization energy in H_3O^+. Contour spacing is 0.5 kcal/mol

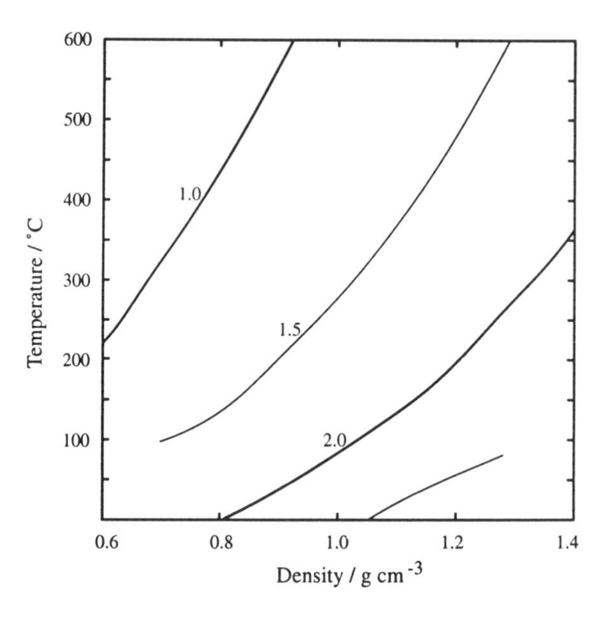

Fig. 2.14. Temperature and density dependence of the electronic reorganization energy in H_2O. Contour spacing is 0.5 kcal/mol

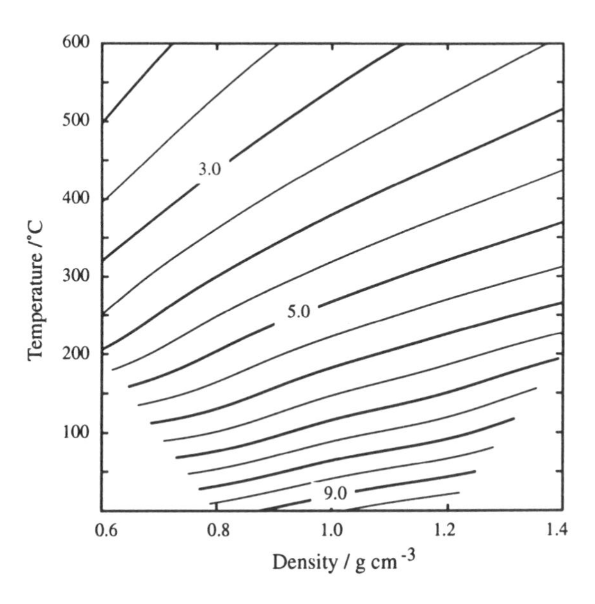

Fig. 2.15. Temperature and density dependence of the electronic reorganization energy in OH^-. Contour spacing is 0.5 kcal/mol

It is noteworthy to make a remark on the features of the contour maps for the three solute species; the spacing of the contours for OH^- is much less than those in H_2O and H_3O^+, indicating the higher sensitivity of the electronic structure of OH^- to density and temperature compared to the other species. As pointed out in a previous study [26], the remarkable sensitivity of the electronic structure to temperature compared to the other species plays an important role in determining the temperature dependence of pK_w at around room temperature.

Solvation Free Energy

Solvation free energies of the three species in solvent water are plotted in Fig. 2.16 (H_3O^+), Fig. 2.17 (H_2O) and Fig. 2.18 (OH^-). Several features can be seen in the figures. The two ionic species obtain much greater stabilization due to solvation compared to H_2O. The free energy increases with increasing temperature as well as density for H_2O, thereby the contours map diagonally from the bottom-left to the top-right corner. In the case of OH^-, the contours map from the bottom-center (1.0 g/cm^3) to both the top-left and -right corners, indicating that the free energy increases as density either increases or decreases at higher temperature. Although it is not so conspicuous, a similar trend can be seen in the case of H_3O^+.

The features seen in the contour maps can be interpreted in terms of the three factors in the solute-solvent interaction, which we have discussed in the previous subsections: the hydrogen bonding, the many-body packing effect, and the solvent polarization around the solutes. The hydrogen bonding stabilizes the solutes, and its disruption gives rise to the positive free energy change. The packing effect contributes to the free energy of cavity formation, and makes a positive contribution.

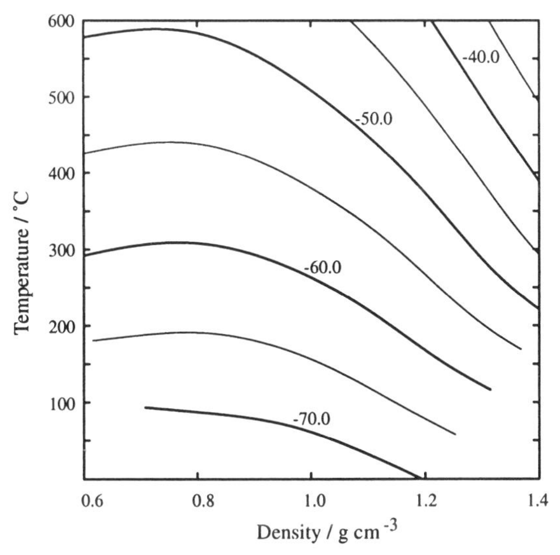

Fig. 2.16. Temperature and density dependence of the solvation free energy of H_3O^+. Contour spacing is 5 kcal/mol

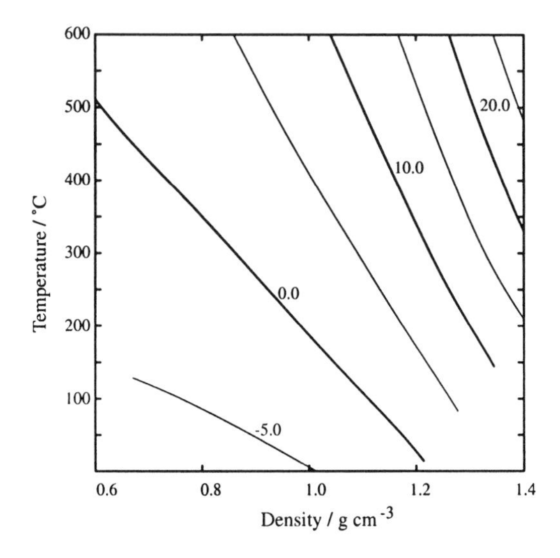

Fig. 2.17. Temperature and density dependence of the solvation free energy of H_2O. Contour spacing is 5 kcal/mol

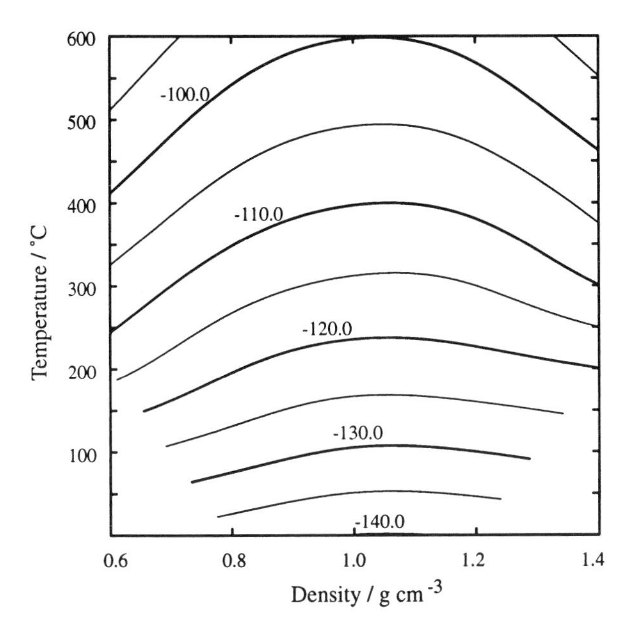

Fig. 2.18. Temperature and density dependence of the solvation free energy of OH^-. Contour spacing is 5 kcal/mol

It therefore becomes larger as density increases. The solvent polarization couples electrostatically with charges on the solutes, and makes a negative contribution to the free energy. In the case of H_2O, the first two effects are dominant and compete against each other. At lower temperature and density, the contribution from the hydrogen bonding dominates, while that from the packing effect is greater at higher temperature and density. Therefore, the contours map diagonally from the bottom left to top right. In the case of OH^-, the three effects stated above contribute with more or less similar strengths, and create the rather complicated pattern in the contour map. An increasing in temperature alone gives rise to positive changes in the free energy. However, either decreasing or increasing density makes positive contributions to the free energy, due respectively to the reduction of coupling between solute charge and solvent polarization, or increase in the energy of cavity formation. This explains the contour map for OH^-. The similar argument applies to the case of H_3O^+. However, contributions from the hydrogen bond and the packing effect are apparently more important than that from coupling between the solute charge and solvent polarization.

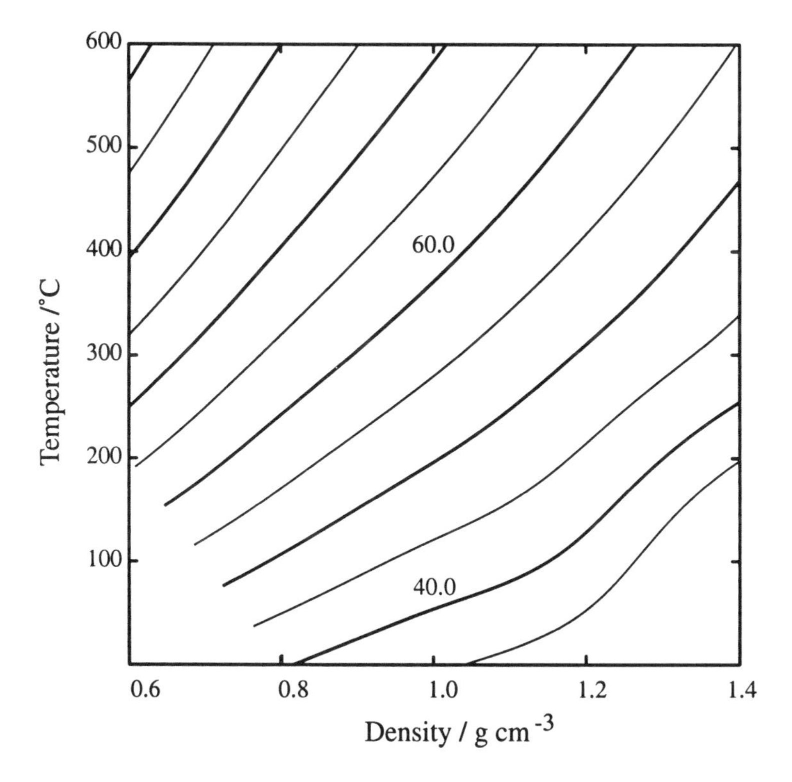

Fig. 2.19. Temperature and density dependence of the total free energy. Contour spacing is 5.0 kcal/mol

Total Free Energy and ΔpK_w

The total free energy (ΔG^{aq}) as the sum of ΔE^{vac}_{elec}, $\Delta\delta E_{reorg}$, $\Delta\delta\mu$ and $\Delta\delta G^{vac}_{kinetic}$ is shown in Fig. 2.19. The contributions compensate for each other and the final profile of free energy surface is rather simple. It monotonically increases as temperature rises and decreases with increasing density.Selected components of ΔpK_w are shown in Fig. 2.19 and Fig. 2.20. As shown in a previous study [26], an interplay of three components, $\Delta pK^{vac}_{w,elec}(\rho,T)$, $\Delta pK_{w,reorg}(\rho,T)$, and $\Delta pK_{w,\mu}(\rho,T)$, determines the final ΔpK_w at around room temperature and normal density. Among these three, contributions from $\Delta pK^{vac}_{w,elec}(\rho,T)$ and $\Delta pK_{w,\mu}(\rho,T)$ are very large, but compensate for each other to make a moderate contribution to the entire profile. It can be readily seen that the sum of these two contributions is very insensitive to temperature change, especially below 200°C.

In contrast, the contribution from $\Delta pK_{w,reorg}(\rho,T)$ shows significant temperature dependence in this region (Fig. 2.21) and becomes progressively less sensitive with rising temperature. The resultant ΔpK_w at around room temperature is determined by a subtle balance between these two contributions. At higher temperature above 200°C, the contribution from electronic reorganization becomes less important. The results of $\Delta \log K_w = -\Delta pK_w$ are plotted as a function of the density (Fig. 2.22), so as to be readily compared with the experimental data [21].

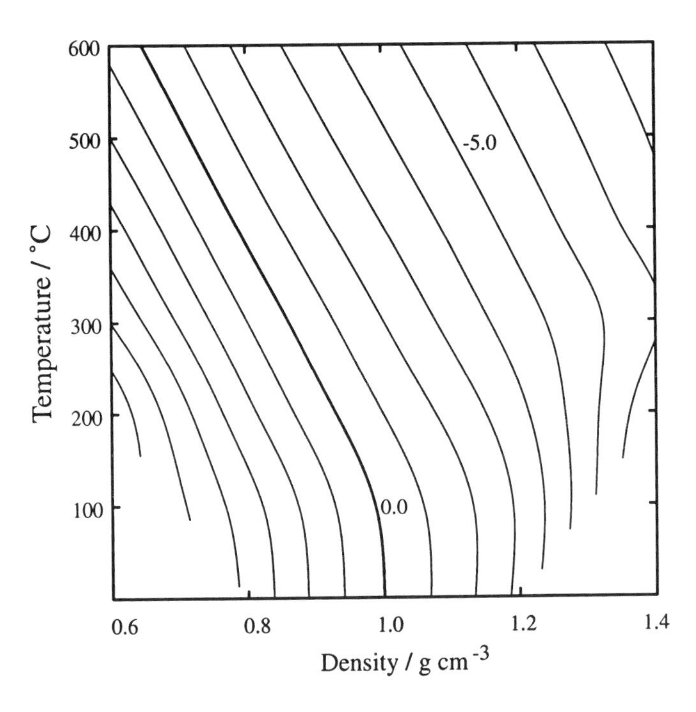

Fig. 2.20. Temperature and density dependence of a ΔpK_w component: the sum of the contributions from the potential energy change associated with the auto-ionization process in an isolated system ($\Delta pK^{vac}_{w,elec}$) and from the free energy change of solvation ($\Delta pK_{w,\mu}$)

The dashed lines in the figure indicate the isobar lines from the literature. The whole profile of density and temperature dependence is in excellent accordance with the experimental data: $\Delta\log K_w$ becomes greater with increasing density and temperature, and the contour spacing at lower temperature is slightly wider than that at higher temperature. As had been discussed, $\Delta\log K_w$ shows a rather monotonic dependence on temperature and density. The inversion experimentally observed in ΔpK_w along the isobar lines can be attributed to the relationship among the state variables, pressure, temperature and density. Essential physics to be clarified lies in the monotonic increase in $\Delta\log K_w$ as density increases at all temperatures.

As mentioned above, at higher temperature, above 200°C, the contribution from electronic reorganization becomes less important and $\Delta pK_{w,\mu}(\rho,T)$, or the change in the solvation free energy of the species associated with auto-ionization, governs the over all result (ΔpK_w). As can be seen, $\Delta\mu(H_2O)$ and $\Delta\mu(H_3O^+)$ exhibit similar monotonic increases with increasing density, which can be essentially attributed to the density dependence of the free energy of cavity formation, as we have discussed earlier. If $\Delta\mu(OH^-)$ behaves in a similar manner, those contributions cancel each other, i.e., $\Delta\mu(\text{total}) = \Delta\mu(H_3O^+) + \Delta\mu(OH^-) - 2\Delta\mu(H_2O)$, and the overall free energy change will not show such a significant dependence on density. However, $\Delta\mu(OH^-)$ shows the opposite density dependence at lower density, and nearly constant at higher density, which gives rise to the observed density dependence for $\Delta\mu(\text{total})$, and in turn for $-\Delta pK_w$. As we have clarified earlier in the present section, the only physical process that gives rise to the negative density dependence on the solvation free energy is the Coulombic coupling of atomic partial charges with solvent polarization.

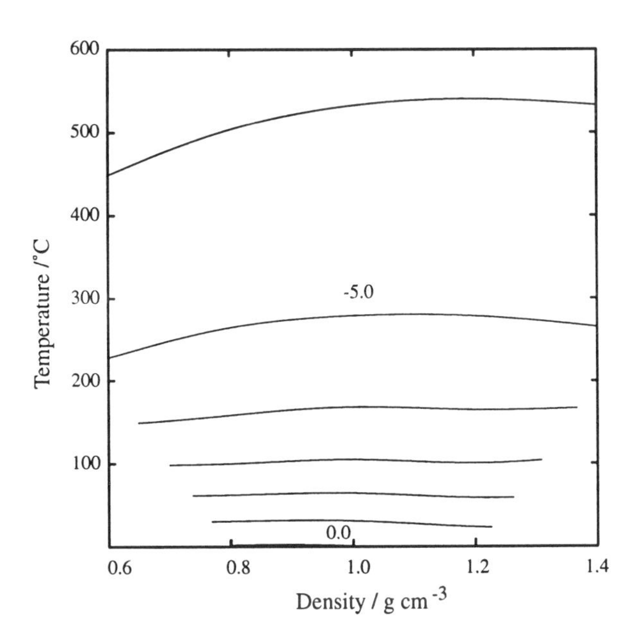

Fig. 2.21. Temperature and density dependence of a ΔpK_w component: the contribution from $\Delta pK_{w,reorg}$

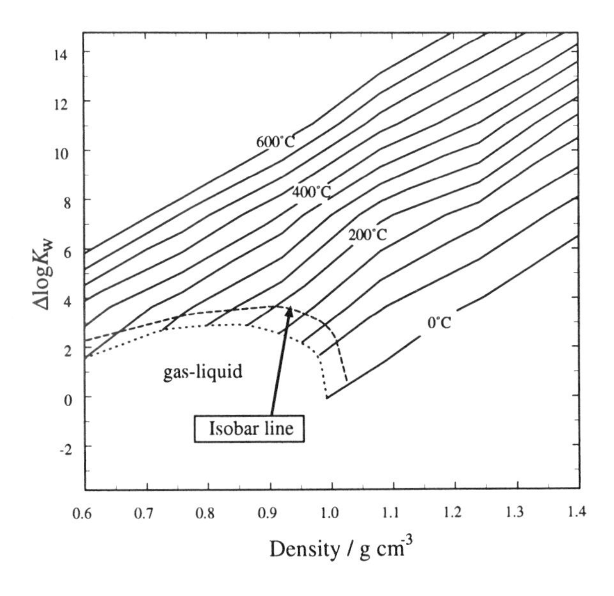

Fig. 2.22. Temperature and density dependence of ΔpK_w

2.4 Conclusions

The electronic and liquid structures of water and its thermodynamic properties have been studied over a wide range of temperature (0 - 600°C) and density (0.6 - 1.4 g/cm^3) based on the *ab initio* molecular orbital theory combined with the integral equation method of liquid.

2.4.1 Electronic and Liquid Structure of Water

The molecular dipole moments and electronic polarization energies decrease with increasing temperature and/or density. This is caused by the reduced reaction field including the hydrogen bond exerted on a molecule from the surrounding molecules due to thermal activation of the system. These theoretical results for dipole moments are in quantitative accord with the experimental data, which has been determined based on the NMR chemical shift coupled with the molecular dynamics simulation.

It was concluded that water largely retains the short-range structure which is characteristic to ice, or the tetrahedral configuration, at normal density (1.0 g/cm^3). The ice-I-like characteristic of water structure disappears to a large extent both at high (1.4 g/cm^3) and low (0.6 g/cm^3) densities for different reasons: at high density due to the packing effect, while at low density due to an essentially entropic cause, or increased configuration space available to a molecule. The distance of closest approach between molecules in water is primarily determined by a balance of two forces, the core repulsion and the hydrogen bond. The core repulsion due to the many-body packing effect becomes important only at high density. This makes separations between the nearest neighbor molecules in water insensitive to the density change compared with those corresponding to the L - J fluid. On the other hand, the distance in L - J fluids is determined essentially by the many-body packing effect, which of course is very sensitive to density. The theoretical results for water are in good qualitative accord with experiments.

The number of hydrogen bonds was calculated from the oxygen - hydrogen pair correlation function using a new definition based on S_{OH}, which enabled us to distinguish the hydrogen-bonded O - H pairs from those just in contact due to packing effect. The number of hydrogen bonds so defined decreases with increasing temperature, while the O - H pairs just in contact increases in accord with the intuitive picture of hydrogen bonds breaking due to activated thermal motions.

2.4.2 The State Dependence of pK_w

The dependence of pK_w on the state variables was calculated based on the thermodynamic cycle which assumes H_2O, OH^- and H_3O^+ ions as solute molecules in aqueous solution.

The electronic structures of the three species involved in the auto-ionization reaction, represented by electronic reorganization (distorsion) energy, relaxed toward those in gas phase with rising temperature and with decreasing density. The change is much sharper for OH^- compared to the other species. The state dependence of the electronic structure was explained by an interplay among the three factors in the solute-solvent and solvent-solvent interactions: the packing effect, the intermolecular hydrogen-bonding, and the electrostatic coupling of solute charges with solvent polarization. The observed density dependence of the effective charge turns out to be dominated by the third factor.

Solvation structures of the three species, represented by the site-site pair correlation functions between the solute and solvent, also exhibited marked dependence on temperature and density. The density dependence of PCF at short separations can be explained in terms of an interplay between essentially two different 'solvation' structures, the contributions of which change as density decreases. The first of those is that characteristic of a simple fluid by which the packing effect dominates the configuration, which is favored at higher density. The other structure is due to hydrogen bonding between solute and solvent species, the contribution of which is greater at lower density. Remarkably high sensitivity to density was observed in the PCF between the O(OH^-) - H(solvent) pair, which can be attributed to the 'softness' of the electronic structure of OH^-.

The results for $\Delta \log K_w$ or $-\Delta pK_w$ obtained from the theory showed a monotonical increase with increasing density at all the temperatures investigated, in good accord with the experimental observation: water becomes more and more acidic as density increases. The behavior is determined essentially by the difference in solvation free energies, $[\Delta\mu(H_3O^+) + \Delta\mu(OH^-) - 2\Delta\mu(H_2O)]$ associated with the reaction ($2H_2O = H_3O^+ + OH^-$). The $\Delta\mu(OH^-)$ shows a density dependence that is entirely different from that of the other species, which gives rise to the observed behavior for $\Delta \log K_w$. The distinct density dependence of $\Delta\mu(OH^-)$ has its origin in its rather 'soft' electronic structure interacting with solvent polarization.

Acknowledgements. We would like to thank Prof. M. Nakahara and Drs. N. Matubayashi, C. Wakai and R. Akiyama for fruitful discussions and invaluable comments. The present study is supported by Grants-in-Aid for Scientific Research from the Ministry of Education, Science, Sports and Culture (MONBUSHO) in Japan.

References

2.1 F. Franks: *Water: A Complehensive Treatise* (Plenum, New York 1972)

2.2 J. D. Bernal, R.H. Fowler, J. Chem. Phys. **1**, 515 (1933)

2.3 W.E. Thiessen, A.H. Narten, J. Chem. Phys. **77**, 2656 (1982)

2.4 J.A. Pople, Proc. R. Soc. A **205**, 163 (1951)

2.5 O.Y. Samoilov: *Structure of aqueous electrolyte solutions and the hydration of ions*, (Consultants Bureau, New York 1965)

2.6 G. Némethy, H.A. Scheraga, J. Chem. Phys. **36**, 3382 (1962) ; *ibid* **41**, 680 (1964)

2.7 A. Ben-Naim, J. Chem. Phys. **52**, 5531 (1970); *ibid* **54**, 3682 (1971)

2.8 F. Hirata, Bull. Chem. Soc. Jpn. **50**, 1032 (1977)

2.9 D. Chandler, H.C. Andersen, J. Chem. Phys. **57**, 1930 (1972) ; F. Hirat, P. J. Rossky, Chem. Phys. Lett. **84**, 329 (1981) ; B. M. Pettitt , P. J. Rossky, J. Chem. Phys. **77**, 1452 (1982)

2.10 G.N. Patey , S. L. Carnie, J. Chem. Phys. **78**, 5183 (1983) ; *ibid* **79**, 4468 (1983)

2.11 T. Ichiye, A.D. J. Haymet, J. Chem. Phys. **89**, 4315 (1988) ; *ibid* **93**, 8954 (1990)

2.12 F. Hirata, Bull. Chem. Soc. Jpn. **71**, 1483 (1998)

2.13 S. Ten-no, F. Hirata, S. Kato, J. Chem. Phys. **100**, 7443 (1994) ; S. Ten-no, F. Hirata , S. Kato, Chem. Phys. Lett. **214**, 391 (1993)

2.14 H. Sato, F. Hirata, S. Kato, J. Chem. Phys. **105**, 1546 (1996)

2.15 S. Maw, H. Sato, S. Ten-no, F. Hirata, Chem. Phys. Lett. **276**, 20 (1997); H. Sato, F. Hirata, J. Chem. Phys.,**111**, 8545 (1999).

2.16 J.K. Gregory, D.C. Clary, K. Liu, M.G. Brown, R.J. Saykally, Science **275**, 814 (1997)

2.17 N. Matubayasi, C. Wakai, M. Nakahara, Phys. Rev. Lett. **78**, 2573 (1997); N. Matubayasi, C. Wakai, M. Nakahara, J. Chem. Phys. **107**, 9133 (1997); N. Matubayasi, C. Wakai, M. Nakahara, J. Chem. Phys. **110**, 8000 (1999)

2.18 M. Nakahara, T. Yamaguchi, H. Ohtaki, *Recent. Res. Dev. Phys. Chem.* **1**, 17 (1997)

2.19 T. Radnai, H. Ohtaki, Mol. Phys. **87**, 103 1996)

2.20 R. D. Mountain, J. Chem. Phys. **90**, 1866 (1989); *ibid* **103**, 3084 (1995)

2.21 W. B. Holzapfel, J. Chem. Phys. **50**, 4424 (1969)

2.22 A. S. Quist, W. L. Marshall, J. Phys. Chem. **69**, 3165 (1965) ; A. S. Quist, *ibid,* **74**, 3396 (1970)

2.23 G. Corongiu, E. Clementi , J. Chem. Phys. **97**, 2030 (1992)

2.24 F. J. Luque, S. R. Gadre, P. K. Bhadane, M. Orozco, Chem. Phys. Lett. **232**, 509 (1995); F. R. Tortonda, J.-L. Pascual-Ahuir, E. Silla, I. Tuñón, J. Phys. Chem. **97**, 11087 (1993); F. R. Tortonda, J.-L. Pascual-Ahuir, E. Silla, I. Tuñón, J. Phys. Chem. **99**, 12525 (1995)

2.25 M. Tuckerman, K. Laasonen, M. Sprik, M. Parinello, J. Phys. Chem. **99**, 5749 (1995)

2.26 H. Sato, F. Hirata, J. Phys. Chem. A **102**, 2603 (1998). ; H. Sato, F. Hirata, J. Phys. Chem. B **103**, 6596 (1999)

3 The Behavior of Proteins Under Extreme Conditions: Physical Concepts and Experimental Approaches

Karel Heremans[1] and László Smeller[2]

[1] Katholieke Universiteit Leuven, Department of Chemistry, 3001 Leuven, Belgium
E-mail: karel.heremans@fys.kuleuven.ac.be
[2] Semmelweis University of Medicine, Institute of Biophysics, 1444 Budapest, Hungary
E-mail: smeller@puskin.sote.hu

Abstract. The behaviour of proteins under the extreme conditions of temperature and pressure is quite unique among the biomacromolecules. This shows up in the stability phase diagram, which has an elliptical shape. This shape results from the temperature and pressure dependence of the partial molar volume changes of the unfolding. At the molecular level this can be interpreted by assuming effects of temperature and pressure on the hydration as well as on the intramolecular cavities.

3.1 Introduction

Proteins are probably unique among the biological macromolecules attracting active interest from various disciplines. For the physicist it is the structure that is interesting, it has characteristics of order as well as of disorder. The chemist is attracted by the unique properties that show up in the catalytic activity of enzymes and in the conversion of chemical into mechanical energy in muscles. Biologists tend to put emphasis on the functional role of proteins. Whereas this chapter emphasizes the pressure effects on proteins, the relation to the effects of cold and heat will be discussed as well.

The pioneering work in high-pressure protein research is that of Bridgman, who observed that a pressure of several 100 MPa will give egg white an appearance similar but not identical to that of a cooked egg. Nowadays it is well known that proteins in solution are marginally stable under conditions of high temperature and pressure. In contrast there is the observation that certain bacteria live under extreme temperature conditions, and it is well known that bacteria can survive in the deepest parts of the ocean. These aspects are discussed in other contributions to this book.

If the conditions for equilibrium or isokineticity are plotted versus temperature and pressure, a phase or stability diagram is obtained with an elliptical shape [1–3]. One of the practical consequences is the stabilization against heat unfolding by low pressures. This has been observed in several proteins and enzymes [4,5]. Interestingly, this also applies to the effect of pressure on the heat gelation of starch [6]. Of special interest is the observation that the inactivation kinetics of microorganisms shows diagrams similar to those of proteins [7].

3.2 Physical Concepts

For a one-component system, the properties of the state depend on the temperature and the pressure. For aqueous solutions the extensive properties are usually not a simple addition of the properties of the various components. To this end, partial molar or partial specific quantities are introduced. Partial quantities are intensive variables, and in addition to the temperature and pressure dependence, they depend on the composition of the system. This may lead to considerable confusion in the molecular interpretation of these thermodynamic quantities. In the following we discuss the volume and its pressure and temperature dependence. For the sake of completeness, we briefly outline the temperature dependence of the partial molar enthalpy, the heat capacity. We also consider the statistical thermodynamic interpretation of these quantities in terms of volume and entropy fluctuations.

3.2.1 Volume and Hydration

From a macroscopic point of view, the volume is a measure of that part of space that is inside a given surface. On the molecular and atomic level there is no well-defined surface and it follows that definition of the volume can use different approaches. The first one, the partial molar volume, is the phenomenological one, and this is used in thermodynamics and in experimental work. The second one defines a surface, such as the van der Waals or any other calculable surface, from which the volume is obtained. This approach is the one that is used in molecular dynamics simulations or other computer calculations. The partial molar volume, V_i, of a solute molecule or ion is defined as the change in volume of the solution by the addition of a small amount of the solute over the number of moles of added solute, keeping the amount of the other components constant:

$$V_i = \left(\frac{\partial V}{\partial n_i} \right)_{n_j, T, p} . \tag{3.1}$$

In this definition V is the volume of the solution and n_i is the number of moles of the solute added. It is not equal to the volume of the molecule or the ions since it also includes the interaction with the solvent. This may be seen from the fact that the partial molar volume for salts such as $MgSO_4$ and Na_2CO_3 is negative at zero concentration because of the strong electrostriction of the solvent around the ions. For the same reason the volume of the uncharged glycolamide is larger (56.2 mL/mol) than that of the amino acid glycine (43.5 mL/mol). For a protein V_i can be written as the sum of three terms:

$$V_i = V_{atoms} + V_{cavities} + \Delta V_{hydration} . \tag{3.2}$$

In this expression V_{atom} and $V_{cavities}$ are the volumes of the atoms and the cavities respectively and $\Delta V_{hydration}$ is the volume change of the solution resulting from the interactions of the protein molecule with the solvent. Care should be taken if quantities derived from the volume (such as compressibility and thermal expansion) are interpreted. The results may depend on the sensitivity range of the method. Global measurements such as ultrasonics detect the whole molar volume, while some local probes may detect only the change of the protein interior volume.

As the volumes of the atoms may be considered, as a first approximation, to be temperature and pressure independent, it follows that both the thermal expansion and the compressibility are composed of two main terms, the cavity and hydration terms. An estimate of the contribution of each factor is not easy to evaluate and relies on assumptions that are not easy to check experimentally. However, the compressibility of amino acids is negative because of changes in the solute–solvent interaction (i.e., the amino acid solution is less compressible than the pure solvent). It follows then that the contribution of cavities compensates this effect so that the compressibility of the protein in solution becomes positive. A more quantitative estimate is possible when one makes assumptions about the compressibility of the water of hydration [8].

The influence of various cosolvents on protein unfolding has been discussed by Timasheff [9]. It has been shown that the interpretation in terms of preferential solvation and excluded volume effects are essentially equivalent [10]. Although pressure studies are limited, it seems that the stabilizing effect of organic cosolvents acting against temperature unfolding is also found acting against pressure unfolding [11]. Kinetic studies under pressure of the folding of staphylococcal nuclease in the presence of xylose show that the sugar effect is primarily on the folding step, suggesting that the transition state, a dry molten globule state, is close to the folded state [12].

3.2.2 Compressibility and Volume Fluctuations

The partial molar isothermal compressibility, β_T, the index i being omitted for clarity, is defined as the relative change of the partial molar volume with pressure:

$$\beta_T = -\frac{1}{V_i}\left(\frac{\partial V_i}{\partial p}\right)_{T,j} . \tag{3.3}$$

As for the volumes of the atoms, the compressibility is composed of two main terms, the cavity and hydration terms. An estimate of the contribution of each factor relies on assumptions that are not easy to check. An estimate of the compressibility of the cavities should be possible with positron annihilation lifetime spectroscopy. This technique has proven to be a useful tool for determining the size of cavities and pores in materials. The lifetime is sensitive to the size of the cavity in which it is localized. A number of empirical relations correlate the distribution of the lifetime and the free volume [13]. Data on the pressure effect on the lifetime are only available for polymers. The results suggest that there may be a considerable contribution of the reduction in cavity size to the compressibility of a protein.

The compressibility is a thermodynamic quantity of interest not only from a static but also a dynamic point of view. Its relevance to the biological function of a protein can be understood through the statistical mechanical relation between the isothermal compressibility, β_T , and volume fluctuations:

$$\langle V - \langle V \rangle \rangle^2 = \beta_T k_B T V . \tag{3.4}$$

The expression between brackets indicates an average for the isothermal-isobaric ensemble. Because of the small size of the protein, the volume fluctuations are relatively large. It seems that the expansion and contraction of the cavities is the only way to generate these volume fluctuations. The biological relevance of the volume, as well as the entropy and volume-entropy fluctuations that will be considered in the following sections, can be illustrated by referring to a number of processes that are related to the dynamical properties of proteins. These include the opening and closing of binding pockets in enzymes, the allosteric effects, the conversion of chemical in conformational energy in muscle contraction, the biological synthesis of proteins and nucleic acids and the transport of molecules through membranes.

3.2.3 Thermal Expansion and Volume-Entropy Fluctuations

The partial molar expansion, α, is defined as the relative change of the partial molar volume with temperature:

$$\alpha = \frac{1}{V_i}\left(\frac{\partial V_i}{\partial T}\right)_{p,j} . \tag{3.5}$$

As for the compressibility, the thermal expansion is considered to be composed of two main terms, the cavity and hydration terms. A quantitative analysis is possible when one makes assumptions about the thermal expansion of the water of hydration [14].

The thermal expansion can be related to the fluctuations of the system. However, this relation is not as widely known as the others mentioned above. The thermal expansion is proportional to the cross-correlation of the volume and entropy fluctuations:

$$\langle SV - \langle S \rangle \langle V \rangle \rangle = k_B T V \alpha . \tag{3.6}$$

This agrees with the intuitive picture that the thermal expansivity characterizes some kind of coupling between the thermal (T, S) and the mechanical (p, V) parameters.

3.2.4 Heat Capacity and Entropy Fluctuations

The partial molar heat capacity, C_p, is defined as:

$$C_p = \left(\frac{\partial H_i}{\partial T}\right)_{p,j} . \tag{3.7}$$

Note that, in contrast to the partial molar volume, this quantity is not a relative one. This follows from the fact that the absolute value of the partial molar enthalpy cannot be determined. In a thermodynamic system with constant T and p, the isobaric heat capacity can be regarded as the measure of the entropy fluctuations of the system:

$$\langle S - \langle S \rangle \rangle^2 = k_B C_p . \tag{3.8}$$

The partial molar heat capacity has been considered to be composed of intrinsic and hydration contributions. The intrinsic component contains contributions from covalent and noncovalent interactions. It has been shown that about 85% of the total heat capacity of the native state of a protein in solution is due to the covalent structure [15]. Changes in the heat capacity upon unfolding are therefore primarily interpreted as due to changes in the hydration. A physical picture of entropy fluctuations means changing the conformation between ordered and less-ordered structures. This can be achieved by hindered internal rotations, low-frequency conformational fluctuations or high-frequency bond stretching and bending modes.

3.2.5 Grüneisen Parameter

For solids and synthetic polymers, the relation between the volume, V, the thermal expansion, α, the compressibility, β_T, and the heat capacity at constant volume, C_V, is given by the Grüneisen parameter, γ:

$$\gamma = \frac{V}{C_V} \frac{\alpha}{\beta_T} . \tag{3.9}$$

This volume-independent parameter can be obtained from pressure-induced vibrational wavenumber shifts in solids or polymers [16]:

$$\gamma_i = -\frac{d \ln \nu_i}{d \ln V} = -\frac{V}{\nu_i} \frac{d\nu_i}{dV} . \tag{3.10}$$

If this parameter is assumed to be the same for all vibrations, one can obtain a bulk thermodynamic definition for γ. The bulk Grüneisen parameter is found to be ca 4 for polymers from the effect of pressure on the velocity of sound. The data suggest that for the heat capacity only the interchain contribution should be taken into account. With this assumption, an order of magnitude calculation shows that the bulk Grüneisen parameter for proteins is of the same order of magnitude as that for polymers. This suggests that the thermal expansion and the compressibility of proteins reflect primarily the movement between the secondary structures. These movements are reflected in the low-frequency part of the vibrational spectrum. Unfortunately, no experimental data are available on the effect of pressure on these vibrations.

3.2.6 Protein Stability and Unfolding

The first experimental evidence for the unusual behavior of proteins with respect to temperature and pressure came from the kinetic studies of Suzuki [1] and thermodynamic studies by Hawley [2]. The mathematical implications and assumptions that are usually made in the analysis of the data have recently been discussed [17]. Of particular interest is the cold unfolding of proteins that was predicted on the basis of the model describing the temperature dependence of the reaction enthalpy accounting for the large increase in heat capacity upon unfolding [18]. As we shall see, the effects of heat, cold and pressure on proteins are interconnected.

Thermodynamic Theory

The elliptic phase diagram on the *p-T* plane is characteristic for proteins. Mathematically, this kind of shape originates from the fact that second-order terms give a significant contribution to ΔG, the change in free energy. Physically these second-order terms are proportional to $\Delta\beta$, ΔC_p, $\Delta\alpha$, the changes in the compressibility, the heat capacity and the thermal expansion respectively between the unfolded (*U*) and the native (*N*) state of the protein.

$$\text{N (Native)} \leftrightarrow \text{U (Unfolded)}. \tag{3.11}$$

The pressure and temperature dependence of ΔG ($\Delta G = G_{\text{Unfolded}} - G_{\text{Native}}$), the difference in free energy between *U* and *N* is given by

$$d(\Delta G) = -\Delta S dT + \Delta V dp. \tag{3.12}$$

Integration at constant pressure, gives the expression for the temperature dependence of the free energy assuming that ΔC_p is temperature independent:

$$\Delta G(p_0,T) = \Delta G_0 - \Delta S_0(T - T_0) + \Delta C_p\left[(T - T_0) - T\ln\frac{T}{T_0}\right]. \tag{3.13}$$

ΔG_0 refers to the conditions p_0 and T_0. The pressure dependence can be taken into account starting from the volume change that is not only pressure dependent but also temperature dependent:

$$\Delta V(p,T) = \Delta V + \Delta\alpha^*(T - T_0) - \Delta\beta^*(p - p_0). \tag{3.14}$$

ΔV refers here to p_0 and T conditions. It follows that the pressure and temperature dependence of the free energy change contains a cross term in temperature and pressure:

$$\begin{aligned}\Delta G(p_0,T) = {} & \Delta G_0 - \Delta S_0(T - T_0) - \Delta C_p\left[(T - T_0) - T\ln\frac{T}{T_0}\right] \\ & + \Delta V_0(p - p_0) - \left(\frac{\Delta\beta^*}{2}\right)(p - p_0)^2 + \Delta\alpha^*(T - T_0)(p - p_0)\end{aligned} \tag{3.15}$$

In these equations $\Delta\beta^*$ is the compressibility factor difference ($\beta^* = \beta V$), and $\Delta\alpha^*$ the difference of the thermal expansion factor ($\alpha^* = \alpha V$) of the denatured and native states of proteins. An important assumption in the derivation of this equation is the temperature and pressure independence of $\Delta\alpha^*$, $\Delta\beta^*$ and ΔC_p. The $\Delta G = 0$ curve is an ellipse on the *p-T* plane, and it describes the equilibrium border between the native and denatured state of the protein. This curve is known as the phase or stability diagram. This is visualized in Fig. 3.1. The diagram illustrates the interconnection between the cold, heat and pressure unfolding of proteins.

An equation similar to the Clausius-Clapeyron equation can be obtained for the slope of the phase boundary:

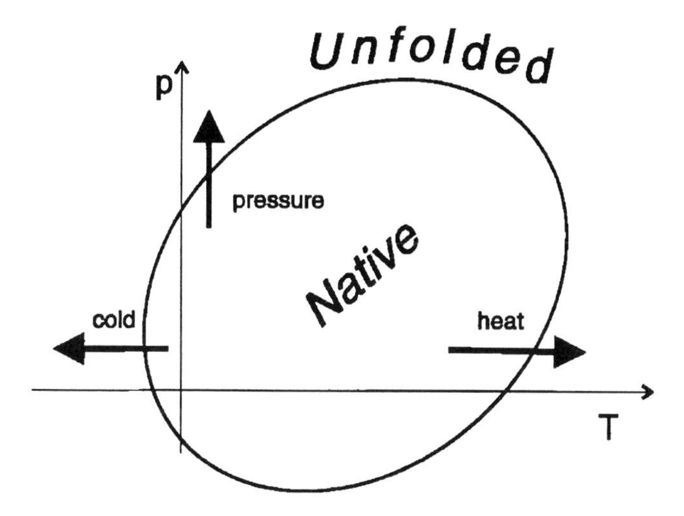

Fig. 3.1. The elliptical temperature-pressure stability phase diagram characteristic for proteins. After Suzuki [1] and Hawley [2]

$$\frac{\partial T}{\partial p} = \frac{\Delta V_0 + \Delta\beta^*(p - p_0) + \Delta\alpha^*(T - T_0)}{\Delta S_0 - \Delta\alpha^*(p - p_0) + \Delta C_p(T - T_0)/T_0} \tag{3.16}$$

This expression reduces to the classical Clausius-Clapeyron equation when the differences in compressibility, thermal expansion and heat capacity vanish, as is observed for most phase transitions in lipids [19]. The shape of the phase diagram for proteins is of considerable interest since, as pointed out previously, it contains information on the volume and entropy fluctuations and on the coupling between volume and entropy fluctuations.

Kinetics and Reversibility of Unfolding

The three common possible ways to denature a protein are heat, pressure, and cold unfolding. According to the above thermodynamic description, the unfolding could be a reversible process. However, in most cases the denatured protein tends to produce a gel stabilized by a network of intermolecular hydrogen bonding, which prevents the refolding to the native structure. Since the equilibrium theory describes the unfolding by treating only two states, the irreversibility is not included in that model. The kinetic theory suggests two denatured states, a metastable state, where the protein is reversibly denatured (*U*), and an irreversibly inactivated (*I*) one:

$$N \leftrightarrow U \rightarrow I\ . \tag{3.17}$$

We should keep in mind that this scheme is especially important in the case of experiments where the sample is subjected to high pressure, but where the amount of inactivation is measured after the pressure treatment on the depressurized sample [20]. There are a few special cases of interest regarding the *rates* of the conformational transitions: When the process $U \rightarrow I$ is slow, then the experiments described above reflect the rate of the kinetics of this slow process. In the opposite case the reversible unfolded protein will be irreversibly captured in an intermolecular network. In this case the kinetic experiment reflects the rate of the first step of the two-step unfolding process. However, in both cases the thermodynamic parameters of the first step will contribute to the sign and the magnitude of the activation parameters.

Recent work in our laboratories has shown that pressure-induced unfolded states play a very important role in the aggregation of proteins. The high-pressure unfolding of horse heart metmyoglobin results in an intermediate form that shows a strong tendency to aggregate *after* pressure release. These aggregates seem to be similar to those usually observed upon temperature denaturation [21].

3.2.7 Glass Transitions

The structure of a protein is the result of a delicate balance between the intramolecular interactions in the polypeptide chain which compete with the solvent interactions. Proteins in the dry state are found to be extremely resistant to temperature and pressure unfolding. The pressure effect on dry proteins may easily be studied in the diamond anvil cell [11]. These experiments emphasize the role of water as a plasticizer in biomacromolecules via the reduction of the glass-rubber transition temperature (T_g). These experiments also suggest that proteins in the dry state are below their glass transition temperature. The fundamental similarities between synthetic polymers and biomacromolecules have stimulated research in this field because of the importance of T_g in all aspects of the physical chemical behaviour. In view of the nonequilibrium nature of the glass transition temperature, this transition is governed by activation rather than equilibrium parameters. Limited studies on the effect of pressure on the glass transitions in polymers suggest a change of ca 22 K/100 MPa for nonhydrogen-bonded systems. The effect of pressure on the hydrogen bonded system sorbitol shows a much weaker dependence of 4 K/100 MPa [22]. This suggests that similar orders of magnitude may be expected for glass transitions in proteins.

3.3 Experimental Approaches

Thermodynamic and structural information on the native and unfolded state s can be obtained from a number of experimental approaches. As well as the specific properties that are being probed, it is important to be aware of the concentration range of the proteins used in the various techniques. This may be of particular interest when one investigates the formation of intermolecular aggregates that may result from the unfolding process.

3.3.1 Thermodynamic Properties

Almost all the data on the compressibility of proteins have been obtained from ultrasonic velocimetry. The velocity of ultrasound, u, is related to the adiabatic compressibility, β_S, and the density, ρ, via the equation:

$$\beta_S = \frac{1}{\rho u^2} . \tag{3.18}$$

Sarvazyan [23] has drawn the attention to the potentialities of ultrasonic velocimetry as a method to obtain information on all molecular aspects related to compressibility. The methodology has been extended to the measurement of the compressibility of proteins as a function of pressure [24, 25].

The compressibility of amino acids in aqueous solution is negative, whereas the compressibility of proteins in the native state is positive. The negative compressibility of the unfolded state is usually interpreted as the consequence of the reduced compressibility of the hydration shell [26]. However, the disappearance of the cavities would also result in a decrease in the compressibility.

The thermal expansion can be obtained from dilatometry, but in most cases it is obtained from the partial molar volume [27]. The interpretation of the changes in the partial molar volume focuses on the inapplicability of the low molecular-weight model systems to the understanding of the behavior of proteins. It is suggested that mutual thermal motions of macromolecules and solvating water molecules involve modes that are absent in small molecules. More specifically, the hydration of nonpolar groups on the surface of proteins is estimated to be different from the low molecular-weight model systems. This topic has also been considered in terms of the "thermal volume", a volume that results from the thermally induced molecular vibrations [27]. For solvents the thermal volume is proportional to the isothermal compressibility.

An estimate of the thermal expansion of the cavities in proteins can be made from X-ray diffraction data as a function of temperature. This will be discussed in section 3.3.5. Positron annihilation lifetime spectroscopy has been used to determine the increase of the free volume of proteins [13]. With this technique, an increase of 0.6 A^3/K has been obtained for lysozyme.

In a seminal paper, Gekko and Hasegawa [8] estimated the contribution of cavities and hydration to the temperature dependence of the adiabatic compressibility of proteins. Bovine serum albumin and lysozyme show an increase in adiabatic compressibility with increasing temperature. The authors attribute this mainly to reduced hydration at high temperature. At low temperature the compressibility of both proteins becomes negative, which is interpreted as a decreased contribution from the cavities.

Tamura and Gekko [28] used precise density and sound-velocity measurements to follow the thermal unfolding of ribonuclease A. The apparent molar volume decreased but the adiabatic compressibility increased. Similar observations have been reported for chymotrypsinogen [29]. The compressibility may thus be used to detect and characterize intermediate states in proteins. More specifically, the compact intermediate state, sometimes called molten globule state, shows an increase in the compressibility compared to the native state.

The change in compressibility between the native and the pressure-induced state can be calculated from the stability diagram as discussed in section 1.2.6. Recently, it has been become possible to measure the adiabatic compressibility *in situ* under pressure [24, 25]. For amino acids the compressibility increases with increasing pressure. The pressure-induced intermediate state of cytochrome c at acid pH shows an increased compressibility compared to the native state. As pointed out by Kharakoz and Bychkova [30], the volume fluctuations for highly hydrated intermediate states cannot be calculated quantitatively without thermodynamic data on the transfer of water into the protein.

3.3.2 Absorption Spectroscopy

Compared to emission spectroscopy, the concentration range used in absorption spectroscopy is much higher, in particular for infrared spectroscopy. Whereas absorption spectroscopy probes changes in the immediate environment of the chromophore, infrared spectroscopy is a useful tool to follow the changes in conformation of the polypeptide chain. The two techniques may therefore complementeach other.

UV/Visible Spectroscopy

In principle it should be possible to obtain the compressibility of a protein matrix surrounding a chromophore from the pressure-induced frequency shifts of the absorption spectrum of that chromophore. This is based on a microscopic model that has been developed for the spectral hole burning technique, which takes into account the long range *induced-dipole—induced-dipole* interactions between the protein matrix and the chromophore. This will be discussed in section 3.3.3.

Jung and coworkers have used this model to calculate the local compressibility of the heme pocket of cytochrome P450-CO complexed with substrate analogues by following the pressure-induced red shift and the broadening of the absorption spectrum in the Soret band at room temperature [31]. The pressure-induced red shift is assumed to result from the interaction between the heme group and the active-site water molecules and/or polar amino acid residues near the heme. In contrast to the observed pressure-induced broadening of the Soret band for cytochrome P450 complexes, a band narrowing is observed for the myoglobin-CO complex. As for the luminescence measurements, this approach gives the local compressibility of the protein interior.

The first systematic studies on pressure- and temperature-induced protein unfolding were performed with UV spectroscopy on ribonuclease A, and chymotrypsinogen and with spectroscopy in the visible region for metmyoglobin. This provided the first systematic thermodynamic description of the stability diagram of a protein as given in section 3.2.6.

Lange and coworkers developed fourth derivative UV spectroscopy as a tool to evaluate changes of the dielectric constant in the vicinity of the aromatic amino acids in proteins which undergo pressure-induced structural changes. Thus, they detected in ribonuclease a pressure-induced intermediate that also occurs in the high-temperature-induced unfolding [32].

Using the same approach, Mombelli *et al.* [33] studied the cold, heat and pressure unfolding of ribonuclease P2 and two mutants from the thermophilic archaebacterium *Sulfolobus solfataricus*. The extreme stability of this protein against pressure and temperature treatment is assumed to be due to a hydrophobic core containing three aromatic acid residues. A strong destabilization takes place when Phe31 is replaced by Ala31. The response of the protein was found to be different towards cold, heat and pressure. The authors suggest that the cold and pressure unfolding may result in intermediate states in which the hydrophobic core is preserved, whereas the outer part of the protein undergoes unfolding. The infrared studies on this protein are discussed in the next section.

Infrared Spectroscopy

Information about the protein secondary structure can be obtained from the amide vibrations. The most characteristic vibrational band is the amide I band (1600−1700 cm^{-1}), which is mainly due to the C=O stretching. Due to hydrogen bonding in the folded protein structures, this band is conformation sensitive. Its frequency is in the range 1600−1700 cm^{-1}, depending on the actual secondary structure. Except for a few high pressure Raman studies [34], most of the vibrational spectroscopy has been done with FTIR (Fourier Transform Infrared) spectroscopy. The infrared spectroscopic experiments must be done in D_2O because of the overlapping of the amide I band with the strong water vibration at 1640 cm^{-1}. The use of D_2O involves the possibility of H/D exchange. While this effect becomes significant at the conformational changes, where new amino acids will be exposed to the solvent, it can also be used to probe the dynamics of proteins by estimating the internal accessibility of the polypeptide chain (see next section).

Several proteins were investigated by Heremans *et al.* [11,35]. The decreasing amide I' frequency was found to be a general feature at ca 500 MPa. An example of the effect of pressure and temperature on the frequency maximum of the amide I' band of sperm whale myoglobin is given in Fig. 3.2. The maximum position of the amide I' band (the index refers to D_2O) has a negative slope in the elastic region according to the figure. One has to mention that the band maximum as such can give only a very rough picture of the changes in secondary structure, because the amide I' band is composed of several overlapping components, and the relative intensity changes can also be observed as frequency shifts of the overall band. A detailed analysis can be made with resolution enhancement and band fitting [36, 37], but in order to perform it properly very good quality spectra are needed.

Elastic effects can also be followed using the ring vibration of the tyrosine side chain. Here the usual blue shift is observed. To correlate the $d\nu/dp$ values with the compressibility is not as straightforward as in the fluorescence and visible/UV spectroscopy as discussed in section 3.3.3.

The first Raman study on the effect of pressure on lysozyme revealed that the high-pressure-induced unfolding of the molecule is irreversible rather than reversible, as observed with fluorescence techniques [34]. A subsequent infrared study on chymotrypsinogen gave similar results [38]. In both cases the high concentrations of protein used (ca 1 mM) showed that the pressure-induced unfolding can lead to intermolecular interactions, thus giving rise to irreversibility. This phenomenon is well known for temperature-induced unfolding of proteins. The temperature-induced aggregation of proteins gives rise to specific bands in the amide I' region of the

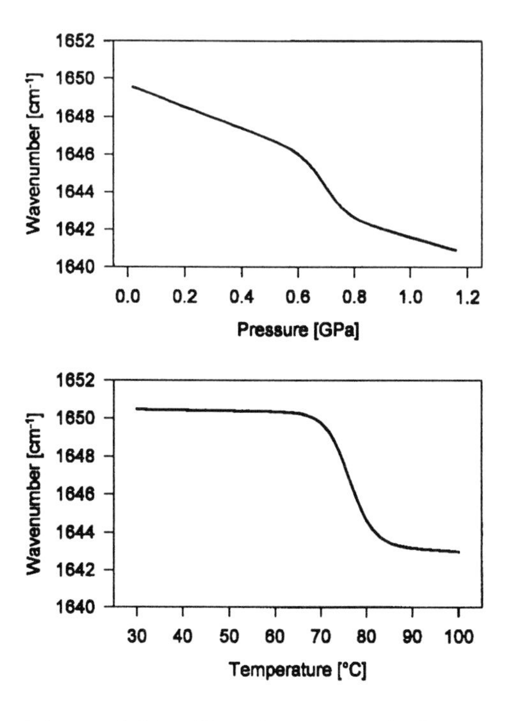

Fig. 3.2. Sperm whale myoglobin: effect of pressure on the frequency maximum of the amide I' band (*top*); Effect of temperature on the frequency maximum of the amide I' band (*bottom*)

infrared spectrum that are characteristic for intermolecular hydrogen-bonded structures [39, 40]. Typical data for the effect of pressure and temperature on the spectrum of horse heart myoglobin are shown in Fig. 3.3. Such bands are, in general, not observed for the pressure-induced unfolding of proteins, suggesting that the degree of unfolding is smaller than in the temperature-induced unfolding. A notable exception is the pressure-induced unfolding of chymotrypsin reversed micelles, where the bands typical for intermolecular hydrogen bonding develop after the pressure is released [41]. Recent data on myoglobin show that the specific side bands characteristic for the aggregation of proteins also appear after pressure unfolding performed at slightly elevated temperatures [21].

An interesting exception to the previous observation is ribonuclease. Takeda *et al.* [42] studied the difference between temperature and pressure-induced unfolding at pH 7 and found no residual secondary structure above 550 MPa and 60°C. Both transitions are reversible, although the pressure-induced transition shows some hysteresis. It is also suggested that the pressure-induced unfolding goes via an intermediate state. These results have received support from the NMR work at pH 2 of Zhang *et al.* [43]. The cold and pressure unfolding of ribonuclease suggests intermediates that are closer to the native state than the unfolded state which is observed in the heat unfolding. This also follows from small angle X-ray scattering experiments as discussed in section 3.3.5.

Infrared studies with high-pressure diamond anvil cell indicate a considerable stability for small proteins. Bovine pancreatic trypsin inhibitor (BPTI) shows pres-

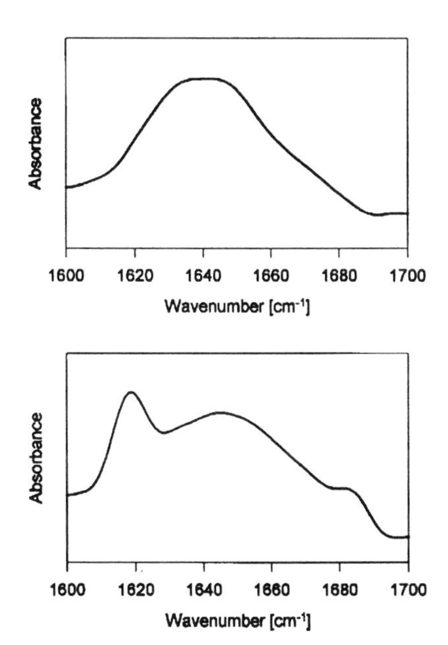

Fig. 3.3. Amide I' band of the infrared spectrum of heat-denatured horse heart myoglobin *(bottom)*. The specific bands characteristic for intermolecular hydrogen bonded structures are seen at 1620 and 1685 cm^{-1}. For comparison, the amide I' band of the pressure-denatured protein is also shown *(top)*

sure-induced changes that are reversible even after compression up to 1.4 GPa [35]. The changes that take place above 1 GPa are reversible and can also be simulated with high-pressure molecular dynamics (see section 3.3.6). Ribonuclease P2 shows no changes in its structure up to 1.4 GPa [44]. On the other hand, the site- directed mutation F31A in the hydrophobic core of the protein destabilizes the protein, giving a reversible pressure-induced unfolding at 400 MPa.

Two-Dimensional Infrared Spectroscopy

Hydrogen/deuterium exchange experiments give important information on the protein dynamics, as shown by the dependence of the accessibility of the protein interior and the pressure and temperature dependence. Even small proteins such as BPTI and ribonuclease P2 show considerable differences in their exchange rates and accessibility under normal conditions. Whereas P2 exchanges all the hydrogen under normal conditions, all hydrogen atoms in BPTI are exchanged after a pressure cycle of 500 MPa. For larger proteins complete unfolding usually brings about a complete exchange of all hydrogen atoms.

It was shown recently [45] that the interaction between H/D exchange and conformational effects can be studied by two-dimensional (2D) infrared spectroscopy [46]. The results for BPTI show that during the pressurization the first step is a small change in secondary structure, which is followed by an increased rate of H/D exchange [45].

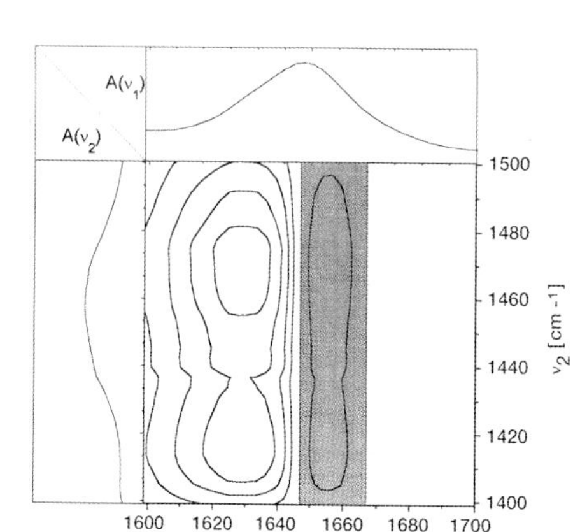

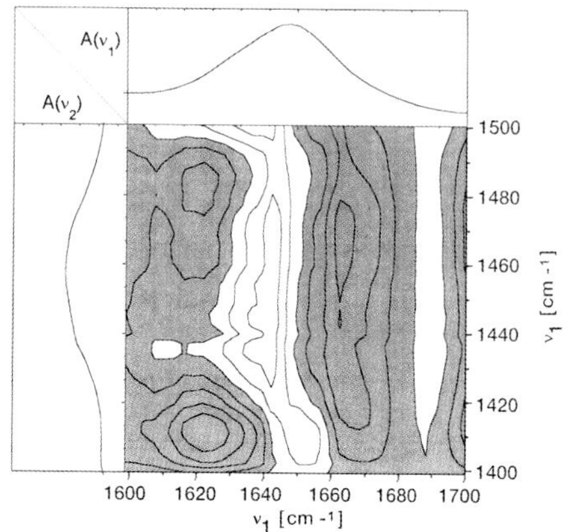

Fig. 3.4. Synchronous (*left*) and asynchronous (*right*) 2D infrared plots of the amide I and II bands of horse heart myoglobin. The plots display changes in the pressure interval 200 500 MPa

Figure 3.4 shows the 2D spectra of the amide I and II band regions of horse heart myoglobin in the pressure range 200–500 MPa. In this region only elastic changes take place. The *synchronous* spectra suggest a positive correlation between the intensities of the amide II' and amide I band. This is in contrast to the helix band, where the intensities are anticorrelated with the degree of exchange. Analysis of the *asynchronous* spectra suggests that the increase in unordered structure happens first and that the increased exchange is the consequence of this structural change.

3.3.3 Emission Spectroscopy

From the discussion in the previous section it is clear that a pressure perturbation can be used to induce reversible (elastic) as well as irreversible (inelastic) effects. One technique that has been found to be especially useful for the study of elastic effects is spectral hole burning. We consider this technique in somewhat more detail than the more classical fluorescence and phosphorescence spectroscopies that can also be used for the determination of the elastic properties. We also discuss the use of these approaches for the study of protein unfolding.

Pressure-Induced Elastic Changes

Marden *et al.* [47] studied the pressure-induced unfolding of several heme proteins from the *fluorescence emission* of the protein tryptophans (Trp). They suggested a method to obtain the compressibility of the protein prior to unfolding from the fluorescence intensity changes with pressure. The decrease in fluorescence yield with increasing pressure suggests an increase in energy transfer due to the protein compression. Neglecting the effect of pressure on the refractive index, it can be shown, on the basis of the Förster theory, that the fractional change in fluorescence intensity is proportional to the difference in compressibility of the protein and the solvent. The observed decrease in intensity suggests that the protein is more compressible than the water. This implies that the observed compressibility is much larger than that observed with ultrasound techniques. However, the fluorescence measurements refer to the Trp-heme distance, whereas the sound velocity technique measures an average protein value.

The *phosphorescence emission* of Trp residues in proteins is a sensitive monitor of the local protein structure. The phosphorescence lifetime technique can be used as an indicator of structural flexibility to investigate pressure effects on monomeric as well as multimeric proteins [48]. At low pressures (<100 MPa), the elastic effects predominate, suggesting a compression effect of the pressure on the protein. At higher pressures, a loosening effect is observed, which is attributed to pressure-induced hydration of the protein. This seems to be consistent with the contribution of internal cavities and hydration of the polypeptide to the compressibility.

Freiberg *et al.* [49] measured the pressure effects on the absorption and *fluorescence emission spectra* of photosynthetic light-harvesting pigment-protein complexes from purple bacteria *Rhodospirillum rubrum* at room temperature and at 77 K. The pressure-induced frequency shifts of the pigments were interpreted in terms of the compressibility of the protein matrix. It was found that compressibility of the proteins surrounding the different pigments is different, thus revealing the local specificity of the elastic properties of the proteins. A compressibility of 25 ± 5Mbar^{-1} is obtained for the bacteriochlorophyl surroundings and 10 ± 2 Mbar^{-1} for the surroundings of the spirilloxanthin pigment.

A more sophisticated luminescence method, *spectral hole burning,* has been applied to measure protein compressibility in the low-pressure region. In this site-selective spectroscopic technique a laser with very narrow bandwidth (10^8 Hz) is used to burn a spectral hole into the inhomogeneously broadened absorption band. The hole arises because the laser light alters photochemically or photophysically a significant population of those molecules with transition frequencies corresponding to the laser (burning) frequency. Although this technique is restricted to chromoproteins and very low temperature conditions (ca 4 K), it is, in theory, possible to extend the method to the aromatic chromophores of the side chains of the proteins.

The compressibility of a protein matrix surrounding the chromophore, can be obtained from the pressure-induced frequency shifts. This is based on a microscopic model which takes into account the long range *induced-dipole induced-dipole* interactions between the protein matrix and the chromophore [50].

$$\Delta\nu = 2(\nu_b - \nu_{vacuum})\beta\Delta p . \tag{3.19}$$

Here $\Delta\nu$ is the shift of the hole frequency caused by the Δp pressure change, ν_b is the frequency where the hole was burnt and ν_{vacuum} is the frequency of the electronic transition of the isolated chromophore. Because of the high spectral resolution, effects of pressures as low as 1 2 MPa can be measured.

It has to be mentioned that all of the above luminescence methods detect the compressibility of the interior of the protein molecule, which is different from that determined by the ultrasonic measurements. By luminescence one measures the changes in the distances of the atoms of the protein from the chromophore. These distances change mainly because of the compression of the voids, while the ultrasonic measurements detect the effect of the hydration as well.

Pressure-Induced Unfolding

Weber and coworkers found the fluorescence emission of Trp and small molecules bound to proteins to follow the pressure-induced unfolding of single-chain proteins [51]. It was observed that the Trp environment becomes more polar at high pressures and that the volume changes are invariable small (<1 %) compared to the volume of the protein itself. Similarly, fluorescence polarization experiments suggest that, upon unfolding at high pressures, there is considerable penetration of water into the protein interior. Although the observed effects are in most cases reversible, Raman and infrared spectroscopy studies on the same proteins, but at higher concentrations, revealed irreversible effects from extensive aggregation of the proteins [34,38].

More recent work has concentrated around the effect of pressure on subunit interactions in multimeric enzymes and proteins. It is invariably observed that pressure dissociates multisubunit proteins into monomers at low pressures (<300 MPa). In many cases the effects are reversible. Of particular interest is the observation that the recovery of the enzyme activity lags behind the reassociation of the subunits. This suggests a 'conformational drift' of the isolated subunits. Dissociation of tetramers and large aggregates, such as viruses, reveals an apparent violation of the law of mass action. This can be understood in terms of deterministic equilibria in the formation of these aggregates in contrast to the classical stochastic equilibria of dimers. The role of deterministic equilibria is even more evident in the case of assembly and disassembly pathway of viruses [52]. The overall conclusion from this work is that pressure affects noncovalent interactions in protein folding, and the assembly of proteins and viruses. Many additional effects can be explained by assuming that pressure induces a number of intermediate conformations between the folded and unfolded state of proteins. This demonstrates the high plasticity of proteins.

3.3.4 NMR Spectroscopy

The pressure-induced effects on the native structure of lysozyme was studied with NMR at 750 MHz [53]. The chemical shifts of 26 protons was followed up to 200 MPa. The main result is a compaction of the hydrophobic core consisting of bulky side chains. By contrast, it was found that the compaction is restricted to the α helical region in the crystal structure (see next section). The pressure effect on the structural dynamics of BPTI was recently estimated by the same research group from the chemical shifts of the individual hydrogen bonds of the peptide bonds [54]. From the linear dependence of the chemical shifts on pressure, a pressure-independent compressibility was assumed up to 200 MPa. This is a rather surprising result. But given the rather small pressure range, the change in compressibility might be with the limit of experimental detection.

The contributions from high NMR spectroscopy to the description of the unfolding behavior of proteins under pressure is reviewed by Jonas in thisbook. In the case of lysozyme it was possible to follow the unfolding from several amino acid residues located in different parts of the protein. As expected this reveals that different parts of the protein show a slightly different pressure response. The cold-induced unfolding of ribonuclease A was studied by Zhang *et al.* [43] with 1D and 2D ^{1}H NMR. In these experiments advantage was taken of the fact that the freezing point of water is depressed by high pressure, thus allowing experiments in the liquid state down to −20°C at 200 MPa. These experiments can therefore be considered to be pressure-assisted cold denaturation. The experiments suggest that the cold- and pressure-denatured protein may contain partially folded structures that are visible as an intermediate in the heat unfolding.

Hydrogen-exchange kinetics of the pressure-assisted cold denaturation of ribonuclease [55] has revealed that this state differs markedly from the temperature- and pressure-denatured state. It is markedly different from a random coil due to patches of residual secondary structure. The situation may be quite different for other proteins, where little secondary structure may be found. Hydrogen exchange studied with infrared spectroscopy was discussed in section 3.3.2.

3.3.5 Diffraction and Scattering Techniques

The compressibility of lysozyme has been obtained from X-ray diffraction by Kundrot and Richards [56]. As expected the compression is nonuniformily distributed. The domain of the α helices is found to be more compressible than the other domain, which is essentially β structure. It should be noted that this compressibility does not include the water of hydration. The thermal expansion of lysozyme is about 1.8% in the range between 100 K and 298 K [57].

Small-angle X-ray scattering (SAXS) is a technique that provides information about the overall shape and size of macromolecules. In this respect it gives complementary information on the unfolded state of a protein to approaches such as infrared and NMR spectroscopy. Kleppinger *et al.* [58] used SAXS and FTIR to follow the pressure-induced changes in ribonuclease. From Kratky plots and distance distribution functions it was concluded that the protein remains compact up to 800 MPa. The radius of gyration increases slightly at 500 MPa, where a pretransition is observed with infrared spectroscopy. The main unfolding step takes place at 700 MPa. For lysozyme small changes in the radius of gyration have been reported to take place at about 400 MPa [59]. Winter and coworkers studied the pressure- and temperature-induced gel formation of β-lactoglobulin with FTIR and SAXS [60]. Under these conditions (80°C/1GPa) the radius of gyration increases dramatically. Since the protein is already highly associated under normal conditions, a quantitative analysis of the data is not a simple task. Further examples are discussed elsewhere in this book.

3.3.6 High-Pressure Computer Simulations

The relation between notions of volumes and surfaces defined in computer calculations and the ones used in thermodynamics of solutions have been discussed in section 3.2.1. A critical discussion of this topic can be found in Paci and Velikson [61]. The Voronoi volume and its related computed compressibility agrees best with the experimental intrinsic compressibility.

Normal Mode Calculations

The compressibility of deoxymyoglobin has been estimated from normal mode analysis calculations up to 100 MPa by Yamato *et al.* [62]. In general, the helices are rigid but the interhelix regions are soft. Interestingly, the large cavities in the hydrophobic clusters do not make these clusters very compressible. Their relative compressibility depends on the size of the cavities, the largest cavities showing the largest compressibility. The distal cavity is also found to be the most compressible.

Molecular Dynamics Simulations

With the exception of the above-mentioned normal mode analysis study all high-pressure computer simulations reported so far have been of the molecular dynamics (MD) type. The first high-pressure MD simulations on BPTI were reported by Kitchen *et al.* [63]. No changes in the conformation were detected at 1 GPa, only the increased hydration of certain amino acids was observed. Subsequent MD simulations up to 2 GPa revealed changes in the secondary structure between 1 and 1.5 GPa [64]. These changes could be correlated with changes in the secondary structure observed with high-pressure infrared studies [35]. Van Gunsteren and coworkers [65] studied the unfolding of lysozyme. No net unfolding was observed at 1 GPa after 210 ps. Infrared studies indicate that the protein unfolds at 500 MPa [34]. Presumably, the origin of the absence of unfolding in the computer experiments might be kinetic. A more extensive discussion of this approach can be found in another chapter in this book.

3.4 Conclusion: Facts and Hypotheses

What is the general picture that emerges from nearly a century of pressure studies on proteins? Although the prediction of the behavior of one particular protein is still far away, a number of simple physical factors that control the pressure behavior is becoming clear.

At low pressures (<100-300 MPa), one may assume that the compression of the cavities and an increased hydration is the main effect. This pressure regime can be studied by a number of experimental approaches. However, the connection between these approaches is not always clear. One of the most promising techniques is ultrasonic velocimetry. Gekko and coworkers have recently shown that changes in the compressibility can be found for protein mutants. Since volume fluctuations are proportional to the isothermal compressibility, changes in the compressibility reflect changes in the dynamic behavior of the mutant protein [66].

The role of the solvent in protein behavior is giving very useful insights into the dynamics. Studies by Priev *et al.* [67] have shown that glycerol decreases the compressibility of the protein interior. These studies suggest a role for water as a lubricant for the conformational flexibility of proteins. In a recent paper Timasheff gives a strong warning against the possible misinterpretations of the role of cosolvents on protein reactions in disperse solutions [68].

Probing the difference between the pressure, heat and cold denaturation and unfolding is not easy in view of the possible side effects that result from the unfolding that make a simple analysis not straightforward. Heat-denatured proteins are extremely prone to aggregation. This is especially clear from the infrared spectra, where the formation of intermolecular hydrogen bonding gives rise to specific bands in the amide I region. As indicated, these specific structures may also occur, under certain conditions and with certain proteins, after compression. It remains to be investigated whether these effects are protein specific or not. It is also of particular interest that certain water-soluble polymers show pressure–temperature phase diagrams similar to those of proteins [69].

Finally, one may ask whether the pressure—temperature behavior of proteins is indeed unique among the biomacromolecules. Hayashi and coworkers have shown that starch also forms a gel by the application of pressure [70]. This suggests that proteins and starch show a similar behavior with regard to temperature and pressure. This raises the question of the presumed role of hydrophobic interactions in the stability of proteins.

Acknowledgements. The material from our research groups presented in this paper was supported by the Research Fund of the K.U.Leuven, F.W.O. Flanders, Belgium, and the European Community. L.S. thanks the Hungarian Academy of Sciences and F.W.O. Flanders for support.

References

3.1 K. Suzuki, Rev. Phys. Chem. Jpn. **29**, 91 (1960)
3.2 S.A. Hawley, Biochemistry **10**, 2436 (1971)
3.3 A. Zipp, W. Kauzmann, Biochemistry **12**, 4217 (1973)
3.4 Y. Taniguchi, K. Suzuki, J. Phys. Chem. **87**, 5185 (1983)
3.5 A. Weingand-Ziadé, F. Renault, P. Masson, Biochim. Biophys. Acta **1340**, 245 (1997)
3.6 J. Thevelein, J.A. Van Assche, K. Heremans, S.Y. Gerlsma, Carbohydrate Res. **93**, 304 (1981)
3.7 C. Hashizume, K. Kimura, R. Hayashi, Biosci. Biotech. Comm. **59**, 1455 (1995)
3.8 K. Gekko, Y. Hasegawa, J. Phys. Chem. **93**, 426 (1989)
3.9 N.S. Timasheff, Annu. Rev. Biophys. Biomol. Struct. **22**, 67 (1993)
3.10 D.J. Winzor, P.R. Wills, in Protein-Solvent Interactions, ed. by R.B. Gregory (Marcel Dekker, New York 1995) pp. 483
3.11 K. Heremans, K. Goossens, L. Smeller, in High pressure effects in molecular biophysics, ed. by J.L. Markley, D.B. Northrop, C.A. Royer (Oxford University Press, New York 1996) pp. 44
3.12 K.J. Frye, C.A. Royer, Protein Sci. **6**, 789 (1997)
3.13 R.B. Gregory, in *Protein-Solvent Interactions*, ed. by R.B. Gregory, (Marcel Dekker, New York 1995) pp.191
3.14 M. Hiebl, R. Maksymiw, Biopolymers **31**, 161, (1991)
3.15 J. Gomez, V.J. Hilser, D. Xie, E. Freire, Proteins: Struct, Funct. Genet **22**, 404 (1995)
3.16 J.J. Flores, E.L. Chronister, J. Raman Spectrosc. **27**, 149 (1996)
3.17 L. Smeller, K. Heremans, in *High Pressure Research in Bioscience and Biotechnology*, ed. by K. Heremans (Leuven University Press, Leuven 1997) pp.55
3.18 P.L. Privalov, Crit. Rev. Biochem. Mol. Biol. **25**, 281 (1990)
3.19 R. Winter, A. Landwehr, T. H. Brauns, J. Erbes, C. Czeslik, O. Reis, in *High Pressure Effects in Molecular Biophysics and Enzymology*, ed. by J. L. Markley, C. Royer, D. Northrup (Oxford University Press, New York 1996) pp.274

3.20 C. Weemaes, S.De Cordt, K. Goossens, L. Ludikhuyze, M. Hendrickx, K. Heremans, P. Tobback, Biotechnology and Bioengineering **50**, 49 (1996)

3.21 L. Smeller, P. Rubens, K. Heremans, Submitted for publication

3.22 T. Atake, C. A. Angell, J. Phys. Chem. **83**, 3218 (1979)

3.23 A.P. Sarvazyan, Annu. Rev. Biophys. Biophys. Chem. **20**, 321 (1991)

3.24 V. Chalikian, A.P. Sarvazyan, Th. Funck, Ch.A. Cain, K.J. Breslauer, J. Phys. Chem. **98**, 321 (1994)

3.25 V.N. Benolenko, T. Chalikian, T. Funck, B. Kankia, A.P. Sarvazyan, in *High Pressure Research in Bioscience and Biotechnology*, ed. by K. Heremans, (Leuven University Press, Leuven 1997) pp.147

3.26 D.P. Kharakoz, Biochemistry **36**, 10276 (1997)

3.27 T.V. Chalikian, M. Totrov, R. Abagyan, K.J. Breslauer, J. Mol. Biol. **260**, 588 (1996)

3.28 Y. Tamura, K. Gekko, Biochemistry **34**, 1878 (1995)

3.29 T.V. Chalikian, J. Völker, D. Anafi, K.J. Breslauer, J. Mol. Biol. **274**, 237 (1997)

3.30 D.P. Kharakoz, V.E. Bychkova, Biochemistry **36**, 1882 (1997)

3.31 C. Jung, G. Hui Bon Hoa, D. Davydov, E. Gill, K. Heremans, Eur. J. Biochem. **233**, 600 (1995)

3.32 R. Lange, N. Bec, V.V. Mozhaev, J. Frank, Eur. Biophys. J. **24**, 284 (1996)

3.33 E. Mombelli, M. Afshar, P. Fusi, M. Mariani, P. Tortora, J.P. Connelly, R. Lange, Biochemistry **36**, 8733 (1997)

3.34 K. Heremans, P. T. T. Wong, Chem Phys. Lett. **118**, 101 (1985)

3.35 K. Goossens, L. Smeller, J. Frank, K. Heremans, Eur. J. Biochem. **236**, 254 (1996)

3.36 L. Smeller, K. Goossens, K. Heremans, Vibrat. Spectrosc. **8**, 199 (1995)

3.37 L. Smeller, K. Goossens, K. Heremans, Appl. Spectrosc. **49**, 1538 (1995)

3.38 P.T.T. Wong, K. Heremans, Biochim. Biophys. Acta **956**, 1 (1989)

3.39 V. Mozhaev, K. Heremans, J. Frank, P. Masson, C. Balny, Proteins: Struct, Funct. Genet. **24**, 81 (1996)

3.40 K. Heremans, J. Van Camp, A. Huyghebaert, in *Food Proteins and Their Applications*, by S. Damodaran, A. Paraf (Marcel Dekker, New York 1997) pp. 473

3.41 G. Vermeulen, K. Heremans, in *High Pressure Research in Bioscience and Biotechnology,* ed. by K. Heremans (Leuven University Press, Leuven 1997) pp.67

3.42 N. Takeda, M. Kato, Y. Taniguchi, Biochemistry **34**, 5980 (1995)

3.43 J. Zhang, X. Peng, A. Jonas, J. Jonas, Biochemistry **34**, 8361 (1995)

3.44 P. Fusi, K. Goossens, R. Consonni, M. Grisa, P. Puricelli, G. Vecchio, M. Vanoni, L. Zetta, K. Heremans, P. Tortora, Proteins: Struct. Funct. Genet. **29**, 381 (1997)

3.45 L. Smeller, K. Heremans, Submitted for publication.

3.46 I. Noda, Appl. Spectros. **47**, 1329 (1993)

3.47 M.C. Marden, G. Hui Bon Hoa, F. Stetzkowski-Marden, Biophys. J. **49**, 619 (1986)

3.48 P. Cioni, G. B. Strambini, J. Mol. Biol. **263**, 789 (1996)

3.49 A. Freiberg, A. Ellervee, P. Kukk, A. Laisaar, M. Tars, K. Timpmann, Chem. Phys. Lett. **214**, 10 (1993)

3.50 M. Koehler, J. Friedrich, J. Fidy, Biochim. Biophys. Acta **1386**, 255 (1998)

3.51 G. Weber, *Protein Interactions*, (Kluwer Academic Publishers. Dondrecht 1992)

3.52 J. L. Silva, D. Foguel, A.T. Da Poian, P. E. Prevelige, Curr. Opin. Struct. Biol. **6**, 166 (1996)

3.53 K. Akasaka, T. Tezuka, H. Yamada, J. Mol. Biol. **271**, 671 (1997)

3.54 H. Li, H. Yamada, K. Akasaka, Biochemistry **37**, 1167 (1998)

3.55 D. Nash, B-S. Lee, J. Jonas, Biochim Biophys Acta **1297**, 40 (1996)

3.56 C.E. Kundrot, F..M. Richards, J. Mol. Biol. **193**, 157 (1987)

3.57 A.C. M. Young, R.F. Tilton, J.C. Dewan, J. Mol. Biol. **235**, 302 (1994)
3.58 R. Kleppinger, K. Goossens, M. Lorenzen, E. Geissler, K. Heremans, in *High Pressure Research in Bioscience and Biotechnology*, ed. by K. Heremans (Leuven University Press, Leuven 1997) pp.135
3.59 M. Kato, T. Fujisawa, Y. Taniguchi, T. Ueki, in *High Pressure Research in Bioscience and Biotechnology*, ed. by K. Heremans (Leuven University Press, Leuven 1997) pp.127
3.60 R. Malessa, G. Panick, R. Winter, K. Heremans, in *High Pressure Research in Bioscience and Biotechnology*, ed. by K. Heremans (Leuven University Press, Leuven 1997) pp.419
3.61 E. Paci , B. Velikson, Biopolymers **41**, 785 (1997)
3.62 T. Yamato, J. Higo, Y. Seno, N. Go, Proteins: Struct. Funct. Genet. **16**, 327 (1993)
3.63 D.B. Kitchen, L.H. Reed, R.M. Levy, Biochemistry **31**, 10083 (1992)
3.64 B. Wroblowski, J.F. Diaz, K. Heremans, Y. Engelborghs, Proteins: Struct. Funct. Genet. **25**, 446 (1996)
3.65 P.H. Hünenberger, A.E. Mark, W.F. van Gunsteren, Proteins: Struct, Funct. Genet. **21**, 196 (1995)
3.66 K. Gekko, Y. Tamura, E. Ohmae, H. Hayashi, H. Kagamiyama, H. Ueno, Protein Sci. **5**, 542 (1996)
3.67 A. Priev, A. Almagor, S. Yedgar, B. Gavish, Biochemistry **35**, 2061 (1996)
3.68 S. N. Timasheff, Proc. Nat. Acad. Sci. USA **95**, 7363 (1998)
3.69 S. Kunugi, K. Takano, N. Tanaka, K. Suwa, M. Akashi, Macromolecules **30**, 4499 (1997)
3.70 S. Ezaki, R. Hayashi, in *High Pressure and Biotechnology*, ed. by C. Balny, R. Hayashi, K. Heremans, P. Masson (John Libbey Eurotext Ltd, Montrouge 1992) pp. 163

4 High-Pressure NMR Spectroscopy of Proteins

Lance Ballard[1] and Jiri Jonas[2]

[1]School of Chemical Sciences, University of Illinois at Urbana-Champaign, 600 South Mathews Avenue, Urbana, IL 61801, USA
E-mail: ballard@scs.uiuc.edu
[2]Beckman Institute for Advanced Science and Technology, University of Illinois at Urbana-Champaign, 405 North Mathews Avenue, Urbana, IL 61801, USA
E-mail: j-jonas@uiuc.edu

Abstract. Advanced high-resolution NMR spectroscopy, including 2D NMR techniques, combined with the high-pressure capability represents a powerful new tool in studies of proteins. Selected results taken from recent studies illustrate the high information content and the range of problems that can be investigated. Design features and performance characteristics of high-sensitivity, high-resolution, variable-temperature NMR probes operating at 500 MHz and at pressures up to 900 MPa are described. The main portion of this chapter deals with an overview of several recent studies or studies in progress from our laboratory using 1D and 2D high-resolution, high-pressure NMR spectroscopy to investigate the pressure-induced reversible unfolding and cold denaturation of proteins. The following proteins were studied: ribonuclease A, lysozyme, apomyoglobin, *arc* repressor, and ubiquitin.

4.1 Introduction

Protein folding, the relationship between the amino acid sequence and the structure and dynamic properties of the native conformation of proteins, represents one of the central problems of biochemistry and biotechnology. The stability of proteins has been extensively studied by temperature and chemical perturbations (for a review, see Dill and Shortle [1]). High temperature changes both the energy content and the volume of a system. Since proteins are flexible polymers folded into three-dimensional structures which are stabilized by interactions of strengths not much larger than the thermal energy, the internal interactions of proteins are changed by temperature in ways that cannot be easily foreseen. Similarly, the effects of denaturants on proteins are difficult to interpret because they also modify the chemical potential of proteins in unpredictable ways by binding to multiple sites with different affinities. The use of hydrostatic pressure to perturb protein structure is more recent and less frequent than the use of heat and chemical agents [2, 3]. However, the effects of pressure are easier to interpret because pressure perturbs internal interactions exclusively by changing the distances between the components, while the total energy of the system remains almost constant. In addition, by taking advantage of the phase behavior of water, high pressure can substantially lower the freezing point of an aqueous protein solution. Therefore, by applying high pressure one can investigate in detail not only pressure-denatured proteins, but also cold-denatured proteins [2] in aqueous solution.

It is well known that pressure affects chemical equilibria and reaction rates. In addition, studies of simple liquids [4] have indicated that quite often volume effects determine the mechanism of a specific dynamic process, whereas temperature only changes the frequency of the motions without actually affecting the mechanism.

It should be noted that the range of pressures used to investigate biochemical systems is from 0.1 MPa (1 bar) to 1 GPa (10 kbar). Pressures within this range change intermolecular distances and affect conformations but do not change covalent bond distances or bond angles. In fact, pressures in excess of 3 GPa (30 kbar) are required to change the electronic structure of a molecule [5]; such pressures are not generally useful for biochemical studies since they modify covalent bonds. There are several excellent general reviews [3, 6] covering the wide spectrum of biochemical problems investigated by high-pressure techniques.

Increasing attention has recently been focused on denatured and partially folded states of proteins since determination of their structure and stability may provide novel information on the mechanisms of protein-folding. The native conformations of hundreds of proteins are known in great detail from structural determinations by X-ray crystallography and, more recently, NMR spectroscopy. However, detailed knowledge of the conformations of denatured and partially folded states is lacking and represents a serious shortcoming in current studies of protein stability and protein-folding pathways [7–10].

From the results obtained so far in our laboratory and elsewhere [3], it is clear that pressure denaturation is more easily controlled and represents a less drastic perturbation of protein structure than thermal or chemical denaturation. The combination of advanced high-resolution NMR techniques with a high-pressure capability represents a powerful new experimental tool in studies of protein folding. Recent advances in superconducting magnets make it possible to attain high homogeneity of the magnetic field over the sample volume so that even without sample spinning one can still obtain high resolution. Achieving an NMR linewidth of 1.5 Hz for a sample diameter of 8 mm at the proton frequency of 500 MHz at 500 MPa (5 kbar) is an impressive feat. The experimental capability of recording high-resolution NMR spectra on dilute spin systems opens new directions for high-pressure NMR spectroscopy on biochemical systems.

The high-resolution, high-pressure NMR studies in our laboratory have concentrated on addressing the following main points:

- Validity of the two-state model of folding
- Structural differences for urea-, pressure-, and heat-denatured states
- Differences in behavior of proteins with and without disulfide bonds
- Relationship between the structure of pressure-assisted, cold-denatured states and the structure of early folding intermediates
- Effects of mutagenesis on pressure, urea, and thermal stability
- Detection of predissociation and intermediate states

This chapter is organized as follows: A brief discussion of the rationale for using high pressure in studies of protein-folding is given in the Introduction. Section 2, which presents an overview of high-pressure, high-resolution NMR instrumentation, is followed by Sect. 3, describing the model proteins used in our NMR studies. The results and discussion section (Sect. 4) gives several selected examples of NMR studies of proteins which illustrate the high information content of combining advanced high-resolution NMR experiments with high-pressure techniques.

4.2 Experimental Methods

When discussing experimental techniques for high-pressure NMR studies of biochemical systems, three criteria must be compared – the pressure range, the spectral resolution, and the sensitivity. While the first two items have a fairly straightforward relationship to NMR probe design, the relationship for sensitivity (*S/N*) is much more complex. Due to the inherently low *S/N* of NMR (compared to optical techniques), as well as the desire in biochemical applications to work with low concentrations (≤1 mM), particular attention should be placed on understanding *S/N* in NMR. Abragam [11] has shown

$$S/N \propto N\sqrt{\eta V_s Q}\,H^{\frac{3}{2}}\,, \tag{4.1}$$

where N is the number of nuclei per unit volume (concentration), η is the filling factor (sample volume/coil volume), V_s is the sample volume, Q is the electronic quality factor for the circuit, and H is the magnetic field strength. The utility of (4.1) will become apparent when comparing different experimental approaches to high-pressure NMR.

Experimental techniques for high-pressure NMR can be divided into three categories – the high-pressure vessel or autoclave, the high-pressure cell, and the diamond anvil cell (DAC). Brief descriptions and representative applications of these variants are discussed in Sect. 4.2.1, while a side-by-side comparison is presented in Table 4.1. Depending on the system under study, each method can offer particular advantages and disadvantages. For most biochemical applications, however, the autoclave approach offers the most versatility and therefore Sect. 4.2.2 describes this method in more detail.

4.2.1 Survey of High-Pressure NMR Techniques

Autoclave or High-Pressure Vessel

The autoclave design historically represents the first type of high-pressure NMR probe [12–14]. This approach places both the NMR sample and the electronic RF circuit inside a high-strength, non-magnetic vessel, with either a gas or a liquid medium used to transmit pressure externally to the glass sample cell. The pressure range for the autoclave is limited only by the strength of the metallic vessel, and is not limited in any way by the strength of the glass sample cell. This fact is advantageous in that it allows one to use a thin-walled glass sample cell and still achieve high pressures. The use of a high-strength metal vessel also contributes some degree of safety to this approach.

Pressure limits for the autoclave design depend both on the material used to construct the vessel and on the vessel dimensions. The highest reported pressure range for an autoclave NMR probe used in biochemical applications is 900 MPa [15]. High spectral resolution ($<3.0 \times 10^{-9}$) in a high-pressure vessel is achieved through the use of thin-walled glass sample cells with precise commercial tolerances. These

Table 4.1. Comparison of high-pressure NMR techniques

	Pressure[a] (MPa)	^{1}H Freq.[b] (MHz)	Sample ϕ[c] (mm)	Res.[d] (x10^{-9})	Rel. *S/N*[e]	Ref. Instrum.	Ref. Biochem.
Autoclave	900	500	5 -10	<3.0	~50%	12-23	2, 24-31
High-pressure cell	250	750	1	<1.1	~10%	32-36	36-40
DAC	>2500	233	n.a.	Poor	Poor	41-46	None

[a]Highest test pressure reported for autoclave vessels and the typical reported working pressures for high-pressure cells.
[b]Highest reported NMR frequency to date.
[c]Tube inner diameter for thick-walled high-pressure cells, and the tube outer diameter for thin-walled cells in autoclave systems.
[d]Best reported resolution values (absolute units), and not necessarily the 'typical' resolution.
[e]Relative *S/N* is the approximate sensitivity relative to a 'typical' commercial 5 mm probe operating at the same frequency. The autoclave value is based on experimental measurements, while the high-pressure cell value is calculated based on filling factor and volume considerations (see text).

thin-walled glass sample cells are also advantageous for *S/N* [see (4.1)] since they allow high filling factors and large sample volumes (5—10 mm diameter sample sizes).

In general, the autoclave approach is well-established [12—23] with an increasing number of biochemical studies being reported [2, 24—31], (Table 4.1). The major limitations of this method are the restrictions imposed on the electronic RF circuit by the metallic pressure vessel. Space limitations in the vessel restrict the complexity of the NMR probe, while the distance between the sample coil and the RF tuning circuitry leads to sensitivity loss and problems in achieving high frequencies. These electronic limitations are reflected in the Q-value of (4.1). Part of these losses, however, can be compensated for by larger sample sizes. For example, in our own 500 MHz autoclave vessel with an 8 mm tube diameter [16, 17], the measured S/N was ~260. This value is ~50% of the measured S/N for a commercial 5 mm NMR probe on the same system. With continuing advances in high-pressure NMR probe design, it is anticipated that the comparison between commercial and high-pressure autoclave NMR probes will continue to show modest improvements.

High-Pressure Cell

A second method for high-pressure NMR is the high-pressure cell approach, where the sample is contained in a high-strength sample cell (usually quartz, sapphire, or fused silica), with pressure applied directly (internally) to the sample [32—35]. The pressure range in this case is limited by the strength of the glass walls. The obvious advantage of the high-pressure cell is that it can be used on any commercial NMR spectrometer without the need to construct special NMR probes. This advantage allows one to achieve high spectral resolution and allows the user ready access to advanced multinuclear NMR techniques on high-field magnets. Yamada and co-workers, for example, have recently studied proteins such as BPTI and lysozyme to 200 MPa on a 750 MHz instrument with the high-pressure cell approach [36, 37].

Yet despite its advantages in implementation, the high-pressure cell faces serious limitations in strength. For example, sapphire tubes generally have a working pressure limit of ~20 MPa [33, 34], while thick-walled quartz pressure tubes have a working pressure limit of ~200 MPa [32, 36]. These modest pressure ranges are adequate for many types of studies, but they do not allow studies of complete protein unfolding in many systems. In addition, the thick sample cell walls necessary for high-pressure applications significantly limit the sensitivity. Yamada et al., for example, used a high-pressure cell in a 5 mm NMR probe with a 1 mm sample diameter [36], limiting both the filling factor and the sample volume. According to (4.1), the *S/N* of such a probe would be ~10% of the measured value for a commercial NMR probe at any given magnetic field. This, in turn, would necessitate that one should use a higher magnetic field strength or a higher sample concentration to compensate for the loss in *S/N*. However, even with these limitations, the ease of implementation for high-pressure cells has contributed to a growing number of biochemical studies performed with this technique [36–40] (Table 4.1).

Diamond Anvil Cell (DAC)

The most recent high-pressure NMR technique is the DAC approach, in which the sample is pressed between two diamonds [41–46]. A metallic ring placed between the diamonds and around the sample cavity provides the necessary high-pressure seal. With this method, an extremely high pressure range (>25 GPa) can be achieved [41].

The high pressure range is indeed the most significant advantage of the DAC, and has led to widespread application of the DAC in biochemical studies using optical techniques. A major problem for NMR, however, is the metallic sealing ring. This ring introduces significant field inhomogeneity and leads to poor spectral resolution (>100 Hz). In addition to the poor resolution, the sample size in the DAC is generally of microliter proportions, which greatly limits the sensitivity. While the DAC-NMR method has been used on some small molecular systems, no biochemical applications of DAC-NMR have been reported to date.

4.2.2 Instrumentation for the Autoclave Approach

As was discussed in Section 4.2.1, both the autoclave approach and the high-pressure cell approach have been used in biochemical studies. With its higher pressure range and higher sensitivity (Table 4.1), though, the authors feel that the autoclave method offers more versatility in studies of biochemical systems. In this section, the specialized instrumentation necessary for the autoclave approach is described in detail, with particular emphasis on the equipment in our own laboratory.

Pressure Generation System

To illustrate the type of equipment required for generating high hydrostatic pressures, a schematic of the 1000 MPa pressurizing system used in our laboratory is provided in Fig. 4.1 [17]. In this setup, an Enerpac hand pump (Butler, WI) generates a pressure that is amplified >20:1 by a pressure intensifier. The pressurizing fluid consists of vacuum pump oil on the low-pressure side, and 90/10 methylcyclohexane/2,2,4-trimethylpentane on the high-pressure side. The high-pressure tubing, fittings, and valves are supplied by Harwood Engineering (Walpole, MA). The pressure is measured directly by a Heise-Bourdon gauge (<700 MPa) supplied by Dresser Industries (Newton, CT) or a manganin cell pressure transducer (to 1300 MPa) supplied by Harwood Engineering. For proton measurements, carbon disulfide (CS_2) is used to transmit pressure to the NMR probe. The corrosive CS_2 is introduced into the system with a special stainless steel loading pump (High Pressure Equipment, Erie, PA) and isolated from the gauges with a pressure separator.

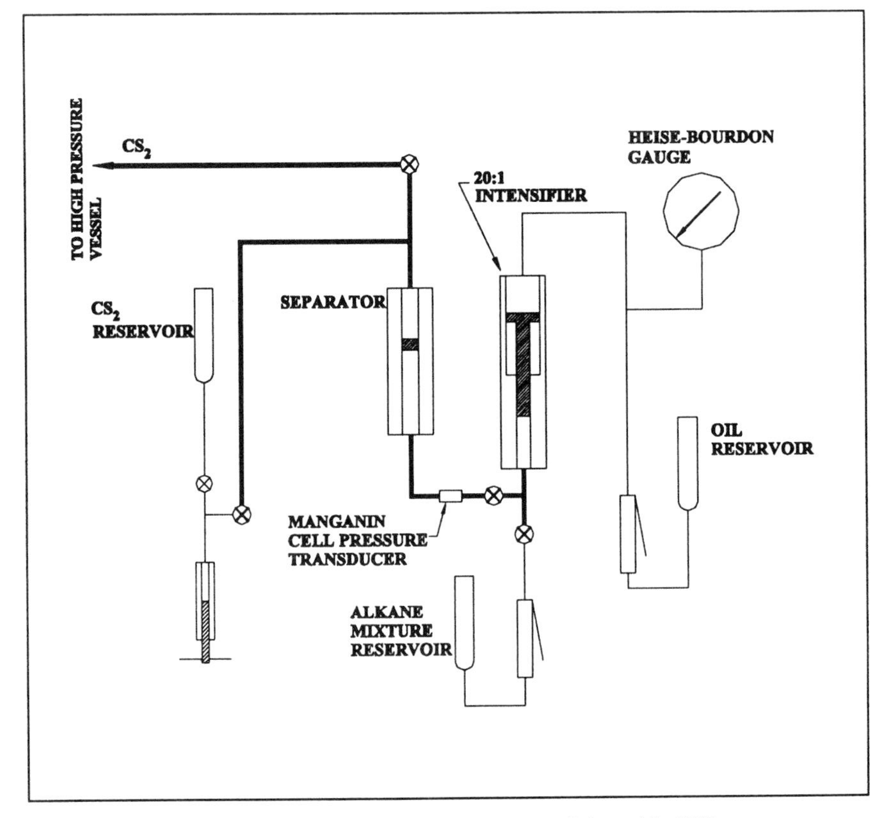

Fig. 4.1. Schematic of 1000 MPa pressurizing table [17]

Spectrometer and Magnet

For our high-pressure, high-resolution NMR studies of proteins, two different systems are used which operate at two different proton Larmor frequencies: 300 and 500 MHz. Both systems have been described previously [15–18]. The 300 MHz system consists of a commercial NMR spectrometer (former GE system with a Tecmag Scorpio interface) and an Oxford wide-bore superconducting magnet (ϕ = 89 mm, 7.4 T). The 500 MHz system consists of a Varian UNITY INOVA spectrometer and an Oxford wide-bore superconducting magnet (ϕ = 89 mm, 11.7 T). The strength of the high-pressure vessel is directly related to the thickness of the metallic walls, meaning that the highest pressures will be achieved with a wide bore magnet. For our own high-pressure systems, this allows pressures up to 900 MPa [15]. It should be noted, though, that autoclave-style probes can also be built for narrow-bore systems, although with lower pressure ranges.

High-Pressure Vessel

The materials used for high-pressure NMR probes must be non-magnetic and of high mechanical strength. Our previous experience with both beryllium copper and titanium alloys has led us to use titanium alloys as much as possible, due to the better corrosion resistance of titanium and its ease of machining [18]. Our two most recent pressure vessels [15–17] have been constructed of the high-strength beta alloy of titanium, 3Al-8V-6Cr-4Zr-4Mo (RMI Titanium, 180 kpsi YS). A schematic of the 500 MPa, 500 MHz vessel is shown in Fig. 4.2. The vessel has a 1.9 cm inner diameter to accommodate NMR sample sizes of up to 10 mm, and has a calculated burst pressure of >1200 MPa at its seal flanges. A 900 MPa, 300 MHz probe has a similar design, with a smaller 1.5 cm inner diameter (8 mm tube) and a calculated burst pressure of >1500 MPa at its seal flanges. For both systems, the bottom of the vessel is closed by a high-pressure screw-in plug with Bridgman-style seals. The top plugs of the vessels are sealed with metallic C-seals tightened by drivers. The bottom plugs contain thermocouples, while the top plugs contain the electrical RF feedthroughs. Channels hollowed into the outer surface of the vessels allow circulation of heating/cooling fluid for temperature control. The temperature range for both vessels is –30 to 80°C. The vessels are supported by aluminum legs connected to an aluminum stand. A complete set of drawings for both vessels and assorted components, as well as more detailed descriptions, has been published separately [17].

High-Pressure NMR Probes and RF Feedthroughs

Our current high-pressure NMR probe design features a single-coil probe double-tuned to both proton and deuterium frequencies. Recent improvements to the design have been the inclusion of a porcelain chip capacitor inside the pressure vessel [15, 17] and the introduction of a single-turn machined copper RF sample coil for high frequencies [7, 16]. Information about the RF circuits has been published in extensive detail [17].

Aside from the RF circuit, the most important feature of the autoclave-style NMR probe is the high-pressure electrical RF feedthrough [15]. Various approaches have been used in the past, but today all of our probes use the high-strength, high-reliability Berylco feedthrough shown in Fig. 4.3. In this design, the short Berylco 25 feedthrough is seated in a matching Vespel SP-1 (DuPont) polyimide plastic insula-

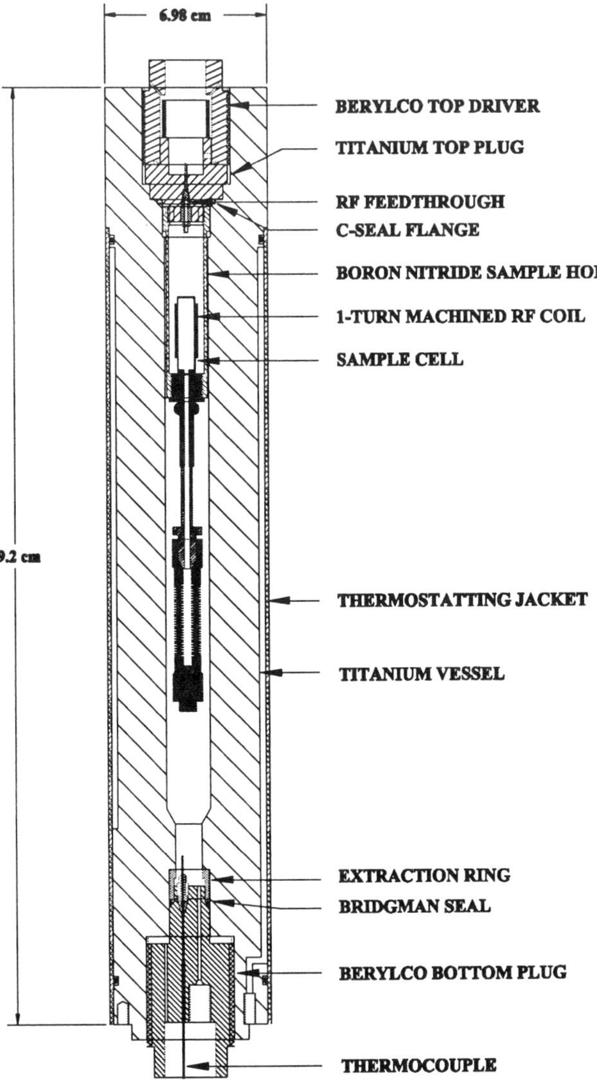

6.98 cm
39.2 cm
BERYLCO TOP DRIVER
TITANIUM TOP PLUG
RF FEEDTHROUGH
C-SEAL FLANGE
BORON NITRIDE SAMPLE HOLDER
1-TURN MACHINED RF COIL
SAMPLE CELL
THERMOSTATTING JACKET
TITANIUM VESSEL
EXTRACTION RING
BRIDGMAN SEAL
BERYLCO BOTTOM PLUG
THERMOCOUPLE

Fig. 4.2. Schematic of the titanium high-pressure vessel used in the wide-bore 11.7 T superconducting magnet. For clarity, only one of the three RF feedthroughs is shown [17]

tor. Copper wires are soldered on to either end of the feedthrough to complete the circuit. The feedthrough has been tested to pressures of 1.0 GPa.

Sample Cells

The type of sample cell used is related to the type of experiment performed. In general, we use two types of high-pressure sample cells [18] that differ in the way the

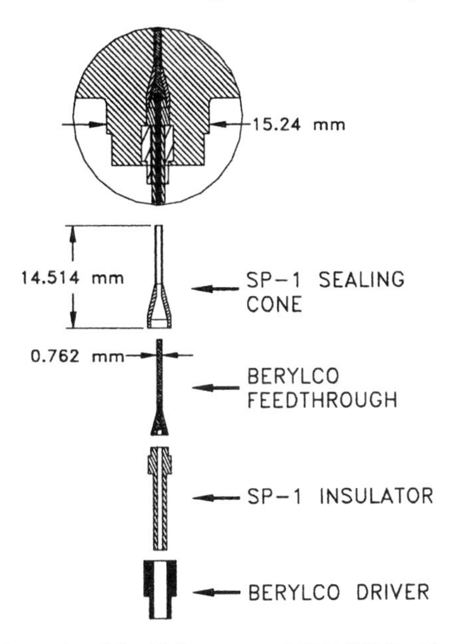

Fig. 4.3. Schematic of the high-pressure NMR RF feedthrough [15]

pressure is transmitted to the sample. One is a piston-based design, while the other uses a bellows system. Both sample cells are shown in Fig. 4.4.

In the piston-style design (Fig. 4.4A), the sample cell consists of a 6.5, 8, or 10 mm NMR tube connected via capillary tubing to precision-bore Pyrex tubing containing the piston. The piston, which is made of Teflon, has a plunger equipped with a driver to provide tighter contact with the glass surface. Two O-rings made of Viton (Parker Seal Group) are used on the piston to limit sample contamination and to provide a good seal.

In the bellow-style design (Fig. 4.4B), a glass (Pyrex) or quartz-to-metal (SS-316) seal (Quartz Scientific, Inc., Palo Alto, CA) connects the 8 or 10 mm glass sample chamber to the SS-316 tubing. Stainless-steel bellows (Mechanized Science Seals, Inc., Los Angeles, CA) accommodate the volume change of the liquid due to compression. Hermetic closure of the sample cell is attained by a copper cone brazed to the SS-316 bellow plug and seated in a flange made of stainless steel.

Each type of sample cell has its own advantages and disadvantages. The bellows-style sample cell represents a hermetically closed system where the danger of contamination of the sample does not exist. The sample can be kept in such a system for a long time and re-measured later. However, the bellows themselves, as well as the glass-metal seals, are quite expensive, and the glass-metal seal is also quite fragile. On the other hand, the piston-style sample cells are not hermetic in repeated decompression runs, and the sample should be replaced once it has been measured. However, we seldom note any serious contamination from use of the piston-style cells, which are much simpler and relatively inexpensive. The piston-style sample cell is used almost exclusively in our studies of biochemical systems.

Each type of sample cell has its own advantages and disadvantages. The bellows-style sample cell represents a hermetically closed system where the danger of contamination of the sample does not exist. The sample can be kept in such a system for

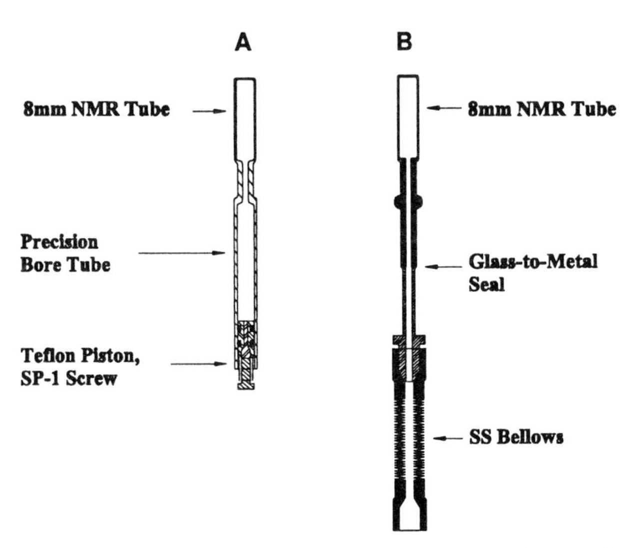

Fig. 4.4. Types of sample cells used in the high-pressure NMR studies. (*A*) The 8 mm piston-type sample cell; (*B*) the 8 mm bellows-type sample cell

a long time and re-measured later. However, the bellows themselves, as well as the glass-metal seals, are quite expensive, and the glass-metal seal is also quite fragile. On the other hand, the piston-style sample cells are not hermetic in repeated decompression runs, and the sample should be replaced once it has been measured. However, we seldom note any serious contamination from use of the piston-style cells, which are much simpler and relatively inexpensive. The piston-style sample cell is used almost exclusively in our studies of biochemical systems.

Performance Characteristics and Applications

Standard tests are routinely performed to assess our NMR pressure probes. This includes a resolution test (1% chloroform in either acetone-*d6* or chloroform-*d*) and a sensitivity test (0.1% ethylbenzene and 0.2% tetramethylsilane in chloroform-*d*). Figure 4.5 contains examples of standard resolution spectra obtained at 300 and 500 MHz [16], clearly demonstrating the excellent resolution attainable without sample spinning –0.7 Hz at 300 MHz (2.3×10^{-9}) and 1.5 Hz at 500 MHz (3.0×10^{-9}). These resolution values are independent of pressure, as we have demonstrated previously [15]. As was noted in Section 4.2.1, the sensitivity of the 8 mm 500 MHz probe is ~50% of the *S/N* for a commercial 5 mm Varian probe in the same system.

To illustrate a practical 1D biochemical NMR application of the 900 MPa, 300 MHz probe, Fig. 4.6 is a stacked plot demonstrating the decrease in native His$^{\varepsilon 1}$-15 intensity and the increase in denatured His$^{\varepsilon 1}$-15 intensity with pressure for a pH* 2.0 lysozyme sample at 37.5°C [15]. With the peak integrals determining the amount of denaturation, one can calculate the ΔV for His$^{\varepsilon 1} - 15$ to be -27 ± 2 cm^3/mol. The utility of the pressure vessel is clearly demonstrated by this calculation and Fig. 4.6, as the large pressure range available was necessary in order for complete denaturation to occur.

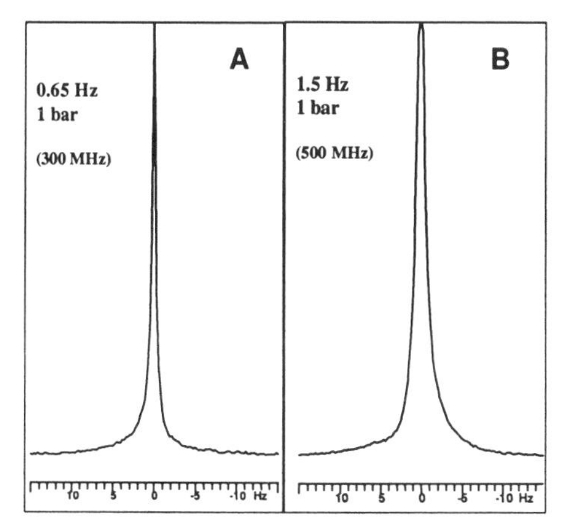

Fig. 4.5. $CHCl_3$ peak of the lineshape sample (1% v/v $CHCl_3$ in acetone-*d6*, 300 MHz; 1% v/v $CHCl_3$ in chloroform-*d*, 500 MHz) measured at 1 bar in an 8 mm sample cell at (A) 300 MHz and (B) 500 MHz [16]

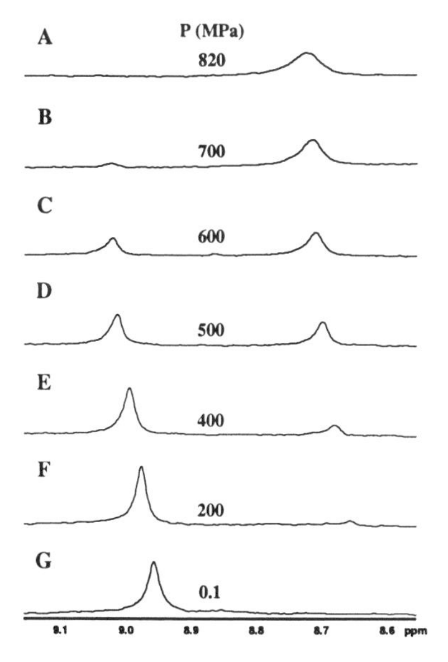

Fig. 4.6. Stacked 1H NMR plots of the lysozyme (pH* 2.2) histidine region at selected pressures and 37.5°C. Note the disappearance of the native Hisε1-15 residue peak (~8.95 ppm) and the appearance of the denatured Hisε1-15 peak (~8.7 ppm) with pressure. Complete denaturation is achieved between 7 and 8.25 kbar [15]

While one can readily appreciate the advantages of higher sensitivity on dilute biochemical studies using 1D NMR techniques, we feel that an even more important extension of this work is in the field of high-pressure 2D NMR. As an example of

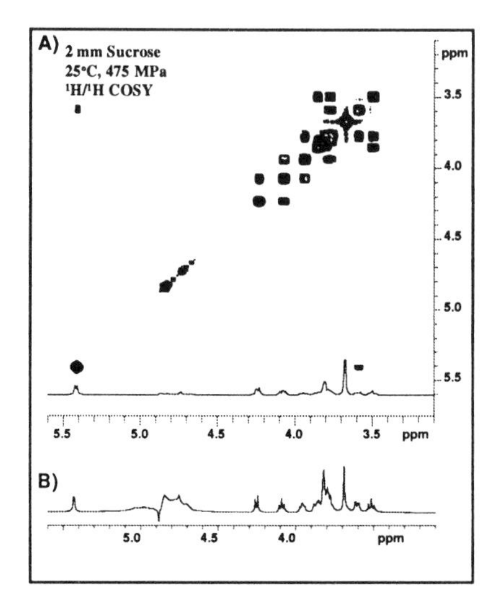

Fig. 4.7. 2D ^{1}H/^{1}H PRESAT COSY spectrum of 2 mM sucrose at 25°C and 4750 bar obtained at 500 MHz. (*A*) The absolute value 2D COSY spectrum; (*B*) the 1D PRESAT spectrum [17]

this capability, we include Figs. 4.7 and 4.8, obtained with the 500 MPa, 500 MHz probe. Figure 4.7 is a 2D ^{1}H/^{1}H COSY spectrum of 2 mM sucrose at 475 MPa, which demonstrates the probe capabilities [17]. Figure 4.8 contains four two-dimensional ^{1}H/^{1}H NOESY spectra of a β-sheet region for two fragments (wildtype and F29W mutant) of 0.6 mM N-domain troponin C, both at 150 and 400 MPa. In this figure, one observes the disappearance of the β-sheet crosspeak for the N-domain mutant at high pressure, which is in contrast to the wildtype protein and indicates a structural difference. As with Fig. 4.7 for the COSY sucrose example, the purpose of this figure is to demonstrate the applicability of the high-pressure autoclave-style probe to biochemical NMR applications.

4.3 Model Proteins

This chapter is restricted to a systematic investigation of the effects of pressure on the structure and dynamics of several model proteins which have been selected using the following criteria: (a) a known structure by NMR and assigned NMR spectra; (b) a reversible unfolding process; and (c) proteins for which thermal and urea denaturation were already studied. The proteins in these high-pressure NMR studies include ribonuclease A, hen lysozyme, *arc* repressor, and apomyoglobin. This section provides background information on these protein systems. Selected high-pressure NMR studies of these proteins are presented in Section 4.4.

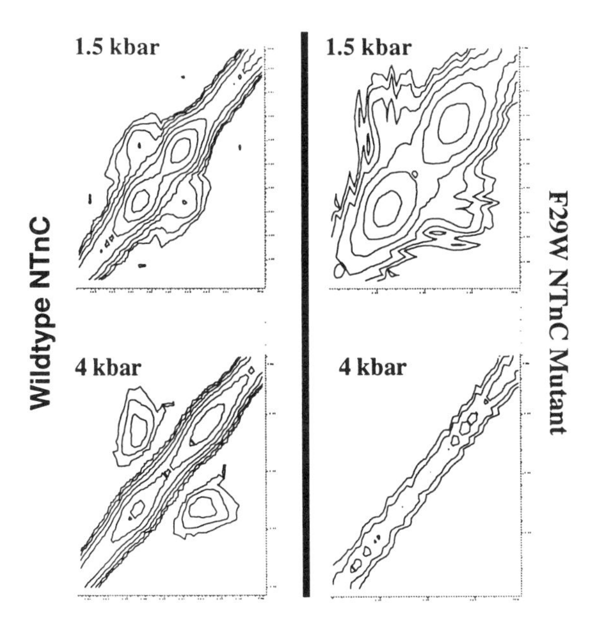

Fig. 4.8. High-pressure 2D $^{1}H/^{1}H$ NOESY spectra of 0.6 mM N-domain troponin C fragments, comparing the β-sheet D36α-D74α crosspeak of the F29W mutant to the wildtype. Loss of the NOESY crosspeak may indicate a decrease in stability for the mutant

4.3.1 Ribonuclease A

Ribonuclease A (RNase A) from bovine pancreas is a model protein that has been extensively investigated in classical protein-folding studies [47]. The kinetics of RNase A refolding are very complex because of the existence of at least five distinct unfolded species that contain different X-Pro peptide bond conformations. An early folding intermediate has been characterized by the pulse-labeling hydrogen exchange technique in a slow refolding pathway [48, 49], and a recent study has identified two folding intermediates in a very rapid conformational pathway [50]. One of the latter intermediates corresponds to a molten globule form [50]. RNase A has also served as a model for the study of disulfide bond formation pathways starting from the unfolded reduced form of the protein [51]. The unfolding of RNase A by guanidine hydrochloride at 1°C using NMR techniques has indicated that the helix containing His-12 is resistant to denaturation [52]. An earlier high-pressure study [53] of the unfolding of RNase A was performed using CD spectroscopy and pressures of 140 MPa. A recent NMR investigation of RNase A at pressures up to 200 MPa suggests that hydration of non-polar side chains upon unfolding determines the pressure dependence of the conformational stability of the protein [40].

4.3.2 Hen Lysozyme

Hen lysozyme is perhaps one of the most studied enzymes. Its X-ray structure in the presence and absence of substrate analogs is known, and its enzymatic mechanism has been thoroughly investigated. An X-ray study by Kundrot and Richards [54] of the crystal structure of lysozyme at a pressure of 100 MPa is relevant for this discussion, particularly the finding that different regions of lysozyme have different compressibilities. Molecular dynamics simulations of lysozyme in the free and substrate-bound states were reported by Karplus and coworkers [55], including extensive analysis of the nature of the average structure and atomic fluctuations. There have been extensive NMR studies of lysozyme structure and folding- unfolding behavior, and detailed sequential assignments of the ^{1}H resonances for lysozyme have been made using 2D NMR methods [56]. Thermal folding-unfolding rate constants have been measured by NMR magnetization transfer and are consistent with a two-state process [57, 58]. However, more recent reports by Dobson and coworkers [59–62] indicate that hen lysozyme in the presence of trifluoroethanol, and human and equine lysozymes, denatures at pH 2, forming molten globule states which retain some of the secondary structure of the native states. Further investigation of the folding pathway of hen lysozyme has revealed that a rapid collapse to a native-like secondary structure occurs first, and it is followed by stabilization of these structures through the formation of domains. Association of these domains finally forms the functional protein. In addition, lysozyme folding may follow two separate kinetic tracks: a rapid and a slow track. The slow track appears to involve a folding intermediate [63]. There have been several high-pressure studies on the denaturation of lysozyme employing fluorescence [64], Raman [65], and quasielastic light scattering [66] methods.

4.3.3 Apomyoglobin

Removal of the heme from myoglobin produces apomyoglobin, which retains a structured hydrophobic core, as is shown by an increase in heat capacity upon denaturation [67]. In addition, the chemical shifts of its backbone resonances are only slightly different from those of myoglobin [68], indicating that much of the structure of apomyoglobin is nearly identical to that of myoglobin. Apomyoglobin denaturation and refolding have been extensively studied, focusing mainly on the ability of apomyoglobin to unfold through an equilibrium intermediate I_1, in which the A, G, and H helices remain intact [69]. Numerous mutants have been constructed to investigate the nature of this intermediate, which has been shown to retain a close-packed, native-like structure [70]. An additional intermediate, I_2, can be created when TCA is added to the I_1 or to the acid-denatured states. Jennings and Wright [71] showed that a refolding intermediate is formed within the first 5 ms which is similar to the I_1 state. By using stopped flow hydrogen exchange, they showed that the apomyoglobin folding pathway is: unfolded → A·G·H→ A·B·G·H→ A·B·C·CD·E·G·H→ native. By analyzing the energy landscape for apomyoglobin, Panchenko et al. [72] have discovered foldons, or regions of apomyoglobin that fold cooperatively and quasi-independently. These foldons are found in the same regions that form equilibrium intermediates. Finally, FTIR measurements, following the relaxation dynamics on the microsecond timescale, have shown that apomyoglobin has three substructures, which denature quasi-independently. The most stable of these substructures is believed to be the AGH core [73].

Myoglobin was one of the first proteins studied under high pressure. Zipp and Kauzmann [74] used changes in the heme visible absorbance of metmyoglobin to monitor pressure denaturation, and constructed a phase diagram of protein stability versus pressure (0.1 MPa to 600 MPa), pH (4 to 13), and temperature (5°C to 80°C). Metmyoglobin was shown to unfold at pH and temperature extremes, as well as at high pressure. Assuming a two-state denaturation, the reaction volume for metmyoglobin denaturation was determined to be −60 to −100 mL/mol. High-pressure apomyoglobin denaturation has also been investigated by fluorescence. In this study, 8-anilino-1-napthalene sulfonate (ANS) was bound into the heme pocket, and frequency domain fluorometry was used to monitor the effects of high pressure, showing a small compaction of apomyoglobin up to 80 MPa, followed by swelling and increasing flexibility at higher pressures [75].

4.3.4 *Arc* Repressor

Arc repressor is a small, DNA-binding dimeric protein consisting of 53 amino acid residues (Mr = 13,000) which represses transcription from the P_{ant} promoter of Salmonella bacteriophage P22 [76−78]. *Arc* repressor belongs to a family of proteins that have an antiparallel β-sheet as the interfacial DNA-binding motif [79-81]. A tertiary structure model for *arc* repressor has been proposed [80] based upon homology between *arc* repressor and the *E. coli met* repressor and on 2D NMR data [82, 83]. This model consists of an intertwined dimer in which residues 8 to 14 of each monomer participate in the formation of an antiparallel β-sheet. *Arc* repressor dimer dissociates reversibly into subunits with increasing pressure at fixed protein concentration, or with dilution at constant pressure [84]. The *arc* repressor monomer obtained by compression is compact and, as measured by its rotational diffusion, has a much smaller hydrodynamic radius than that of *arc* repressor denatured by urea. The properties of the *arc* repressor dissociated by pressure agree with the characteristic properties of a molten globule. Fluorescence studies [84] clearly show that the pressure-dissociated form of the *arc* repressor retains significant three-dimensional structure, in contrast to the urea-denatured or temperature-denatured forms of the protein.

4.4 Results and Discussion

4.4.1 Determination of the Activation Volume of the Uncatalyzed Hydrogen Exchange Reaction Between N-Methylacetamide and Water

In connection with our experiments on pressure denaturation and pressure-assisted cold denaturation, we have become aware of the scarcity of data dealing with hydrogen exchange at high pressure. The overall hydrogen exchange rate constant (k) for each amide proton in peptides and proteins is pseudo first order and can be written as a sum of three terms:

$$k = k_{H^+}\left[H^+\right] + k_{OH^-}\left[OH^-\right] + k_W, \qquad (4.2)$$

where k_{H^+} is the rate constant of the acid-catalyzed exchange reaction, k_{OH^-} is the rate constant of the base-catalyzed exchange reaction, and k_w is the contribution due to the pH-independent exchange reaction, or uncatalyzed exchange reaction, assuming constant water concentration, which is typical in peptide and protein studies. The amide-water hydrogen exchange rates are very sensitive to the local environment of the protons. Thus, fluctuations in the local conformations of proteins can be probed by measuring the hydrogen exchange rates of the individual backbone amide protons. The effect of the local structure on the hydrogen exchange rate constants is typically quantified in terms of the protection factor (P), which is the ratio of the observed rate constant of each amide proton in the protein to a calculated rate constant for the amide proton in an analogous unfolded structure. The calculated unfolded rate constants are determined from the rate constant for the model compound, polyalanine, by factoring in side-chain effects on the acid-, base-, and water-catalyzed exchange rate constants and by using the activation energies of exchange to correct the rate constants for temperature differences [85]. In our laboratory, the amide-water hydrogen exchange rate constants of the individual amide residues of proteins are used to probe the local structure surrounding the residues in cold- and pressure-denatured proteins. Thus, in addition to side-chain and temperature corrections, the effect of pressure on the rate constants must be included. Currently, the data analysis is difficult since information on the activation volume of the uncatalyzed amide-water hydrogen exchange reaction is lacking. The activation volumes of exchange for the acid- and base-catalyzed exchange reactions have been determined by Carter et al. [86], who investigated the effect of pressures up to 250 MPa on hydrogen exchange in polypeptides and proteins using tritium-hydrogen exchange experiments. However, their study was incomplete in that it neglected the contribution of the uncatalyzed hydrogen exchange reaction to the rate constant.

Therefore, we investigated hydrogen exchange [87] between N-methylacetamide (NMA) and water since it serves as a model for the amide-water hydrogen exchange in peptides and proteins. The rate constants for amide-water hydrogen exchange between 16 mol % NMA and water at 66°C were measured at pressures ranging from 0.1 MPa to 500 MPa at three different pH values using the NMR magnetization transfer technique, and the apparent activation volumes of hydrogen exchange were obtained at each pH. The activation volume of the acid-catalyzed hydrogen exchange reaction was calculated to be +1.7 $cm^3\ mol^{-1}$, and the activation volume of the base-catalyzed hydrogen exchange reaction was calculated to be +11.0 $cm^3\ mol^{-1}$. The hydrogen exchange rate constants were also measured as a function of concentration and pH at ambient pressure at 66°C so that the rate constants of the acid-, base-, and uncatalyzed hydrogen exchange reactions could be found. With these values, the activation volume of the uncatalyzed hydrogen exchange reaction was calculated to be equal to −9.0 $cm^3\ mol^{-1}$. This quantity has not been previously determined and is required in hydrogen exchange studies of peptides and proteins at high pressures.

4.4.2 Cold, Heat, and Pressure Unfolding of Ribonuclease A

This work [26, 27] was a continuation of our systematic studies [2, 24] of pressure-induced unfolding of proteins and had the following specific objectives: (a) to investigate the pressure unfolding of RNase A; (b) to characterize the structure of the pressure-denatured protein; and (c) to compare the unfolded structures of RNase A produced by cold, heat, and pressure denaturation.

The reversible pressure-denaturation experiments in the pressure range from 0.1 MPa to 500 MPa were carried out at pH* 2.0 and 10°C. The cold denaturation was carried out at 300 Mpa, at which pressure the protein solution can be cooled down to –25°C without freezing, according to the pressure-temperature phase diagram of water. Including heat-denaturation experiments, the experimental data obtained allowed us to construct the pressure-temperature phase diagram of RNase A shown in Fig. 4.9 [26].

The experimental results suggested that pressure denaturation leads to non-cooperative unfolding of this protein. The appearance of a new histidine resonance in the cold-denatured and pressure-denatured RNase A spectra, compared to the absence of this resonance in the heat-denatured state, indicated that the pressure-denatured state and the cold-denatured state may contain partially folded structures, which are similar to that of the early folding intermediate found in the temperature-jump experiment reported by Blum et al. [47]. A hydrogen-exchange experiment was performed to confirm the presence of partially folded structures in the pressure-denatured state. Stable hydrogen-bonded structures protecting the backbone amide hydrogens from solvent exchange were observed in the pressure-denatured state. These experimental results suggested that pressure-denatured RNase A displays the characteristics of a molten globule. The cold-, heat-, and pressure-denaturation experiments on the complex of RNase A with the inhibitor 3'-UMP showed, as expected, that the RNase A-inhibitor complex was more stable in comparison to RNase A without the inhibitor.

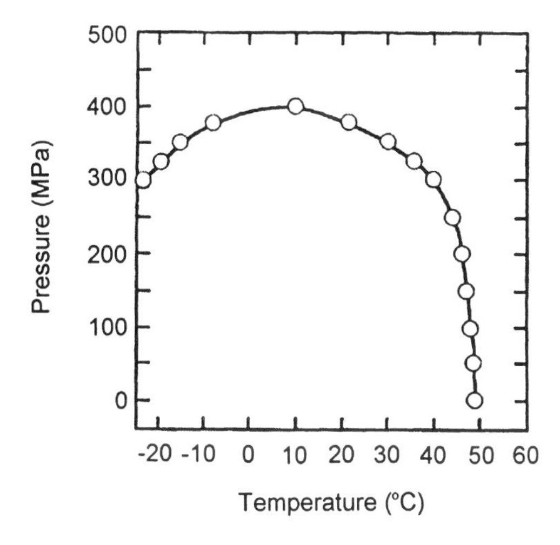

Fig. 4.9. Pressure-temperature phase diagram of ribonuclease A at pH* 2.0 [26]

Figure 4.10. [27] compares the protection factors (P) as a function of residue position for pressure-assisted cold-denatured RNase A, pressure-denatured RNase A, and heat-denatured RNase A. Clearly, the cold-denatured state differs markedly from the thermally denatured state.

4.4.3 Pressure-Assisted, Cold-Denatured Lysozyme Structure and Comparison with Lysozyme Folding Intermediates

In view of the importance of the structure of partially denatured states [26, 27] to the understanding of protein folding pathways, and our results dealing with RNase A, we decided to expand our hydrogen-exchange experiments on pressure-assisted, cold-denatured states to other proteins [88]. We chose lysozyme for these studies because its structure and folding had been studied extensively, but information about

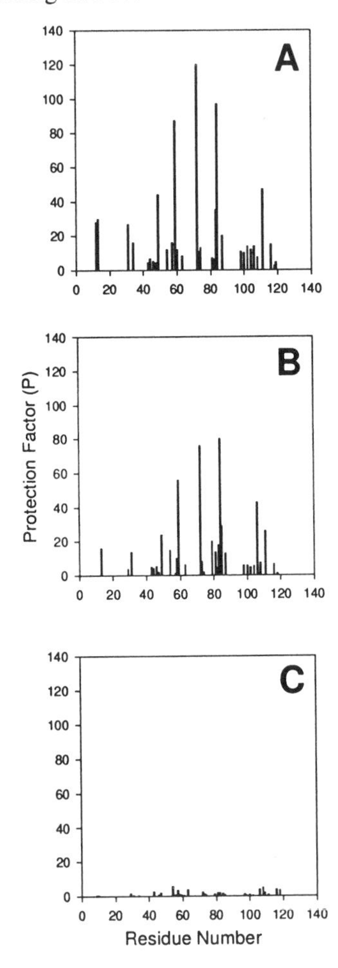

Fig. 4.10. Comparison of the protection factors (P) obtained as a function of residue number for (A) cold-denatured, (B) pressure-denatured, and (C) heat-denatured RNase A [27]

its cold denaturation was lacking. From our hydrogen exchange rate data on lysozyme in the cold-denatured state, it is clear that many regions of lysozyme are significantly protected from exchange, with protection (*P*) values exceeding 10. For example, *P* factors ranged from 1.19 for F3 to 71.0 for R114. Except for a partially folded form observed in 50% trifluoroethanol by Buck and co-workers [89], cold-denatured lysozyme is the only form of denatured lysozyme that shows appreciable protection from exchange. Data for heat-, urea-, and acid-denatured $CM^{6\text{-}127}$ lysozyme, analyzed using the same temperature and side chain correction methods used in this study, show few residues with $P > 5$ [89]. It should be made clear that the pressure-assisted, cold-denatured state is not a 'pure' cold-denatured state, since the pressure affects the protein directly. Furthermore, the *P* factors are low and incompatible with anything more than small or transiently formed regions of structure. Nevertheless, any effect that pressure and low temperature have is clearly very different from that induced by high temperature or chemical denaturants such as urea.

Using a variant of conventional pulsed-labeling hydrogen-exchange studies, Gladwin and Evans [90] observed relatively early stages in the folding of lysozyme by confining the exchange to the dead time of their instrumentation (~3.5 ms). Slowing of exchange measured during this time can be quantified by comparing the measured exchange rate to that for a random coil, as in other hydrogen-exchange studies. The resulting quantity, though similar to a *P* factor, is determined from a rate that changes considerably as the protein refolds; as a result, it is best referred to as a 'dead-time inhibition factor' (I_D) rather than a true *P* factor. Nevertheless, I_D values provide information similar to that provided by *P* values in equilibrium-denatured states. During the first 3.5 ms of folding, lysozyme shows moderate degrees of protection ($I_D > 5$), not only in the α-helices and the C-terminal 3_{10} helix, but also in the loop region from residues 60−65, as well as at residue 78. The early stages of lysozyme refolding thus show a marked resemblance to our cold-denatured state. We have thus observed a strong parallel between a stable, denatured form of lysozyme and the transient species observed during folding. Figure 4.11 compares the inhibition of hydrogen exchange observed during the first 3.5 ms of folding by Gladwin and Evans [90] to the *P* factors obtained for pressure-assisted, cold-denatured lysozyme.

In the case of lysozyme and RNase A, the patterns of protection against hydrogen exchange are similar to those observed in early (refolding time < 10 ms) folding intermediates for these proteins, leading to the idea that the cold-denatured state is structurally similar to such intermediates. To help test this idea, ubiquitin, which has folding kinetics that are markedly different from those of either RNase A or lysozyme, was investigated [91]. In particular, ubiquitin shows much less evidence of early structure formation than lysozyme; it more closely resembles a random coil.

Cold-denatured ubiquitin ($P = 225$ MPa, $T = -16°C$) shows little deviation from a random coil in its hydrogen exchange kinetics, with no *P* values above 5 and most below 2. As shown in Fig. 4.11, these values are typical of highly denatured proteins such as the urea-denatured state of lysozyme. Under other circumstances, however, such as the A state produced by a 60% methanol solution at a pH of 2.0, ubiquitin shows significantly more protection from exchange. It is important to point out that in ubiquitin, the same dead-time inhibition study [90] showed no evidence for protection from exchange, with the largest dead-time inhibition factors being ~2 (Fig. 4.11). Moreover, there was no correlation of even these modestly elevated factors within significant elements of secondary structure as there were in lysozyme.

The similarity of cold-denatured state P factors to the protection observed in protein refolding studies has led to the idea that the cold-denatured state is populated by species comprising elements of secondary and perhaps tertiary structure that are comparatively stable. The similarities between the cold-denatured state and the early folding states suggest that these structural elements would be expected to form first during folding. In the case of ubiquitin, the folding reaction appears to proceed in one highly cooperative step, with little of the multiphasic behavior observed in proteins like lysozyme or RNase A. Although a possible intermediate for ubiquitin refolding has been characterized by other means [92], the intermediate thus identified shows no protection from hydrogen exchange. The results for the cold-denatured state of ubiquitin parallel these results: there is no partially folded state that is stable enough to be characterized by hydrogen-exchange methods when ubiquitin is cold-denatured.

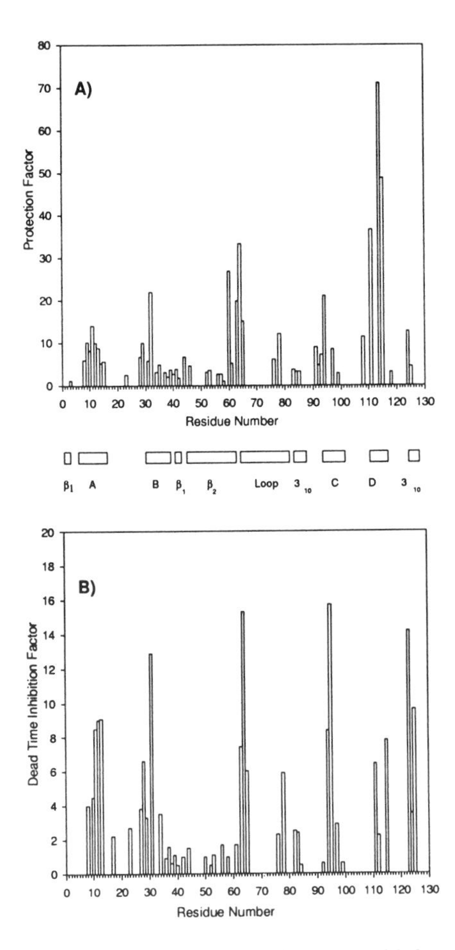

Fig. 4.11. Comparison of protection factors observed in (*A*) cold-denatured lysozyme to (*B*) the dead-time inhibition factors obtained for the first 3.5 ms of lysozyme folding by Gladwin and Evans [88, 90]

4.4.4 Denaturation of Apomyoglobin Mutants by High Pressure

To study the effects of mutagenesis on protein stability, the pressure-denaturation behavior of recombinant sperm whale apomyoglobin and several mutants of this protein was investigated [93]. Apomyoglobin is a good target for these studies for three reasons: (a) the pH and some urea denaturation has already been studied for wildtype and mutant proteins; (b) intermediates have been identified for apomyoglobin folding by Baldwin and co-workers [69, 70]; and (c) apomyoglobin contains no proline or cysteine residues to complicate the folding process.

Three kinds of mutants were studied which alter the hydrophobic core that remains in apomyoglobin after removal of the heme. The first set (A130K, S108K and F123K) introduces a positive charge into the hydrophobic core. The second set (F123W and S108L), increases the size and hydrophobicity of a residue. This kind of disruptive mutation can potentially cause steric clashes and a general expansion of the surrounding area to accommodate the extra volume. The third type of mutation (F123G) decreases the size of a residue in the core and creates a cavity. The goal of these experiments [93] in progress was to understand which kind of mutation is the most destabilizing to hydrostatic pressure. Also, having mutations at different positions can show which sites in the protein can accommodate the mutations without exposing more of the hydrophobic core, thus resulting in a higher ΔV. So far, we have carried out high-pressure NMR experiments on 4 of the 6 mutants and on wildtype apomyoglobin. From the changes in the intensity of His proton peaks with pressure, we obtained the denaturation curves shown in Fig. 4.12. All mutants are significantly destabilized to pressure denaturation relative to the wildtype protein. The wildtype apomyoglobin has a sigmoidal denaturation curve and its pressure denaturation does not begin until 80 MPa. The ΔV of wildtype apomyoglobin is high at –90 mL/mol, which is the result of partial exposure of the interior of the protein upon heme removal. For the mutants which have introduced a charge, denaturation starts immediately at low pressures and both mutants are denatured by 100 MPa. The A130K mutant denatures so readily that its reaction volume is roughly

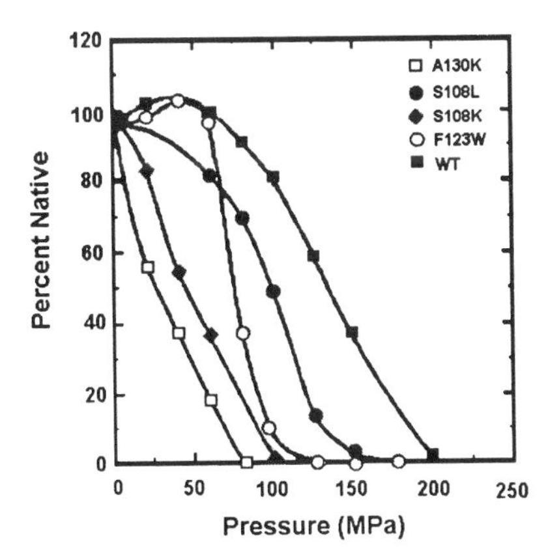

Fig. 4.12. Pressure denaturation of wildtype apomyoglobin (WT) and four destabilizing mutants. The percentage of native form was obtained from the intensity changes of His resonances between 8.0 and 8.8 ppm

−300 mL/mol. We believe these mutants are in the native state at 0.1 MPa because the denatured His peaks at about 8.5 ppm have not yet appeared. The second group of mutants, F123W and S108L, increase the hydrophobicity and size of core residues. These mutants are more stable under pressure than the first group, but the denaturation of the F123W mutant appears to be much more cooperative than that of the other forms of the protein. These results remain to be confirmed and expanded; however, compared to the pH denaturation [69] and the urea denaturation [70] of some of the same mutants, the patterns of pressure denaturation are quite distinct.

4.4.5 High-Pressure NMR Study of the Dissociation of the *Arc* Repressor

In view of the important observation that a pressure-induced predissociated state exists in a dimeric protein, we mention the study of *arc* repressor [28, 30]. Our high-resolution NMR studies of the pressure dissociation and denaturation of *arc* repressor showed that pressure denaturation leads to a more controlled, less drastic perturbation of the protein structure than temperature or chemical denaturation. In fact, these studies demonstrated that pressure produces molten globule or compact denatured states of *arc* repressor.

Different denatured states of *arc* repressor were characterized by 1D and 2D NMR and by fluorescence spectroscopy. Increasing pressure promoted sequential changes in the structure of *arc* repressor: from the native dimer through a predissociated state to a denaturated molten globule monomer. A compact state (molten globule) of *arc* repressor was obtained in the dissociation of *arc* repressor by pressure, whereas high temperature and urea induced dissociation and unfolding to less-structured conformations. The NMR spectra of the monomer under pressure (up to 500 MPa) were typical of a molten globule, and they were considerably different from those of the native dimer and the thermally or chemically denatured monomer. The substantial line broadening and overlap of many resonances in the NMR spectra at high pressures indicated that there was interconversion between a number of different conformations of the molten globule at an intermediate exchange rate. The 2D NOE spectra showed that the pressure-denatured monomer retains substantial secondary structure. The presence of NOEs in the β-sheet region in the dissociated state suggested that the intersubunit β-sheet (residues 6−14) in the native dimer was replaced by an intramonomer β-sheet. The detection of the existence of a pressure-induced predissociated state of *arc* repressor represents an excellent example of the power of advanced high-resolution 2D NMR techniques at high pressure.

Figure 4.13 shows the proposed β-sheet structure of the *arc* repressor in the native state, the predissociated state, and the dissociated molten globule state. The coordinates of the native dimer state were provided by R. Kaptein (University of Utrecht, The Netherlands).

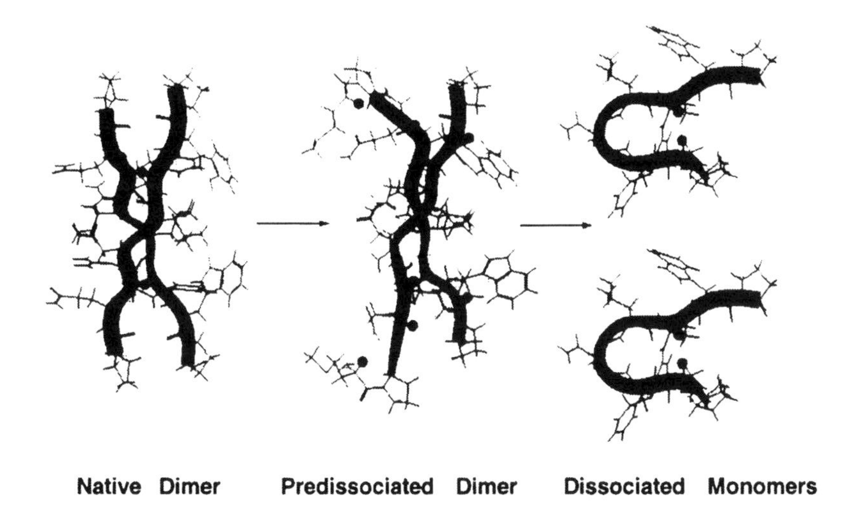

Fig. 4.13. Proposed β-sheet of the *arc* repressor in the native state, the predissociated state, and the dissociated molten globule state. The coordinates of the native dimer state were provided by R. Kaptein [28]

4.5 Conclusions

On the basis of recent NMR experiments on proteins at high pressure, we can formulate the following main conclusions:

- Pressure and pressure-assisted cold-denaturated states exhibit significant residual secondary structure
- The cold-denaturated state is different from the thermally denaturated state
- Pressure allows stabilization of intermediate states
- Pressure-assisted cold-denaturated states have residual secondary structure resembling that of early folding intermediates
- Mutations in the core of proteins affect their pressure stability differently than their thermal stability or resistance to chemical denaturation

Acknowledgements. This work was supported in part by the National Science Foundation under grant NSF CHE 95-26237 and by the National Institutes of Health under grant PHS 5 R01 GM42452-09.

References

4.1 K.A. Dill, D. Shortle: Annu. Rev. Biochem. **60**, 795 (1991)
4.2 J. Jonas, A. Jonas: Annu. Rev. Biophys. Biomol. Struct. **23**, 287 (1994)

4.3 V.V. Mozhaev, K. Heremans, J. Frank, P. Masson, C. Balny, Proteins: Struct. Funct. Genet. **24**, 81 (1996)
4.4 I. Artaki, J. Jonas, J. Chem. Phys. **82**, 3360 (1985)
4.5 H.G. Drickamer, C.W. Frank, *Electronic Transition and the High Pressure Chemistry and Physics of Solids* (Chapman and Hall, London 1973)
4.6 M. Gross, R. Jaenicke, Eur. J. Biochem. **221**, 617 (1994)
4.7 D. Shortle, Curr. Opin. Struct. Biol. **3**, 66, (1993)
4.8 D.R. Shortle, Curr. Opin. Struct. Biol. **6**, 24 (1996)
4.9 D. Shortle, FASEB J. **10**, 27 (1996)
4.10 S.M. V. Freund, K.B. Wong, A.R. Fersht, Proc. Natl. Acad. Sci. USA **93**, 10600 (1996)
4.11 Abragam, *The Principles of Nuclear Magnetism* (Oxford University, Oxford 1961)
4.12 G.B. Benedek, E.M. Purcell, J. Chem. Phys. **22**, 2003 (1954)
4.13 J. Powles, M. Gough, Mol. Phys. **16**, 349 (1969)
4.14 J. Jonas, Rev. Sci. Instrum. **43**, 643 (1972)
4.15 L. Ballard, C. Reiner, J. Jonas, J. Magn Reson. **123A**, 81 (1996)
4.16 L. Ballard, A. Yu, C. Reiner, J. Jonas, J. Magn. Reson. **133**, 190 (1998)
4.17 L. Ballard, Ph.D. thesis, University of Illinois, 1997
4.18 J. Jonas, P. Koziol, X. Peng, C. Reiner, D.M. Campbell, J. Magn. Reson. **102B**, 299 (1993)
4.19 Zahl, A. Neubrand, S. Aygen, R. van Eldik, Rev. Sci Instrum. **65**, 882 (1994)
4.20 U. Frey, L. Helm, A. Merbach, High Press. Res. **2**, 237 (1990)
4.21 F. Bachl, H.D. Lüdemann, High Press. Res. **6**, 91 (1990)
4.22 K. Woelk, J.W. Rathke, R.J. Klingler, J. Magn. Reson. **109A**, 137 (1994)
4.23 L. Ballard, J. Jonas, in Annu. Rep. NMR Spectrosc., ed. by G.A. Webb, **33**, 115, 1997.
4.24 J. Jonas, L. Ballard, D. Nash, Biophys. J. **75**, 445 (1998)
4.25 K.E. Prehoda, E.S. Mooberry, J.L. Markley, Biochemistry **37**, 5785 (1998)
4.26 J. Zhang, X. Peng, A. Jonas, J. Jonas, Biochemistry **34**, 8631 (1995)
4.27 D. Nash, B.S. Lee, J. Jonas, Biochim. Biophys. Acta **1297,** 40 (1996)
4.28 X. Peng, J. Jonas, J. Silva, Biochemistry **33**, 8323 (1994)
4.29 C.A. Royer, A.P. Hinck, S.N. Loh, K.E. Prehoda, X. Peng, J. Jonas, J.L. Markley, Biochemistry **32**, 5222 (1993)
4.30 X. Peng, J. Jonas, J. Silva, Proc. Natl. Acad. Sci. USA **90**, 1776 (1993)
4.31 S.D. Samarasinghe, D.M. Campbell, A. Jonas, J. Jonas, Biochemistry **31**, 7773 (1992)
4.32 H. Yamada, K. Kubo, I. Kakihara, A. Sera, in *High Pressure Liquids and Solutions*, ed. by Y. Taniguchi, M. Senoo, and K. Hara (Elsevier Science, Amsterdam, **49** 1994)
4.33 D.C. Roe, J. Magn. Reson. **63**, 388 (1985)
4.34 S. Bai, C.M. Taylor, C.L. Mayne, R.J. Pugmire, D.M. Grant, Rev. Sci. Instrum. **67**, 240 (1996)
4.35 C.R. Yonker, T.S. Zemanian, S.L. Wallen, J.C. Linehan, J.A. Franz, J. Magn. Reson. **113A**, 102 (1995)
4.36 H. Li, H. Yamada, K. Akasaka, Biochemistry **37**, 1167 (1998)
4.37 K. Akasaka, T. Tezuka, H. Yamada, J. Mol. Biol. **271**, 671 (1997)
4.38 J.L. Urbauer, M.R. Ehrhardt, R.J. Bieber, P.F. Flynn, J.A. Wand, J. Am. Chem. Soc. **118**, 11329 (1996)
4.39 T. Yamaguchi, H. Yamada, K. Akasaka, Prog. Biotechnol. **13**, 141 (1996)
4.40 T. Yamaguchi, H. Yamada, K. Akasaka, J. Mol. Biol. **250**, 689 (1995)
4.41 M.G. Pravica, I.F. Silvera, Rev. Sci. Instrum. **69**, 479 (1998)

4.42 J.L. Yarger, R.A. Nieman, G.H. Wolf, R.F. Marzke, J. Magn. Reson. **114A**, 255 (1995)
4.43 S.G. Goo. S.H. Lee, Ungyong Mulli **7**, 477 (1994)
4.44 R.F. Marzke, D.P. Raffaelle, K.E. Halvorson, G.H. Wolf, J. Non-Cryst. Solids **172-174**, 401 (1994)
4.45 S.H. Lee, M.S. Conradi, and R.E. Norber, Rev. Sci. Instrum. **63**, 3674 (1992)
4.46 R. Bertani, M. Mali, J. Roos, D. Brinkmann, Rev. Sci. Instrum. **63**, 3303 (1992)
4.47 A.D. Blum, S.H. Smallcombe, R.L. Baldwin, J. Mol. Biol. **118**, 305 (1978)
4.48 J.B. Udgaonkar, R.L. Baldwin, Nature **335**, 694 (1988)
4.49 J.B. Udgaonkar, R.L. Baldwin, Proc. Natl. Acad. Sci. USA **87**, 8197 (1990)
4.50 W.A. Houry, H.A. Scheraga, Biochemistry **35**, 11734 (1996)
4.51 D.M. Rothwarf, H.A. Scheraga, Biochemistry **32**, 2671 (1993)
4.52 A. Bierzynski , R.L. Baldwin, J. Mol. Biol. **162**, 173 (1982)
4.53 S.J. Gill, R.L. Glogovsky, J. Phys. Chem. **69**, 1515 (1965)
4.54 C.E. Kundrot, F.M. Richards, J. Mol. Biol. **193**, 157 (1987)
4.55 C.B. Post, B.R. Brooks, M. Karplus, C.M. Dobson, P. J. Artymiuk, J.C. Cheetham, and D.C. Phillips, J. Mol. Biol. **190**, 455 (1986)
4.56 C. Redfield and C. M. Dobson, Biochemistry **27**, 122, 1988.
4.57 C.M. Dobson, P.A. Evans, R.O. Fox, in *Structure and Motion: Membranes, Nucleic Acids and Proteins*, ed. by E. Clementi, G. Corongiu, M.H. Sarma, R.H. Sarma (Adenine Press, Guilderland, N. Y. 1985) pp.265
4.58 C.M. Dobson, P.A. Evans, in *Structure and Dynamics of Nucleic Acids, Proteins, and Membranes*, ed. by E. Clementi, S. Chin (Plenum Press, New York 1986) pp.127
4.59 M. Buck, S.E. Radford, C.M. Dobson, Biochemistry **32**, 669 (1993)
4.60 P. Haezebrouck, M. Joniau, H.Van Dael, S.D. Hooke, N.D. Woodruff, C.M. Dobson, J. Mol. Biol. **246**, 382 (1995)
4.61 L.A. Morozova, D.T. Haynie, C. Arico-Muendel, H.Van Dael, C.M. Dobson, Nature Struct. Biol. **2**, 871 (1995)
4.62 L.A. Morozova-Roche, C.C. Arico-Muendel, D.T. Haynie, V.I. Emelyanenko, H.Van Dael, C.M. Dobson, J. Mol. Biol **268**, 903 (1997)
4.63 A. Matagne, S.E. Radford, C.M. Dobson, J. Mol. Biol. **267**, 1068 (1997)
4.64 T.M. Li, J.W. Hook III, H.G. Drickamer, G. Weber, Biochemistry **15**, 5571 (1976)
4.65 K. Heremans, P.T.T. Wong, Chem. Phys. Lett. **118**, 101 (1985)
4.66 B. Nyström, J. Roots, Makromol. Chem. **185**, 1441 (1984)
4.67 Y.V. Griko, P.L. Privalov, S.Y. Venyaminov, V.P. Kutyshenko, J. Mol. Biol. **202**, 127 (1988)
4.68 J.T.J. Lecomte, Y.H. Kao, M.J. Cocco, Proteins: Struct. Funct. Genet. **25**, 267 (1996)
4.69 F. M. Hughson, P.E. Wright, R.L. Baldwin, Science **249**, 1544 (1990)
4.70 M.S. Kay, R.L. Baldwin, Nature Struct. Biol. **3**, 439 (1996)
4.71 P.A. Jennings, P.E. Wright, Science **262**, 892 (1993)
4.72 A.R. Panchenko, Z. Luthey-Schulten, P.G. Wolynes, Proc. Natl. Acad. Sci. USA **93**, 2008 (1996)
4.73 R. Gilmanshin, S. Williams, R.H. Callender, W.H. Woodruff, R.B. Dyer, Proc. Natl. Acad. Sci. USA **94**, 3709 (1997)
4.74 A. Zipp, W. Kauzmann, Biochemistry **12**, 4217 (1973)
4.75 E. Bismuto, G. Irace, I. Sirangelo, E. Gratton, Protein Science **5**, 121 (1996)
4.76 M. M. Susskind, J. Mol. Biol. **138**, 685 (1980)
4.77 R.T. Sauer, W. Krovatin, J. DeAnda, P. Youderian, M.M. Susskind, J. Mol. Biol. **168**, 699 (1983)
4.78 A.K. Vershon, J.U. Bowie, T.M. Karplus, R.T. Sauer, Proteins: Struct. Funct. Genet. **1**, 302 (1986)

4.79 K.L. Knight, R.T. Sauer, Proc. Natl. Acad. Sci. USA **86**, 797 (1989)
4.80 J.N. Breg, J.H.J. van Opheusden, M.J.M. Burgering, R. Boelens, R. Kaptein, Nature **346**, 586 (1990)
4.81 S.E. V. Phillips, Curr. Opin. Struct. Biol. **1**, 89 (1991)
4.82 J.N. Breg, R. Boelens, A.V.E. George, R. Kaptein, Biochemistry **28**, 9826 (1989)
4.83 M.G. Zagorski, J.U. Bowie, A.K. Vershon, R.T. Sauer, D.J. Patel, Biochemistry **28**, 9813 (1989)
4.84 J.L. Silva, C.F. Silveira, A. Correia Jr., L. Pontes, J. Mol. Biol. **223**, 545 (1992)
4.85 Y. Bai, J.S. Milne, L. Mayne, S.W. Englander, Proteins: Struct. Funct. Genet. **17**, 75 (1993)
4.86 J.V. Carter, D.G. Knox, A. Rosenberg, J. Biol. Chem. **253**, 1947 (1978)
4.87 S.A. Mabry, B.S. Lee, T. Zheng, J. Jonas, J. Am. Chem. Soc. **118**, 8887 (1996)
4.88 D.P. Nash, J. Jonas, Biochemistry **36**, 14375 (1997)
4.89 M. Buck, S.E. Radford, C.M. Dobson, J. Mol. Biol. **237**, 247 (1994)
4.90 S.T. Gladwin, P.A. Evans, Folding Des. **1**, 407 (1996)
4.91 D.P. Nash, J. Jonas, Biochem. Biophys. Res. Commun. **238**, 289 (1997)
4.92 S. Khorasanizedeh, I.D. Peters, H. Roder, Nature Struct. Biol. **3**, 193 (1996)
4.93 S. Bondos, S. Sligar, J. Jonas, unpublished results.

5 Pressure-Induced Secondary Structural Changes of Proteins Studied by FTIR Spectroscopy

Yoshihiro Taniguchi[1] and Naohiro Takeda[2]

[1]Department of Applied Chemistry, Faculty of Science and Engineering, Ritsumeikan University, 1-1-1, Noji-Higashi, Kusatsu, Shiga 525-8577, Japan
E-mail: taniguti@se.ritsumei.ac.jp
[2]Ashigara Research Institute, Fujifilm Co. Ltd., 210, Nakanuma, Minamiashigara, Kanagawa 250-0123, Japan
E-mail: takeda@ashikenn.fujifilm.co.jp

Abstract. The pressure-induced secondary structural changes of ribonuclease A (RNase A), ribonuclease S (RNase S), and bovine pancreatic trypsin inhibitor (BPTI) were studied by Fourier transform infrared (FTIR) spectroscopy combined with a diamond anvil cell up to about 1000 MPa. From the studies of pressure-induced changes in the secondary structure and pressure effects on the hydrogen–deuterium exchange, pressure-denatured RNase A and RNase S did not have any secondary structure at 840 MPa and 670 MPa, respectively. The poor pressure stability of RNase S compared with that of RNase A is due to the larger structural fluctuation of the peptide backbone. On the other hand, the polypeptide backbone of BPTI was not fully unfolded even above 1000 MPa, and pressure induced the structural rearrangements of the larger fluctuated local parts in native form.

5.1 Introduction

For globular proteins, thermodynamic studies have revealed that the native state is only marginally stable relative to the denatured state [1, 2]. It is well known that the reversible denaturation of proteins in aqueous solution can be caused by application of high pressure as well as by an increase in temperature. Many high pressure experimental techniques have been used: ultraviolet and visible spectroscopy [3–5]; fluorescence spectroscopy [6, 7]; NMR[1] spectroscopy [7–9]; and enzyme activity [10]. However, the structural features created during the pressure-induced denaturation of proteins remain uncertain.

On the basis of red-shifts in the intrinsic fluorescence emission spectra and binding of a hydrophobic dye, ANS, the hydrophobic residues of lysozyme and chymotrypsinogen, which are buried in the native state, are exposed to aqueous medium upon the pressure-induced denaturation [6]. The ^{1}H NMR spectra of pressure-denatured *Arc* repressor monomer are considerably different from those of the thermally or urea-denatured monomer, while the three denatured states are not distinguishable from one another using the tryptophan fluorescence spectra [9]. Only a difference in the microenvironment of amino acid residues, used as a probe to compare between native and pressure-denatured proteins, has been distinguished. There is very little information directly concerning structural features of the polypeptide backbone of proteins at high pressure. Disruption of tight packing interactions between the side-chains is necessary for protein denaturation, which does not necessarily lead to extensive unfolding of the polypeptide backbone. Indeed, study of BPTI by MD simulation demonstrates no evidence for pressure-induced unfolding of the polypeptide backbone even at 1000 MPa [11]. In addition, an increasing number of partially folded states of proteins stable at equilibrium have been recently identified. The molten globule state, a compact denatured state with a significantly native-like secondary structure but a largely flexible and disordered tertiary structure, is particularly interesting [12, 13]. Pressure-denatured proteins might be also characterized by such structural features.

Fourier transform infrared (FTIR) spectroscopy is one of the most powerful techniques for determining the secondary structure of proteins in aqueous solution [14–19]. Of all the amide vibrational modes of the backbone peptide groups, the most widely used one is the amide I mode (designated amide I' in the deuterated peptide group). This vibrational mode originates from C=O stretching vibration of the peptide group, which is weakly coupled with in-plane N–H bending and C–N stretching vibration, and gives rise to an infrared band in the region between approximately 1600 and 1700 cm^{-1} [20]. The major factors responsible for conformational sensitivity of the amide I vibrational frequency are the hydrogen-bonding pattern in the polypeptide backbone and the geometrical arrangement of the peptide groups. However, proteins usually contain different types of secondary structure elements such as α-helices, β-sheets, turns, and non-ordered structures. Since each of these conformational entities contributes to the infrared spectrum, the observed amide I band contour consists of many overlapping component bands related to different structural elements. The bandwidth of the contributing component bands is usually greater than the separation between the maxima of adjacent peaks. Consequently, in order to extract structural information from the amide I band, extensive mathematical manipulation of the experimentally measured infrared spectra is required. Resolution enhancement techniques such as second-derivative [21, 22] and Fourier self- deconvolution [23] can be used to allow increased separation and thus better identification of the overlapping amide I component bands.

Another advantage of the application of FTIR spectroscopy to proteins is that the hydrogen–deuterium exchange for each secondary structure element can be monitored. The reason for this is that shifts in amide I vibrational frequency due to hydrogen–deuterium exchange are always to lower frequency. In general, the backbone amide protons involved in the α-helices or β-sheets in native proteins are strongly protected against exchange with solvent deuterons. The protection patterns found in the two secondary-structure elements are complex, reflecting tertiary as well as secondary structure. Measurements of the hydrogen–deuterium exchange are suitable for detection of persistent secondary-structure and destabilization of native tertiary interactions in partially folded proteins [24–26].

In the present study, FTIR spectroscopy combined with resolution enhancement techniques has been used to probe the pressure-induced structural changes in RNase A, RNase S, and BPTI in aqueous solution. This method provides a sensitive diagnostic tool for monitoring the nature of changes in the conformation of the protein backbone. The structural features at high pressure are additionally described by the behavior of the hydrogen–deuterium exchange when compressed. Such detailed structural characterization is essential to understanding of the factors determining the pressure stability of proteins and the mechanism of pressure-induced unfolding.

5.2 Experimental Methods

5.2.1 Sample and Solutions

Bovine pancreatic ribonuclease A (type XII-A), bovine pancreatic ribonuclease S (grade XII-S), and bovine pancreatic trypsin inhibitor (type I-P) were purchased from Sigma Chemical Co. and poly (L-glutamic acid) sodium salt (PLGA-Na) was purchased from Peptide Institute Inc., and they were used without further purification. Sample solutions of RNase A, RNase S, BPTI, and PLGA-Na were prepared by dissolving samples in 0.05 M Tris-DCl D_2O buffer, pD 7.0. The proteins or polypeptide concentration was 50 mg/mL. The pD was read directly from a pH meter and no adjustments were made for isotope effects.

5.2.2 Deuterated Solutions

The completely deuterated protein solutions were prepared by means of cooling rapidly in an ice/water bath after incubating the solution at 62℃ for 20 min in RNase A, at 45℃ for 20 min in RNase S, and at 85℃ for 12 min in BPTI. The completion of hydrogen–deuterium exchange was confirmed by no further changes in the amide II band. This amide band in the frequency region around 1550 cm^{-1} shifts to around 1450 cm^{-1} as a result of deuteration of the backbone amide protons. All sample solutions were prepared immediately prior to the infrared measurements. It took about 1 h to record the first infrared spectrum after sample preparation.

5.2.3 High-Pressure FTIR Measurements

For the high-pressure experiments, the sample solutions were placed together with a small amount of powdered α-quartz in a 1.0 mm diameter hole of a 0.05 mm thick anti-corrosive steel (SUS 304 or Hasteloy C-276) gasket mounted on a diamond anvil cell thermostated at 25℃ [BPTI] or 30℃ [RNase A, RNase S]. The α-quartz was used as an internal pressure calibrator [27]. Infrared spectra of the samples were recorded by using a Perkin-Elmer 1725X Fourier transform infrared spectrometer equipped with a liquid-nitrogen cooled MCT (HgCdTel) detector. The infrared beam was condensed by a zinc selenide lens system onto the sample in the diamond anvil cell as shown in Fig. 5.1. For each spectrum, 1000 interferograms were co-added and Fourier transformed to give a spectral resolution of 2 cm^{-1}. Thirty minutes were allowed to equilibrate the sample at the chosen pressure prior to each infrared measurement, which itself took 16 min. Pressure was increased or released at the average rate of about 150 MPa/h. In order to eliminate spectral contributions of atmospheric water vapor, the spectrometer and sample chamber were continuously purged with dry air. Reference spectra were recorded in the same cell and under identical conditions with only the medium in which the proteins were dissolved. Each infrared spectrum of proteins dissolved in D_2O buffer was obtained by digitally subtracting the appropriate reference spectrum from the spectrum of each sample solution. The bands originating from water vapor in the second-derivative infrared spectrum were subtracted until the absorption-free region of the amide I(I') band above 1700 cm^{-1} was featureless. Second-derivative spectra were generated by using a 9-data-point (9 cm^{-1}) Savitzky-Golay function [21]available from Perkin-Elmer software, IRDM2.

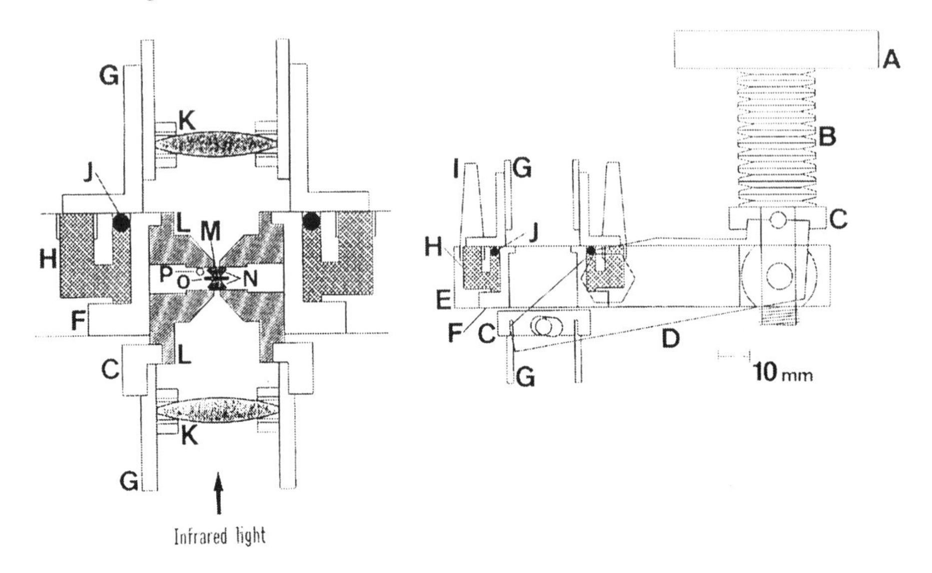

Fig. 5.1. Diamond anvil cell for the high pressure FTIR measurements. (A) Screw, (B) Spring washer, (C) Pressure plate, (D) Lever arm, (E) Frame, (F) Cylinder, (G) Lens holder, (H) Water jacket, (I) Nozzle, (J) O-ring, (K) Lens (ZnSe), (L) Piston, (M) Sample, (N) Diamond anvil, (O) Gasket (stainless steel), (P) Thermistor

5.3 Results and Discussion

5.3.1 Ribonuclease A

Pressure-induced Changes in the Secondary Structure of RNase A. Figures 5.2A and B show the second-derivative infrared spectra of completely deuterated RNase A upon pressure increase and release, respectively. The observed changes in the amide I' band shown in Figs. 5.2 directly indicate the pressure-induced changes in the secondary structure of RNase A, because the hydrogen–deuterium exchange has been already completed. The second-derivative analysis permits the direct separation of the amide I band into its components. Absorbance bands in the original spectrum are revealed as negative bands in the second-derivative spectrum. On the basis of the results previously reported by FT-IR spectroscopic studies [28–30] for completely deuterated RNase A, the component bands at 1632, 1650, 1663, 1673, and 1681 cm^{-1} 0.1 MPa are due to the amide I' vibrational mode of peptide segments in the β-sheet, α-helices, turns, turns, and β-sheet, respectively (Fig. 5.2A at 0.1 MPa). Another band at 1609 cm^{-1} arises from amino acid side-chain absorption, primarily from the tyrosine residues [31, 32]. The relative intensity of the amide I' component band at 1632 cm^{-1} is strongest, which suggests that the β-sheet predominates over other secondary structure elements in RNase A in aqueous solution. This result is in agreement with those of both X-ray crystallographic studies [33] and 2D-NMR spectroscopic studies [34, 35]. Figure 5.3 shows ribbon representation of the polypeptide backbone of RNase A.

As pressure is increased, little frequency shift in all of the five amide I' component bands and no significant changes in the amide I' band contour are observed up to 570 MPa. A further increase in pressure induces the cooperative disappearance of the amide I' component bands characteristic of native RNase A. These spectral changes, which occur over the pressure range between 570 and 1030 MPa, suggest the pressure-induced unfolding of RNase A. The spectrum at 1030 MPa after the pressure-induced unfolding is complete exhibits only a broad, nearly featureless amide I' band contour centered at 1639 cm^{-1} with a shoulder around 1670 cm^{-1}. PLGA in neutral pH range is frequently used as a random coil conformation model of polypeptide in aqueous solution [36]. The amide I' band contour shown in Fig. 5.2A at 1030 MPa is well identified with that of PLGA. Pressure-denatured RNase A does not have any residual secondary structure elements. The bands at 1601, 1609, and 1615 cm^{-1} appear at 770 MPa, although they were already detectable at 370 MPa. However, the two band assignments are unknown. As soon as pressure is further increased up to 1240 MPa, pressure-unfolded RNase A is precipitated, and then the inside of the diamond anvil cell gradually turns clear within 10 min. Water freezes at room temperature above about 1200 MPa; however, the protein solution does not freeze. This is indicated by the lack of change in the water infrared absorption arising from uncoupled O—H stretching vibrational mode. On the basis of the fact that α-quartz does not move at all in the cell, it is confirmed that fluidity of the protein solution is considerably reduced. Thus, such a phenomenon as a pressure-induced gelation is identified. Surprisingly, the observed amide I' band contour of the clear gel is similar to that of the protein solution at 1030 MPa. However, the contour is broadened and the negative maximum shifts to 1637 cm^{-1}. The clear gel changes quickly back into solution by pressure release to 1060 MPa, while the polypeptide backbone of RNase A remains fully unfolded, as shown in Fig. 5.2. At 780 MPa, the broad bands at 1667 and 1673 cm^{-1}, generally assigned to turns, are observed prior to the appearance of the bands characteristic of native RNase A. It is doubtful whether

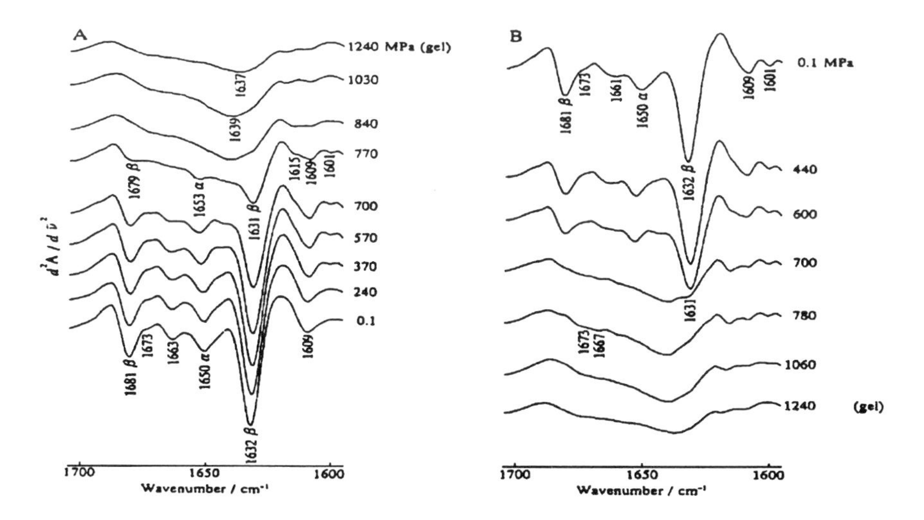

Fig. 5.2. Second-derivative infrared spectra in the amide I' region of completely deuterated RNase A upon pressure increase (*A*) and release (*B*) at 30℃. Solution conditions: RNase A is dissolved in 0.05 M Tris-DCl D_2O buffer, pD 7.0; the protein concentration is 50 mg/mL

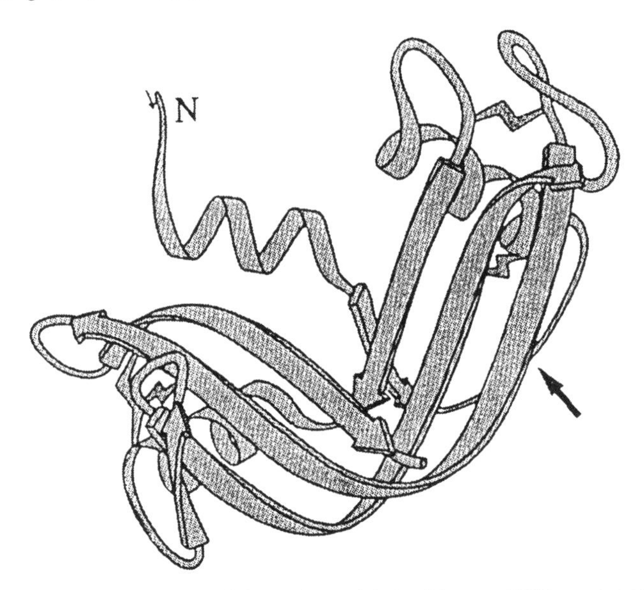

Fig. 5.3. Ribbon representation of the polypeptide backbone of RNase A. RNase A, a small monomeric enzyme, consists of 124 amino acid residues. The structural feature of the polypeptide backbone is an N-terminal α-helix and two shorter α-helices packed against a central, V-shaped antiparallel β-sheet. Various turns join a α-helix to a β-strand, or are located at a hairpin connection between antiparallel β-strands. An arrow represents the site where RNase A may be cleaved with subtilisin to give the two fragments, S-peptide (residues 1–20) and S-protein (residues 21–124)

these bands are similar to those identified in the native state. The dramatic spectral changes occur over the pressure range between 780 and 440 MPa. The transition upon pressure release is found in a lower pressure range, at about 150 MPa, than that upon pressure increase. The amide I' band contour after pressure is released to 0.1 MPa is identical to that before pressure is applied, except for the bands at 1601, 1609, and 1661 cm^{-1}. The pressure-induced changes in the secondary structure of RNase A, at least in the α-helices and β-sheet, are completely reversible. On the other hand, the turns and the microenvironment of tyrosine residues are different between before when pressure is applied and after when pressure is released. It is interesting to note that two of the six tyrosine residues of RNase A are located in the turns [37].

Pressure Effects on the Hydrogen—Deuterium Exchange of RNase A. Figure 5.4 shows the second-derivative infrared spectra of partially deuterated RNase A upon pressure increase. At 0.1 MPa, the component bands at 1638, 1658, and 1688 cm^{-1} are assigned to the β-sheet, α-helices, and β-sheet, respectively [28, 29]. These band frequencies are 6—8 cm^{-1} higher than those corresponding to completely deuterated RNase A. This results from the overlapping of the amide I vibrational mode of both deuterated and undeuterated peptide segments in the α-helices and β-sheet. Shifts in amide I vibrational frequency due to hydrogen–deuterium exchange are always to lower frequency. The backbone amide protons involved in the α-helices and β-sheet

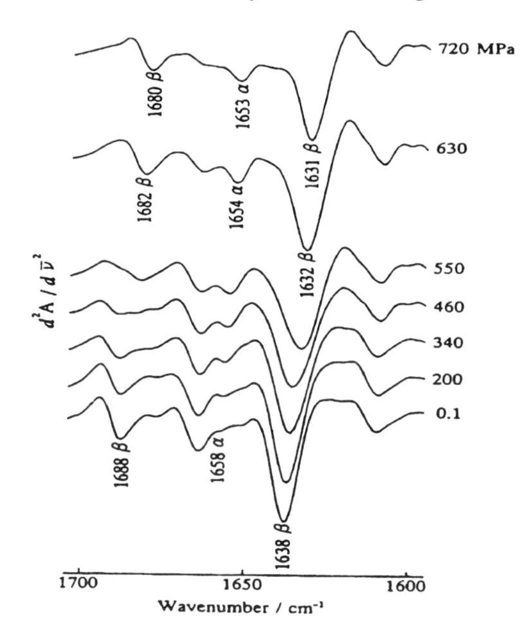

Fig. 5.4. Second-derivative infrared spectra in the amide I(I') region of partially deuterated RNase A upon pressure increase at 30 ℃. It takes about 1 h to record the first infrared spectrum after the sample preparation. 30 min are allowed to equilibrate the sample solution at the chosen pressure prior to each infrared measurement, which itself takes 16 min

of RNase A are strongly protected against exchanging with solvent deuterons [28, 29, 34, 35]. Under the present conditions, the amide I(I') component bands, assigned to the α-helices and β-sheet, shift little toward low frequency after 3 days incubation at 30 ℃ and 0.1 MPa. As pressure is increased from 0.1 MPa to 720 MPa, especially between 550 and 630 MPa, the dramatic spectral changes are observed. The amide I(I') band contour at 630 MPa identifies well with that shown in Fig. 5.2A at 570 MPa. This suggests that most of the protected amide protons involved in the α-helices and β-sheet are exchanged with solvent deuterons upon pressure increase. It is noteworthy that the hydrogen–deuterium exchange is practically completed by application of high pressure where the two secondary structure elements are identified as virtually intact.

Pressure-Induced Denaturation of RNase A. Pressure-denatured RNase A does not have any secondary structure elements such as α-helices, β-sheets and reverse turns. The amide I(I') vibrational mode principally originates from the C=O stretching vibration of peptide group [38]. It is expected that the stronger the hydrogen bonding at the C=O site of peptide group, the lower the observed frequency of amide I vibrational mode. The amide I' bands shown in Fig. 5.2A at 1030 MPa have a broad maximum at 1639 and 1645 cm^{-1}. From the results of the hydrogen–deuterium exchange, it is clear that the backbone peptide segments in the unfolded states are easily accessible to D_2O molecules. The backbone peptide segments of pressure-unfolded

RNase A are hydrated in a stronger manner. Interestingly, MD simulation studies have suggested that the average number of water molecules hydrated to the backbone peptide groups of BPTI increases with increasing pressure, while the average bond length between hydrated water molecules and the backbone peptide groups does not change [11].

It is interesting to note that the broad amide I' component bands at 1667 and 1673 cm^{-1} appear suddenly at 780 MPa in the refolding from the pressure-unfolded state (see Fig. 5.2B). Both the bands are generally assigned to turns, judging from their band frequencies [14, 17, 38, 39]. They are reproducibly observed prior to the obvious appearance of bands assigned to other secondary structure elements such as α-helices or β-sheets. It seems unlikely that the bands at 1667 and 1673 cm^{-1} are identical with those due to the turns in native RNase A. The native turns join the α-helix to the β-strand or are located at the hairpin connection between the antiparallel β-strands [33−35]. If they were identical, the refolding of not only the turns but also the α-helices and β-sheet would occur cooperatively. In an initial stage during protein folding, 'local' interactions should predominate over 'non-local' ones. ('local' interactions are those among residues that are neighbors or near in the sequence. 'Non-local' refers to interactions among residues that are significantly apart in the sequence.) The use of short linear peptides as model systems to detect the formation of such local structural elements is reasonable. Recently, the following conclusions have been drawn by using 2D-NMR spectroscopy: (a) The key structural element involved in the initiation of protein folding is probably a turn-like conformation; and (b) initial folding events do not necessarily give rise to the structures which are retained in the final folded protein [40, 41]. The broad bands at 1667 and 1673 cm^{-1} shown in Fig. 5.2B and at 780 MPa possibly indicate that there are such non-native turns in equilibrium. The formation of local structures in certain regions of the polypeptide chain could be an initiating step in protein folding. Even if the non-native turns are only marginally stable, they would efficiently induce the steric restriction of the polypeptide chain, thus directing the course of protein folding. Indeed, the appearance of the amide I' component band at 1631 cm^{-1} is induced by pressure release from 780 to 700 MPa, which indicates the refolding of the β-sheet (see Fig. 5.2B). The structural features of the polypeptide backbone in the initiation of protein folding might be first characterized by high pressure FT-IR spectroscopy.

Interpretation of the Pressure Effect on the Hydrogen–deuterium Exchange of RNase A. The major factor responsible for exchange retardation is undoubtedly due to the hydrogen bonds which are formed in α-helices and β-sheets. However, the protection patterns found in the two secondary-structure elements are complex, reflecting tertiary as well as secondary-structure. The role of tertiary (side-chain) interactions in stabilizing secondary-structure elements against the exchange is difficult to define. Tertiary interactions may influence the exchange by burial of peptide segments in the protein interior, by distortion of secondary structure in hydrophobic regions, or by local restraints on dynamics [42–44]. These complicated contributions preclude a rigorous and definitive interpretation of the exchange. The following three exchange mechanisms have been suggested: (a) global unfolding of the protein; (b) local unfolding in certain regions of the protein; (c) solvent penetration into internal regions of protein. In any mechanism, the backbone amide protons must be transiently exposed to solvent, permitting hydrogen–deuterium exchange to occur [45−47]. These exchange mechanisms depend on the protein stability. Under conditions favoring the denatured state, the dominant mechanism of exchange involves global unfolding. Under strongly native conditions, more localized structural fluctuations predominate.

An increase in pressure also induces the exchange completion for RNase A prior to the pressure-induced structural transition of the polypeptide backbone (see Figs. 5.2A and 5.4). Localized structural fluctuations are probably dampened with increasing pressure, which is suggested by MD simulation studies of BPTI [11, 48]. Consequently, it is expected that the exchange for RNase A at high pressure occurs primarily via global unfolding. This interpretation is applicable to chymotrypsinogen A. That is, the exchange completion at high pressure occurs only when the pressure-induced denaturation of chymotrypsinogen A is completed [49]. However, the correlation between the exchange and the secondary-structure disruption for RNase A is much poorer at high pressure than that for chymotrypsinogen A. It would seem that consideration of other contributions is required in order to understand the pressure-induced exchange completion for RNase A. Recently, an X-ray crystallographic study at a hydrostatic pressure of 100 MPa has reported the non-uniform distribution of compressibility in the lysozyme molecule [50]. If proteins were non-uniformly compressed in aqueous solution, the application of high pressure where proteins were not extensively unfolded might be enough to form new sites for either water penetration or local unfolding as the first step in hydrogen–deuterium exchange. Binding of a kind of hydrophobic dye, ANS, to lysozyme, which is associated with the protein unfolding, occurs above 600 MPa, while exposure of the tryptophan residues to aqueous medium is inceased at much lower pressure [6]. This should also indicate minor conformational changes of protein at high pressure, as have been previously described. Moreover, it is very interesting to note structural features in the molten globule state recognized as the partially unfolded states of various proteins. The key structural features characterizing the molten globule state are the following: (a) compactness close to the native state; (b) presence of native-like secondary structure; (c) little or no evidence for specific tertiary structure [12, 13]. In spite of the existence of secondary structure comparable to the native state, the amide protons involved in the α-helices and β-sheets in the molten globule state are considerably destabilized compared with those in the native state [14, 25, 26]. The drastic destabilization of native tertiary interactions may enable solvent to be highly accessible to the internal regions. Therefore, there is a possibility that such a partially unfolded state is induced by application of high pressure up to 570 MPa to RNase A.

5.3.2 Ribonuclease S

Pressure-Induced Changes in the Secondary Structure of RNase S. Figure 5.5A and B show the second-derivative infrared spectra of completely deuterated RNase S upon pressure increase and release, respectively. The observed changes in the amide I' band shown in Fig. 5.5 directly indicate the pressure-induced changes in the secondary structure of RNase S. At 0.1 MPa, a visual comparison of the second-derivative infrared spectra between completely deuterated RNase A and RNase S in aqueous solution clearly indicates the absence of major changes in the secondary structure due to the proteolytic cleavage. This result is in agreement with that demonstrated by the X-ray crystal structure of the two proteins [51]. The second-derivative infrared spectrum of RNase S at 0.1 MPa allows the identification of

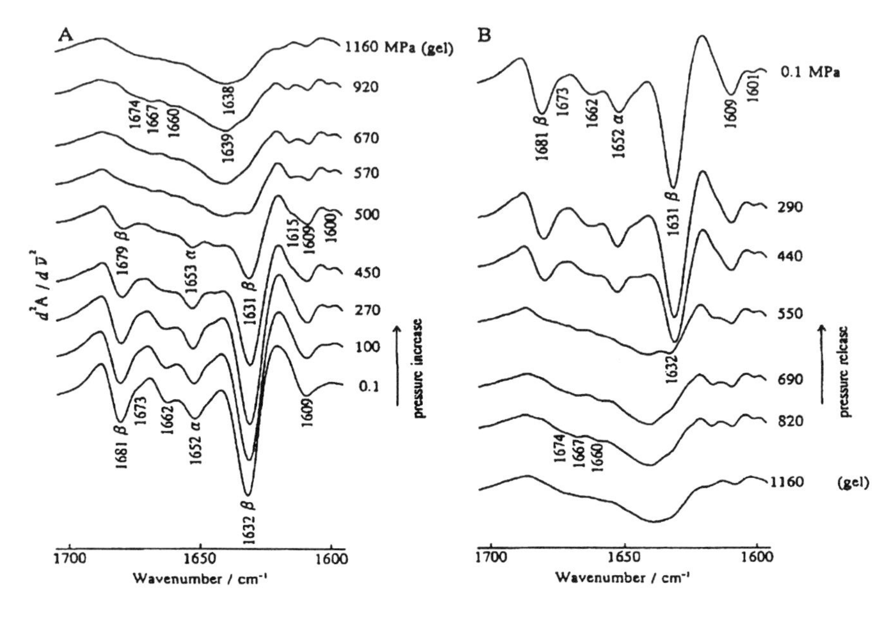

Fig. 5.5. Second-derivative infrared spectra in the amide I' region of completely deuterated RNase S upon pressure increase (A) and release (B) at 30 ℃. Solution conditions: RNase S is dissolved in 0.05 M Tris-DCl D_2O buffer, pD 7.0; the protein concentration is 50 mg/mL

five amide I' component bands. On the basis of the results previously reported by FT-IR spectroscopic studies [29, 30], the component bands at 1632, 1652, 1662, 1673, and 1681 cm^{-1} are assigned to the β-sheet, α-helices, turns, turns, and β-sheet, respectively. Another band at 1609 cm^{-1} arises primarily from side-chain absorption of the tyrosine residues [31, 32].

The amide I' band contours of RNase S upon pressure increase from 0.1 MPa up to 270 MPa are very similar to one another, indicating no or merely minor changes in the secondary structure over this pressure range. A further increase in pressure dramatically induces the cooperative disappearance of the amide I' component bands characteristic of native RNase S. These spectral changes, which occur over the pressure range between 270 and 670 MPa, suggest the pressure-induced unfolding of RNase S. This transition range is lower than that of RNase A by about 300 MPa (see Fig. 5.2A). The proteolytic cleavage between residues Ala20 and Ser21 considerably decreases the pressure stability of RNase A. The amide I' band contour at 920 MPa after the pressure-induced unfolding is complete indicates no evidence of the secondary-structure elements characteristic of native RNase S. However, the contour is not as featureless as that of pressure-unfolded RNase A. The broad bands at 1660, 1667, and 1674 cm^{-1} are observed, although their assignments are uncertain. As is the case for the RNase A D_2O solution, the RNase S D_2O solution reversibly becomes a gel with the polypeptide backbone unfolded by application of pressure above 1000 MPa. The spectral transition upon pressure release occurs over the same pressure range as that upon pressure increase. The amide I' band contour after pressure is released to 0.1 MPa is identical with that before pressure is applied, except for the bands at 1601 and

1609 cm^{-1}. The microenvironment of tyrosine residues are different between before when pressure is applied and after when pressure is released. In contrast, the difference in the turns (the amide I' component band at 1662 cm^{-1}) is not as obvious as that of RNase A. The pressure-induced changes in the secondary structure of RNase S are completely reversible.

Behavior of the Hydrogen–Deuterium Exchange of RNase S at High Pressure. Figure 5.6 shows the second-derivative infrared spectra of partially deuterated RNase S upon pressure increase. At 0.1 MPa, the component bands at 1636, 1658, and 1684 cm^{-1} are assigned to the β-sheet, α-helices, and β-sheet, respectively[29]. These band frequencies are 4–6 cm^{-1} higher than those corresponding to completely deuterated RNase S (see Fig. 5.6A at 0.1 MPa). This indicates that a lot of backbone amide protons involved in the α-helices and β-sheet are protected against exchange with solvent deuterons. However, the bands assigned to the β-sheet of RNase S are broader and found in a frequency region lower than those corresponding to RNase A, which suggests that RNase S is more flexible than RNase A. This is probably responsible for the proteolytic cleavage of the polypeptide chain which lies across the three β-strands. Under the present conditions, the amide I(I') component bands, assigned to the α-helices and β-sheet, shift little toward low- frequency after 24 h incubation at 30 ℃ and 0.1 MPa. On the basis of low frequency shifts in the amide I(I') component bands, assigned to the α-helices and β-sheet, the protected backbone amide protons in RNase S are exchanged with solvent deuterons upon pressure increase. The hydrogen–deuterium exchange is practically completed at 250 MPa, where the two secondary-structure elements are identified as virtually intact.

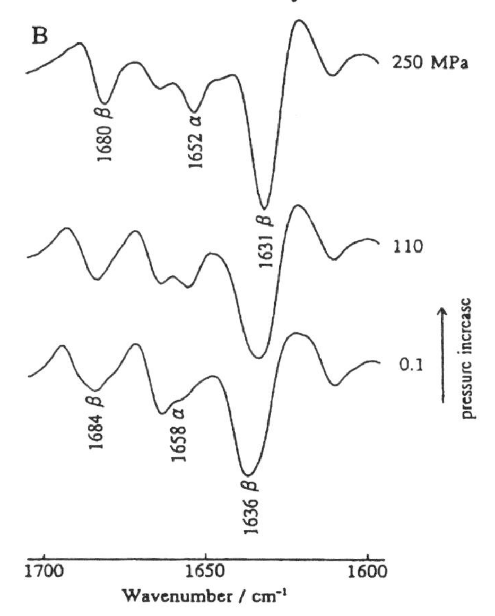

Fig. 5.6. Second-derivative infrared spectra in the amide I(I') region of partially deuterated RNase S upon pressure increase at 30℃. It takes about 1 h to record the first infrared spectrum after the sample preparation. 30 min are allowed to equilibrate the sample solution at the chosen pressure prior to each infrared measurement, which itself takes 16 min

Difference in the Structural Features Between Pressure-Unfolded RNase A and RNase S. The amide I' band contour of RNase A after the pressure-induced unfolding is complete identifies well with that of PLGA at high pressure. PLGA in the neutral pH range is frequently used as a random coil conformation model of polypeptide [36]. It is evident that pressure-denatured RNase A does not have any residual secondary-structure elements. In contrast, the amide I' band contour of RNase S after the pressure-induced unfolding is complete is not as featureless as that of RNase A. Only for RNase S are the broad amide I' component bands at 1660, 1667, and 1674 cm^{-1} observed. The three bands suggest the formation of such nonnative turns as are described in Sect. 5.3.1. Despite the same amino acid sequence, a distinct difference in the structural features of the polypeptide backbone between pressure-unfolded RNase A and RNase S is found. This seems to result from a difference in the polypeptide chain length [52]. There is a strong possibility that RNase S dissociates into the two fragments after the pressure-induced unfolding.

Difference in the Pressure Stability Between RNase A and RNase S. In the native state, the infrared spectra of completely deuterated RNase A and RNase S hardly distinguish any significant differences in secondary structure between the two proteins. Measurements of hydrogen–deuterium exchange were originally introduced as a way to probe the dynamic properties of proteins. As the result of a visual comparison of the infrared spectra between partially deuterated RNase A and RNase S, it is concluded that a considerable decrease in the pressure stability of RNase S results from more extensive structural fluctuations in the native state, which is induced by the proteolytic cleavage. For staphylococcal nuclease and its mutants, however, there seems to be little correlation between the structural fluctuations assessed by time-resolved fluorescence spectroscopy and the pressure stability [7]. This may reflect that the location of the single tryptophan residue probed by fluorescence spectroscopy in the protein molecules is not suitable for identifying the dynamic properties. It is interesting to note the behavior of hydrogen–deuterium exchange at high pressure. In any exchange mechanism, the peptide segments in protein must be transiently exposed to solvent, permitting exchange to occur. As shown in Figs. 5.4 and 5.6, it is clear that the peptide segments, which are buried in the interior, of both RNase A and RNase S are more accessible to solvent at high pressure before their pressure-induced unfolding starts. Application of high pressure to protein aqueous solution promotes any structural change that brings about a decrease in the volume of the system. A probable contribution to the volume change arising from solvent access to protein internal regions is the decrease in partial molar volume upon the exposure of buried hydrophobic surfaces to water [10]. On the other hand, the further pressure is increased, the harder tightening protein internal regions becomes [53]. The reason for this is that the internal cavities which result from imperfect packing of the amino acid residues are very compressible [54]. The increase in solvent accessibility to protein internal regions may easily induce the replacement of intramolecular interactions between the amino acid residues by intermolecular interaction between the amino acid residues and water molecules. This process could provoke extensive unfolding of the polypeptide backbone of proteins. The solvent accessibility to protein internal regions should provide valuable insights into the factors determining the pressure stability of proteins and the mechanism of pressure-induced unfolding.

5.3.3 Bovine Pancreatic Trypsin Inhibitor

Assignments of the Amide I(I') Component Bands for Completely and Partially Deuterated BPTI. The conformational sensitive amide I region of the infrared spectra of completely and partially deuterated BPTI dissolved in D_2O buffer under high pressure are illustrated in Figs. 5.7 and 5.9, respectively. The second-derivative infrared spectrum of completely and partially deuterated BPTI allows the identification of four and six component bands as shown in Figs. 5.7A and 5.9, respectively. The assignments of these component bands are based on theoretical calculations [20] and on correlation experimentally established between amide I(I') component bands and specific types of secondary structures [14, 15, 17, 39].

Figure 5.8 shows the ribbon representation of the polypeptide backbone of BPTI. In addition to the known X-ray crystal structure of BPTI [55, 56], the different behavior of individual amide I component bands upon hydrogen–deuterium exchange facilitates their band assignments. In general, the backbone amide protons involved in the α-helix and β-sheet are strongly protected against hydrogen-deutrium exchange. NMR studies have revealed that BPTI is no exception [57 – 59]. The amide I component bands which should be assigned to such secondary structure elements shift to lower frequency only after lengthy incubation times or heating in D_2O. The bands which shift quickly to lower frequency after dissolution in D_2O can be assigned to other secondary-structure elements. On the basis of the criteria described above, the band assignments of completely and partially deuterated BPTI to specific types of secondary structures are as follows: the amide I' component bands at 1638, 1654, 1669, and 1682 cm^{-1} for completely deuterated BPTI are assigned to the β-sheet,

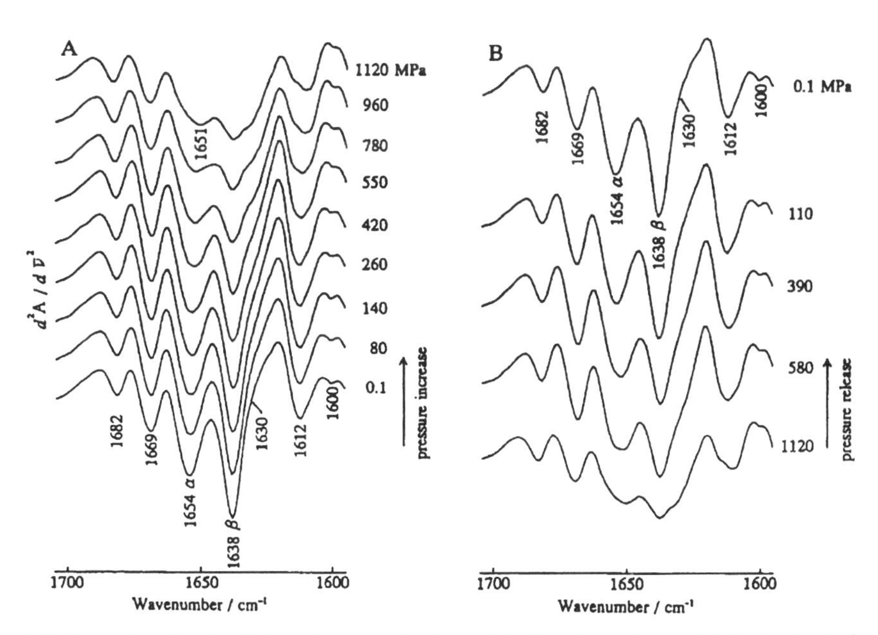

Fig. 5.7. Second-derivative infrared spectra in the amide I' region of completely deuterated BPTI upon pressure increase (*A*) and release (*B*) at 25℃. Solution conditions: BPTI is dissolved in 0.05 M Tris-DCl D_2O buffer, pD 7.0; the protein concentration is 50 mg/mL

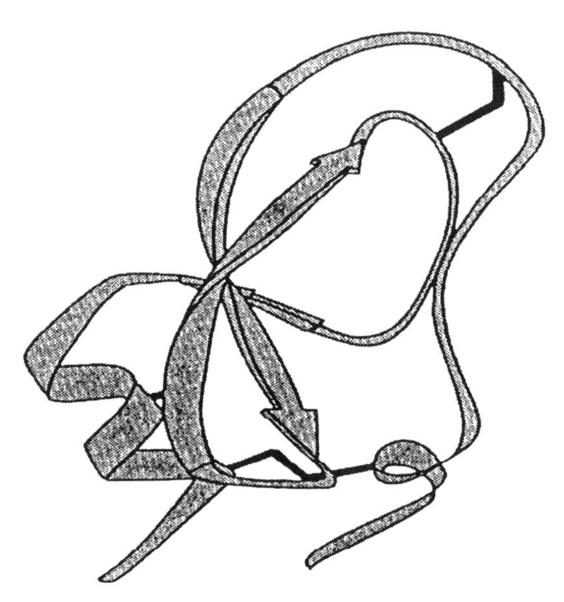

Fig. 5.8. Ribbon representation of the polypeptide backbone of BPTI. BPTI consists of 58 amino acid residues. The most conspicuous feature of the polypeptide backbone is a central, twisted β-sheet composed of two antiparallel strands (residues 18—24 and 29—35) and a very short strand (residues 44—45). In a hairpin-like arrangement, the two longer antiparallel strands are connected by a type I β-turn (residues 25—28). Two helices, a 3_{10}-helix at the N-terminus (residues 3—7) and an α-helix at the C-terminus (residues 47—56), are packed at the closed end side of the hairpin. Finally, two long and overlapping loops (residues 9—17 and 36—43) are contained in BPTI molecule

α-helix, β-turn and/or 3_{10}-helix, and β-sheet, respectively; the component bands at 1642, 1656, 1669 and 1677, and 1688 cm^{-1} for partially deuterated BPTI originate primarily from the b-sheet, a-helix, b-turn and/or 310-helix, and b-sheet, respectively. The component band at 1630 cm-1 for partially deuterated BPTI is difficult to assign. Because the band does not shift to lower frequency after the deuteration of BPTI is complete, it is clear that such a band originates from the peptide segments in which hydrogen-deutrium exchange is rapidly completed. The bands between 1600 and 1612 cm^{-1} shown in Fig. 5.5A at 0.1 MPa arise primarily from side-chain absorption of the tyrosine residues [31, 32].

Pressure-Induced Changes in the Secondary Structure of BPTI. As pressure is increased up to 550 MPa, no frequency shift is observed in all of the four amide I' component bands, as shown in Fig. 5.7. An amide I' component band centered around 1630 cm^{-1} appears, and an increase in the absorbance at 1630 cm^{-1} is observed by the difference spectra in the same pressure range (see Fig. 5.10). BPTI dissolved in D_2O buffer is precipitated above 550 MPa under the present experimental conditions. The pressure-induced precipitation is found with the naked eye. However, no additional bands characteristic of the aggregates of many other thermally denatured proteins are observed in the amide I' region [60—63]. This suggests that the aggregation mecha-

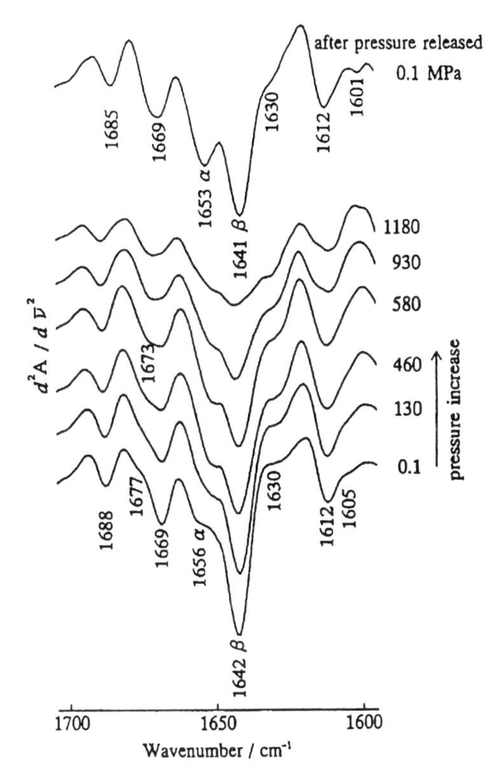

Fig. 5.9. Second-derivative infrared spectra in the amide I(I') region of partially deuterated BPTI upon pressure increase at 25℃. It takes about 1 h to record the first infrared spectrum after the sample preparation. 30 min are allowed to equilibrate the sample solution at the chosen pressure prior to each infrared measurement, which itself takes 16 min

nism at high pressure is different from that at elevated temperature. The band at 1614 cm^{-1} appears above 550 MPa. This band appearance might suggest the changes in the microenvironment of the tyrosine residues due to pressure-induced precipitation. Judging from the amide I' band contour at 1120 MPa shown in Fig. 5.7A, it is evident that the polypeptide backbone of BPTI is not fully unfolded even above 1000 MPa. Precipitation undoubtedly indicates the formation of specific or non-specific interactions between BPTI molecules, which may prevent it from being completely unfolded by application of high pressure. Precipitated BPTI vanishes at 580 MPa upon pressure release. There is no difference in the amide I' band contour between before when pressure is applied and after when pressure is released to 0.1 MPa.

Figure 5.10 shows the difference spectra of completely deuterated BPTI up to 1120 MPa. Only the absorbance at 1630 cm^{-1} increases strikingly with increasing pressure up to 550 MPa. On the other hand, the absorbance at 1638 cm^{-1} for the β-sheet, at 1655 cm^{-1} for the α-helix, and at 1667 cm^{-1} for the β-turn and/or 3_{10}-helix dramatically decrease, and the band at 1630 cm^{-1} shifts lower to 1619 cm^{-1} with a small decrease in absorbance above 550 MPa.

Behavior of the Hydrogen–Deuterium Exchange of BPTI at High Pressure. Figure 5.9 shows the second-derivative infrared spectra of partially deuterated BPTI in the amide I(I') region upon pressure increase. The component bands at 1642 and 1688 cm^{-1}, assigned to the β-sheet, do not shift to lower frequency at all in the pressure range where BPTI is not precipitated (up to 580 MPa). The component bands at 1641 and 1685 cm^{-1} after pressure is released to 0.1 MPa are located between the bands at 1642 and 1688 cm^{-1} before pressure is applied and the bands at 1638 and 1682 cm^{-1} for completely deuterated BPTI, respectively. This result suggests that at least half of the protected backbone amide protons in the β-sheet remain intact up to 580 MPa. The component band around 1656 cm^{-1}, assigned to the α-helix, shifts to lower frequency in the relatively-low-pressure range, at least below 460 MPa. The component band at 1653 cm^{-1} after pressure is released to 0.1 MPa indicates that all of the protected backbone amide protons in the α-helix are exchanged with solvent deuterons by application of high pressure. The gradual changes in the bands between 1660 and 1680 cm^{-1} are observed up to 580 MPa. The component band at 1669 cm^{-1} after pressure is released to 0.1 MPa is more asymmetric than that for completely deuterated BPTI. It seems that some of the protected backbone amide protons in the β-turn and/or 3_{10}-helix are exchanged, and the others remain intact.

Pressure-Induced Structural Rearrangements in BPTI. As shown in Fig. 5.10, the absorbance at/around 1638, 1654, 1669, and 1682 cm^{-1} scarcely changes in the pressure range where BPTI is not precipitated. The absorbance centered around 1630 cm^{-1} increases enormously, although it is unknown what type of structural element corresponds to its frequency. The pressure-induced structural disruption in the β-sheet, α-helix, and β-turn and 3_{10}-helix is probably not correlated with the increase in absorbance around 1630 cm^{-1}. However, the observed changes in the amide I' band under high pressure undoubtedly indicate the pressure-induced changes which occur somewhere in the polypeptide backbone of BPTI. It is interesting to compare the results demonstrated by MD simulation of BPTI at high pressure (1000 MPa). The average RMS deviations of the backbone atoms between the simulated structure at high pressure and the X-ray crystal structure are largestin the loop regions [11]. Such structural rearrangements in the two overlapping loops might contribute to an increase in the absorbance around 1630 cm^{-1}.

By assessing from the amide I' band contours, it is evident that BPTI is hardly unfolded in the pressure range where the protein is not precipitated. An increase in the absorbance around 1630 cm-1 implies that minor structural changes in the certain regions of polypeptide chain in BPTI are induced by application of high pressure. Unless BPTI is precipitated by an increase in pressure, such minor structural changes in the polypeptide backbone may develop into global unfolding with increasing pressure. Precipitation also suggests conformational changes which the hydrophobic residues buried in the native state expose to solvent, D_2O [64]. In addition, it is important to note the behavior of hydrogen-deutrium exchange in the individual secondary-structure elements under high pressure. Most of the protected backbone amide protons involved in the β-sheet remain intact up to 580 MPa, while those involved in the α-helix and β-turn and/or 3_{10}-helix are exchanged with solvent deuterons. The core of BPTI is composed of the antiparallel β-sheet structure. The α-helix, β-turn, and 310-helix are located at the molecular surface. The new sites where either water penetration or local unfolding occurs as the first step in hydrogen-deutrium exchange

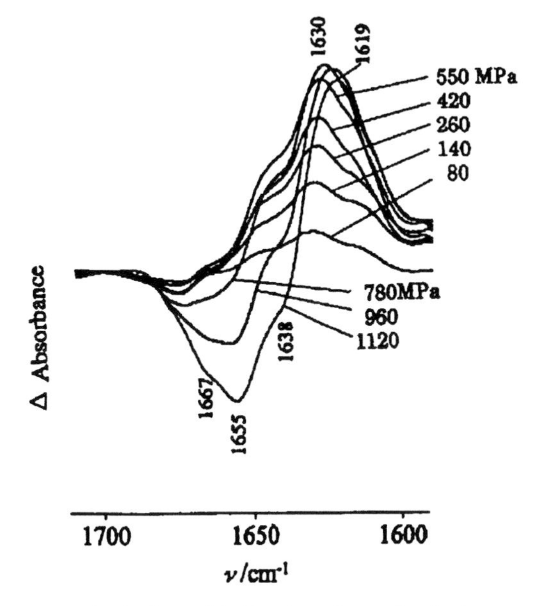

Fig. 5.10. The difference spectra in the amide I' region of completely deuterated BPTI shown in Fig. 5.7A

are formed at the surface regions by application of high pressure and then may be propagated to the internal regions.

5.4 Conclusions

The amide I(I') band analysis provides detailed information directly concerning the structural features of the polypeptide backbone of not only native but also pressure-denatured protein in aqueous solution. The dramatic structural transition of RNase A occurs over the pressure range between 570 and 1030 MPa. Pressure-denatured RNase A, as well as thermally denatured RNase A, does not have any secondary-structure elements such as α-helices, β-sheets, and reverse turns. Although the secondary structure of RNase S in the native state is hardly distinguishable from that of RNase A by the infrared spectra, the structural transition range of RNase S is lower than that of RNase A by about 300 MPa. The proteolytic cleavage between residues Ala20 and Ser21 considerably decreases the pressure stability of RNase A. Pressure-denatured RNase S, as well as RNase A, does not have any secondary-structure elements characteristic of the native state, while a slight but distinct difference in their structural features of pressure-unfolded polypeptide backbone is found. Non-native turns are formed in pressure-unfolded RNase S. For RNase A, non-native turns appear only in the refolding from the pressure-unfolded state prior to the formation of the secondary-structure elements characteristic of the native state. The formation of non-native turns might be correlated with the hysteresis of reversible pressure-induced unfolding observed only for RNase A. On the other hand, the polypeptide backbone of BPTI is not fully unfolded even above 1000 MPa. BPTI is precipitated above 550 MPa, which likely prevents it from being completely unfolded. However, the polypeptide backbone is partly rearranged in the pressure range where BPTI is not precipitated. The pressure-induced structural changes in proteins may be local and minor before the polypeptide backbone is extensively unfolded.

The behavior of hydrogen–deuterium exchange under high pressure is important in the understanding of the factors determining the pressure stability of proteins and the mechanism of pressure-induced unfolding. Most of the protected backbone amide protons in the native state, which are involved in the α-helices and β-sheet of both RNase A and RNase S, are exchanged with solvent deuterons before their pressure-induced unfolding starts. Different as the extent is, similar behavior of hydrogen-deutrium exchange is observed for BPTI. The application of high pressure where proteins are not extensively unfolded must be enough to form new sites where either water penetration or local unfolding occurs as the first step in hydrogen–deuterium exchange. There is a possibility that application of high pressure to proteins induces a partially unfolded state which has native-like secondary-structure elements but permits solvent to be highly accessible to the internal regions. In any mechanism of hydrogen–deuterium exchange, the peptide segments in protein must be transiently exposed to solvent, D_2O. Application of high pressure to a protein aqueous solution promotes any structural change that results in a decrease in the volume of the system. Since the partial molar volume of proteins decreases upon the exposure of buried hydrophobic surfaces to water, the volume change of the system arising from solvent access to protein internal regions is negative. The increase in solvent accessibility to protein internal regions under high pressure may easily induce the replacement of intramolecular interactions between the amino acid residues by intermolecular interaction between the amino acid residues and water molecules. This process should provoke extensive unfolding of the polypeptide backbone of proteins with increasing pressure, because packing the internal regions of proteins is substantially completed at relatively low pressure (< 200 MPa). Therefore, it seems that the easiness of solvent accessibility to protein internal regions is strongly correlated with the pressure stability of proteins.

References

5.1 P. L. Privalov: Adv. Protein Chem. **33**, 167 (1979)

5.2 K. A. Dill, Biochemistry **29**, 7133 (1990)

5.3 J. F. Brandts, R. J. Oliveira, C.Westort, Biochemistry **9** 1038 (1970)

5.4 S. A. Hawley, Biochemistry **10**, 2436 (1971)

5.5 A.Zipp and W. Kauzmann, Biochemistry **12**, 4217 (1973)

5.6 T. M. Li, J. W. Hook III, H. G. Drickamer, G. Weber, Biochemistry **15**, 5571 (1976)

5.7 C. A. Royer, A. P. Hinck, S. N. Loh, , K. E. Prehoda, X. Peng, , J. Jonas, J. L. Markley, Biochemistry **32,** 5222 (1993)

5.8 S. D. Samarasinghe, D. M. Campbell, A. Jonas, J. Jonas, Biochemistry **31**, 7773 (1992)

5.9 X. Peng, J. Jonas, J. L. Silva, Biochemistry **33**, 8323 (1994)

5.10 Y. Taniguchi, K. Suzuki, J. Phys. Chem. **87**, 5185 (1983)

5.11 D. B. Kitchen, L. H. Reed, R. M. Levy, Biochemistry **31**, 10083 (1992)
5.12 O. B. Ptitsyn, J. Protein Chem. **6**, 273 (1987)
5.13 K. Kuwajima, Proteins: Struct. Funct. Genet. **6**, 87 (1989)
5.14 D. M. Byler, H. Susi, Biopolymers **25**, 469 (1986)
5.15 W. K. Surewicz, H. H. Mantsch, Biochim. Biophys. Acta **952**, 115 (1988)
5.16 M. Jackson, P. I. Haris, D. Chapman, J. Mol. Struct. **214**, 329 (1989)
5.17 A. Dong, P. Huang, W. S. Caughey, Biochemistry **29**, 3303 (1990)
5.18 D. C. Lee, P. I. Haris, D. Chapman, R. C. Mitchell, Biochemistry **29**, 9185 (1990)
5.19 W. K. Surewicz, H. H. Mantsch, D. Chapman, Biochemistry **32**, 389 (1993)
5.20 S. Krimm, J. Bandekar, Adv. Protein Chem. **38**, 181 (1986)
5.21 A. Savitzky, J. E. Golay, Anal. Chem. **36**, 1628 (1964)
5.22 D. G.Cameron, D. J. Moffatt, Appl. Spectrosc. **41**, 539 (1987)
5.23 J. K. Kauppinen, D. J. Moffatt, H. H. Mantsch, Appl. Spectrosc. **35**, 271 (1981)
5.24 F. M. Hughson, P. E. Wright, R. L. Baldwin, Science **249**, 1544 (1990)
5.25 M.-F. Jeng, S. W. Englander, G. A. Elve, A. J. Wand, H. Roder, Biochemistry **29**, 10433 (1990)
5.26 C.-L. Chyan, C. Wormald, C. M. Dobson, P. A. Evans, J. Baum, Biochemistry **32**, 5681 (1993)
5.27 P. T. T. Wong, D. J. Moffatt, F. L. Baudais, Appl. Spectrosc. **39**, 733 (1985)
5.28 J. M. Olinger, D. M. Hill, R. J. Jakobsen, R. S. Brody, Biochim. Biophys. Acta **869**, 89 (1986)
5.29 P. I. Haris, D. C. Lee, D. Chapman, Biochim. Biophys. Acta **874**, 255 (1986)
5.30 T. Yamamoto, M. Tasumi, J. Mol. Struct. **242**, 235 (1991)
5.31 Y. N. Chirgadze, O. V. Fedorov, N. P. Trushina, Biopolymers **14**, 679 (1975)
5.32 H. Matsuura, K. Hasegawa, T. Miyazawa, Spectrochim. Acta **42A**, 1181 (1986)
5.33 A. Wlodawer, N. Borkakoti, D. S. Moss, B. Howlin, Acta Crystallogr. **B42**, 379 (1986)
5.34 A. D. Robertson, E. O. Purisima, M. A. Eastman, H. A. Scheraga, Biochemistry **28**, 5930 (1989)
5.35 M. Rico, M. Bruix, J. Santoro, C. Gonzalez, J. L. Neira, J. L. Nieto, J. Herranz, Eur. J. Biochem. **183**, 623 (1989)
5.36 M. Jackson, P. I. Haris, D. Chapman, Biochim. Biophys. Acta **998**, 75 (1989)
5.37 A. Wlodawer, R. Bott, L. Sj^lin, J. Biol. Chem. **257**, 1325 (1982)
5.38 S. Krimm, J. Bandekar, Adv. Protein Chem. **38**, 181 (1986)
5.39 D. F. Kennedy, M. Crisma, C. Toniolo, D. Chapman, Biochemistry **30**, 6541 (1991)
5.40 H. J. Dyson, M. Rance, R. A. Houghton, R. A. Lerner, P. E. Wright, J. Mol. Biol. **201**, 161 (1988)
5.41 P. E. Wright, H.-J. Dyson, R. A. Lerner, Biochemistry **27**, 7167 (1988)
5.42 E. M. Goodman, P. S. Kim, Biochemistry **30**, 11615 (1991)
5.43 S. E. Radford, M. Buck, K. D. Topping, P. A. Evans, C. M. Dobson, Proteins: Struct. Funct. Genet. **14**, 237 (1992)
5.44 S. E. Radford, C. M. Dobson, P. A. Evans, Nature **358**, 302 (1992)
5.45 C. Woodward, I. Simon, E. T¸chsen, Mol. Cell. Biochem. **48**, 135 (1982)
5.46 G., Q Wagner: Rev. Biophys. **16**, 1 (1983)
5.47 S. W. Englander, N. R. Q. Kallenbach, Rev. Biophys. **16**, 521 (1984)
5.48 R. M. Brunne, W. F. van Gunsteren, FEBS Lett. **323**, 215 (1993)
5.49 P. T. T. Wong, K. Heremans, Biochim. Biophys. Acta **956**, 1 (1988)
5.50 C. E. Kundrot, F. M. Richards, J. Mol. Biol. **193**, 157 (1987)

5.51 E. E. Kim, R. Varadarajan, H. W. Wyckoff, F. M. Richards, Biochemistry **31**, 12304 1992)
5.52 D. O. V. Alonso, K. A. Dill, D. Stigter, Biopolymers **31**, 1631 (1991)
5.53 P. Cioni, G. B. Strambini, J. Mol. Biol. **242**, 291 (1994)
5.54 K. Gekko, Y. Hasegawa, Biochemistry **25**, 6563 (1986)
5.55 A. Wlodawer, J. Deisenhofer, H. Huber, J. Mol. Biol. **193**, 145 (1987)
5.56 A. Wlodawer, J. Nachman, G. L. Gilliland, W.Gallagher, and C. Woodward, J. Mol. Biol. **198**, 469 (1987)
5.57 H. Roder, G. Wagner, K. Wüthrich, Biochemistry **24**, 7396 (1985)
5.58 W. Gallagher, F. Tao, C. Woodward, Biochemistry **31**, 4673 (1992)
5.59 K. S. Kim, J. A. Fuchs, C. K. Woodward, Biochemistry **32**, 9600 (1993)
5.60 W. K. Surewicz, J. J. Leddy, H. H. Mantsch, Biochemistry **29**, 8106 (1990)
5.61 A. Muga, H. H. Mantsch, W. K. Surewicz, Biochemistry **30**, 7219 (1991)
5.62 A. A. Ismail, H. H. Mantsch, P. T. T. Wong, Biochim. Biophys. Acta **1121**, 183 (1992)
5.63 D. Naumann, C. Schultz, U. G^rne-Tschelnokow, F. Hucho, Biochemistry **32**, 3162 (1993)
5.64 L. R. De Young, K. A. Dill, A. L. Fink, Biochemistry **32**, 3877 (1993)

6 The Small-Angle X-Ray Scattering from Proteins Under Pressure

Tetsuro Fujisawa[1] and Minoru Kato[2]

[1] Structural Biochemistry Laboratory, RIKEN Harima Institute/ Spring-8, 1-1-1 Kouto, Mikazuki, Sayo, Hyogo 679-5148, JAPAN
E-mail: fujisawa@sp8sun.spring8.or.jp
[2] Department of Chemistry, Faculty of Science and Engineering, Ritsumeikan University, 1-1-1 Noji-Higashi, Kusatsu, Shiga 525-8577, JAPAN
E-mail: kato-m@se.ritsumei.ac.jp

Abstract. The overall solution structures of proteins under pressure were studied by using the synchrotron small-angle X-ray scattering (SAXS) technique. The measurements were made with a hydrostatic pressure cell with diamond windows, which enabled quantitative analysis of scattering profiles. Two applications were demonstrated: the pressure-induced denaturation of a monomeric protein, myoglobin, and the dissociation of an oligomeric protein, lactate dehydrogenase (LDH). Myoglobin showed sigmoid transition with a mid-point at 180 MPa. Denatured myoglobin at 300 MPa gave 20.9 ± 0.9Å for the radius of gyration (*Rg*), which enabled quantitative comparison with different unfolding states. Pressure-denatured myoglobin at acidic pH was much more compact than myoglobin in urea-induced unfolded state and even more compact than myoglobin in a molten globule state. LDH consists of four identical subunits. Forward scattering, *I(0)/C,* which is proportional to molecular weight, showed that LDH dissociated into not monomers but dimer subunits with applied pressure. The conformational drift was confirmed by the value of *Rg*. There is little structural difference between native and drifted LDH. The presence of five scattering peaks in the medium-angle region indicates the dissociated dimer does not take a molten-globule-like structure but instead a structure in which the core is retained. The analysis of SAXS profiles under high pressure enabled modeling of the orientation of dimer subunits, which seems to be chosen so as to reduce the volume without disrupting the core structure.

6.1 Introduction

Since the small-angle scattering (SAS) method is very powerful for detecting volume change, many studies have been performed on the lipid membrane system, especially using neutron scattering [1, 2]. The SAS from protein solutions, however, has only recently been investigated because of difficulties associated with small signals from protein solutions compared with those from lipids as well as the selection of good materials for cell windows. By using a combination of an intense synchrotron X-ray source and a hydrostatic high-pressure cell with diamond windows, we have been able to quantitatively study proteins in solution [3]. In this chapter we confine the application of a high-pressure cell to the protein-folding problem.

6.1.1 Protein Folding Under Pressure Related to SAXS

Since the pioneering work done by Prof. Suzuki [4] on the reversible denaturation of a protein, many thermodynamic features have been studied by stoichiometric or kinetic methods using UV or fluorescence spectroscopy [5–8] under high pressure. In contrast, microscopic studies started only a decade ago: high-pressure FTIR spectroscopy [9] and NMR techniques [10] have been used to prove changes in the secondary and local tertiary structures, respectively.

At atmospheric pressure the use of the small-angle X-ray scattering (SAXS) technique on molten globular and various denatured states of proteins is very successful in characterizing the size and compactness of each state, the results for which were unique compared to other techniques [11]. The quantitative results on the size and compactness of various denatured states gave new insight into the protein-folding problem. By *Le Chatlier*'s principle, which states that at equilibrium a system tends to minimize the effect of any external factor by which it is perturbed, an increase in pressure reduces the volume of a system, consequently producing a change in the SAS. SAXS capable of detecting volume change accurately should be an essential tool for studying pressure effects on proteins. However, there have been few high-pressure studies using SAS because of experimental difficulty. There have been preliminary reports on lysozyme [12], on ribonuclease [13], and on ATCase [14]. Recently, the pressure-denatured state of staphylococcal nuclease was investigated using SAXS together with FTIR spectroscopy [15]; this showed the possibility of performing a kinetic experiment on the time scale of several minutes.

In this chapter we deal with two applications: pressure-induced denaturation of a small protein and the dissociation of an oligomeric protein. For the former experiments, myoglobin (MW 17 kDa) was used. The thermodynamic denaturation features under high pressure have been already characterized [7]. Further, at atmospheric pressure the various non-native structures have been characterized using the SAXS technique [16]. Determination of the radius of gyration (Rg) for pressure-denatured myoglobin enabled quantitative comparison with other denatured states. For the latter experiments, lactate dehydrogenase (LDH; MW 140 kDa) was used. It consists of four identical subunits. G. Weber's group characterized the pressure-induced dissociation of the four subunits into single subunits by using fluorescence intensity change [17, 18]. They showed a 'hysteresis effect', or 'conformational drift', that is, the dissociation and association cycle of subunits took different pathways. This is not common for other structural perturbations such as temperature or denaturants. While the detection of size change with a probe of fluorescence intensity is indirect, we show a direct measure for the hysteresis effect of a pressurizing and depressurizing cycle. Making the best use of the scattering profile, the dissociation pathway of LDH will be discussed [19].

6.1.2 Information Available from SAXS

The analysis of SAXS profiles is made by applying them in various plots based on theorem. Here, we shortly summarize the fundamental equations for SAXS.

Fundamental Formulas for SAXS. Scattering intensity of X-rays, $I(S)$, where S is the reciprocal vector $|S|=\sin 2\theta/\lambda$, in which λ is the wavelength and 2θ is the scattering angle, can be written as follows [20, 21]:

$$I(S) = NF(S) \cdot F^{*}(S) \tag{6.1}$$

$$F(S) = \int_{volume} \rho(r)\exp(2\pi i S \cdot r)dr \quad , \tag{6.2}$$

$\rho(\mathbf{r})$: excess scattering density at **r**,
F* : complex conjugate of F,
N : number of the particle in the radiation volume.

In SAXS, particles can take any orientation with equal probability. Then, the spatial average of $\exp(2\pi i\mathbf{S}\mathbf{r})$ is equal to $\sin(2\pi Sr)/(2\pi Sr)$ and $I(S)$ yields the isotropic intensity function, $I(S)$.

$$I(S) = N \iint_{volume} \rho(r_1)\rho(r_2)\frac{\sin(2\pi\ S\ r)}{2\pi\ S\ r}dr_1 dr_2 \quad , \tag{6.3}$$

where r is defined as $r = |\mathbf{r}_1 - \mathbf{r}_2|$
In the very-small-angle region the term $\sin(2\pi Sr)/(2\pi Sr)$ can be approximated as $\exp(-(2\pi rS)^2/3)$ so,

$$I(S)=I(0)\exp[-(4\pi^2/3)Rg^2S^2] \quad , \tag{6.4}$$

where

$$I(0) = N \iint_{volume} \rho(r_1)\rho(r_2)\, dr_1\, dr_2 \tag{6.5}$$

$$Rg^2 = \frac{\int_{volume} \rho(r)r^2 dr}{\int_{volume} \rho(r)dr} \quad . \tag{6.6}$$

The logarithm of (6.4) gives

$$\ln[I(S)] = \ln[I(0)] - \frac{4\pi^2}{3}Rg^2S^2 \quad . \tag{6.7}$$

The plot of $\ln\ [I(S)]$ versus S^2 is called the Guinier plot and from (6.4), (6.5) and (6.6), the radius of gyration, Rg, and forward scattering, $I(0)$, are determined.

The other important equations are from (6.3):

$$I(S)=\int_0^{\infty} P(r)\frac{\sin(2\pi Sr)}{2\pi Sr}dr \quad , \tag{6.8}$$

where

$$P(r)=Nr^2\int_0^{4\pi}\left(\int_{volume}\rho(r_1)\rho(r+r_1)\right)d\omega \quad . \tag{6.9}$$

P(r) in (6.9) is called the distance-distribution function, which shows the distribution of distances between two points inside the particle. For example, the *P(r)* function of a solid sphere becomes symmetrical. Its first zero cross point gives the maximum dimension of a particle, i.e., the diameter of a sphere in this case. This function is one of the important functions that is obtainable from the inverse Fourier transform of *I(S)* without introducing assumptions.

6.2 Experimental Methods

There are two major difficulties in collecting SAXS data under pressure: (a) There is high absorption and parasitic scattering from cell-window materials. (b) It is very difficult to maintain the same conditions between the sample solution and the buffer solution. By applying pressure, it is very easy to change the sample thickness irreversibly and to destroy parallelism between the two windows.

In this chapter we consider a practical method of SAXS data collection under pressure.

6.2.1 High-Pressure Cell for SAXS

For SAXS measurement under pressure, the diamond anvil cell (DAC) has often been used. This type of cell is very compact and it is easy to achieve high pressures; thus, most SAXS diffraction work employs this type of cell. The DAC, however, is not suitable for the protein-solution system, which requires identical conditions between sample and buffer solutions [3]. Figure 6.1 shows the high-pressure cell used in our experiment. The cell employs two synthetic (type Ib) diamonds (Sumicrystal, Sumitomo Denco Co.), each of 1.0 mm thickness and 5.0 mm diameter. Beryllium is often used for the window material because of its low absorption; however, it has parasitic scattering in the small-angle region [14]. Its use is not recommended. The exchange of sample was done through a hole connected to a high-pressure tubing plug without touching the windows, which is a necessity for keeping identical conditions between sample and buffer solutions.

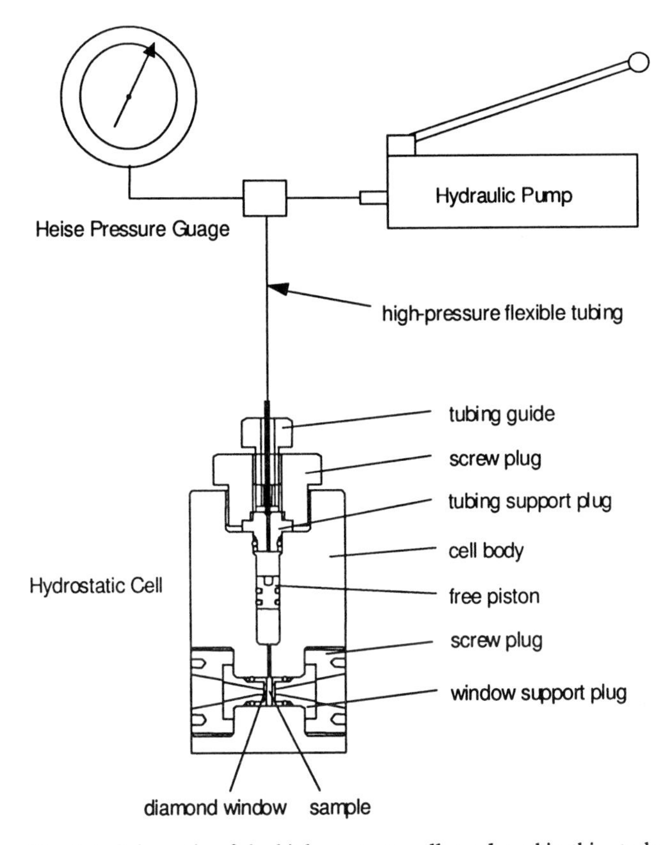

Fig. 6.1. Schematic of the high-pressure cell employed in this study

6.2.2 Absorption of X-Rays

The absorption of X-rays can be expressed as follows [20]:

$$I = I_0 \exp\left[\left(\frac{\mu}{\rho}\right)\rho x\right] , \tag{6.10}$$

I_0 : incident intensity,
I : transmitted intensity,
μ / ρ : mass absorption coefficient (cm^2/g),
ρ : density (g/cm^3),
x : path length (cm).

Equation (6.10) indicates two important things. First, as the mass absorption coefficient (μ / ρ) depends on the wavelength of the X-rays, the choice of the wavelength of the X-rays is critical in reducing the absorption by the window materials. For example, the linear absorption coefficient (μ), the product of the density and the mass absorption coefficient, decreases from 14.9 cm^{-1} to 4.91 cm^{-1} for 1.5 Å and 1.0 Å X-ray wavelengths, respectively. It corresponds to a 7-fold decrease in case of a 1 mm thick diamond window. Second, the absorption will change when pressure is applied. In (6.10) μ/ρ is generally constant irrespective of the status of material. At a fixed wavelength, therefore, because the density increases with increasing pressure, the absorption also increases. Besides this principle effect, the cell reversibly deforms a few tens of microns, so the apparent scattering profile practically decreases more than specified by (6.10). The decrease in transmission at 400 MPa was 37 % in our case [3]. This absorption effect can be corrected by measuring direct intensity. The corrected scattering profiles of water under various pressures coincide with each other within 2.8 % irrespective of pressure [3].

6.2.3 Contrast Effect by Pressure

As implicated by (6.3) scattering phenomena come from the electron-density difference between protein and bulk water, more strictly, the excess electron density of the scattered particle and its associated hydration shell. The isothermal compressibility of protein (about 10×10^{12} cm^2/dyn) is much smaller than water (45×10^{12} cm^2/dyn) [22] so that the excess electron density of protein and its associated hydration shell decreases with increasing pressure. The change with pressure especially affects the forward scattering. According to (6.5), *I(0)* is proportional to the square of the product of the mean excess electron density, $\Delta\rho$, and the volume, *V*.

$$I(0) \propto (\Delta\rho \cdot V)^2 \quad . \tag{6.11}$$

In case of lysozyme *I(0)* decreases by 9 % after correcting for the absorption effect. This decrease includes the decreases in both $\Delta\rho$ and *V*. If we assume this decrease comes only from contrast ($\Delta\rho$), a 9 % decrease in *I(0)* corresponds to an electron-density difference of 0.06 e/Å^3, estimated from a comparison with the decrease in *I(0)* found in contrast variation experiments. In this range of contrast, a change in *Rg* is so small that we can interpret hydrodynamic parameters under pressure as practically the same as those at ambient pressure. However it should always be kept in mind that this rule of thumb depends on the system and pressure region.

6.2.4 High-Pressure SAXS Experiments at a Synchrotron Facility

The SAXS experiments were performed at beamline 10C at Photon Factory, Tsukuba, Japan [23]. With a fix-exit, double-crystal monochromator Si(111), the X-ray wavelength was tuned to 1.3 Å. The beam size was reduced to 1.2 mm × 1.2 mm by guard slits. The SAXS profile was recorded with a one-dimensional position-sensitive proportional counter (PSPC) with a 10 mm slit. The incident X-ray intensity was scaled using the ion chamber current established before sampling. The distance between the sample position and the detector was 55 cm. The X-ray absorption was measured as the direct beam intensity attenuated by 1.3 mm Al foil at each pressure.

In the high-pressure experiments any series of measurements where transmission between the solution and the buffer differed was rejected. From the definition of scattering intensity, the scattering from protein should always be positive (6.1), which means that the raw sample profile is always larger than that of the buffer in all scattering regions. We did not accept any series of experiments where the buffer profile was larger than that of the sample in some region. The temperature of the cell was maintained by circulating temperature-controlled water in a cell jacket. After changing the pressure we waited for at least 5 min for temperature to equilibrate.

Pressure-Induced Unfolding of Metmyoglobin. Sperm whale myoglobin was purchased from Biozyme Co. Myoglobin was purified using ion-exchange column chromatography, PM52, Whatman Co. The protein was dissolved in 50 mM acetate buffer (pH 4.4). The protein concentration was determined spectrophotometrically using $E^{409}_{(mM)} = 159$. The SAXS data collection time was 600 s. No radiation damage was detected during any of the measurements. Experiments were done with a sequential series of compression, that is, starting from ambient pressure an increasing to 300 MPa, with a 75 MPa interval. Since it took more than 1 h from ambient pressure to equilibrate the denature state at 300 MPa, further denaturation from the slow transition to the denaturation state was negligible. Pressure release recovers the initial values of *I(0)* and *Rg*, which certifies that the pressure-induced denaturation of myoglobin is reversible up to 300 MPa.

Pressure-Induced Dissociation of LDH. Pig heart LDH was purchased from Boehringer Mannheim Co. The protein was passed through a desalting column, P6-DG (Bio-Rad Co.), in order to replace buffer media as well as to remove small particles. The buffer was 50 mM Tris-HCl pH 7.5, 1 mM EDTA, and 1 mM DTT containing 8% glycerol. The protein concentration was determined spectrophotometrically using $A_{280}^{1\%} = 14.9$. The data collection time was 900 s. No radiation damage was detected during the compression and decompression cycle.

Measurement of the ligand-induced structural change of LDH at ambient pressure was done in a buffer, containing 100 mM K-phosphate (pH 7.0), 1 mM EDTA, 0.1mM DTT, and with/without 10 mM NAD. A series of different concentrations (1–7 mg/mL) was used to eliminate inter-particle interference. Two camera lengths, 80 cm and 200 cm, were used. The data collection time was 600 s.

6.2.5 Data Analysis of SAXS Profiles

Subtraction of the buffer profile, determination of *Rg* from the Guinier plot, and the extrapolation to an infinitely dilute protein concentration were done at NEC PC9800 [24]. Calculation of the *P(r)* function was done using both Moore's algorithm [25] and the GNOM package [26]. Intensity functions of the crystal structure were calculated using the Debye formula, *i.e.*, the following extension of (6.3):

$$I(S) = \sum \sum f_i f_j \frac{\sin(2\pi S r_{ij})}{2\pi S r_{ij}} , \tag{6.12}$$

where f_i and f_j are the structure factors of the ith and the jth atoms, respectively. Equation (6.12) does not take into account the exclusion volume or the hydration shell, so that only a qualitative comparison between crystal and solution structures can be made. Values of Rg for model structures were calculated using the CRYSOL program package [27]. CRYSOL is a program which evaluates solution scattering with the hydration shell from atomic coordinates by approximating the envelope particle with a spherical harmonic function. The hydration shell width was taken as 3 Å, and the maximum harmonics number was 12.

The surface area and volume were calculated with Voronoi polyhedra using Gerstein's surface and volume calculation program library [28, 29].

6.3 Results and Discussion

6.3.1 Pressure Denaturation of Metmyoglobin

The Change of Rg with Pressure. Figure 6.2a shows the Guinier plot {*ln[I(S)]* vs S^2 plot} of myoglobin (16.7 mg/ml) up to 300 MPa. No upward curvature in the inner scattering region guarantees that there was no significant aggregation in this work. This plot clearly indicates the increase in Rg as a consequence of applying pressure. Figure 6.2b shows the apparent values of Rg change according to pressure. The transition curve indicates that the expansion follows a sigmoid curve. The transition point is around 180 MPa (~900 μM, 25℃),which is compatible with that of 150 MPa at pH 4.5 (~30 μM, 20℃) obtained by studying the spectral change in the visible light region [7] when temperature and protein concentration differences are considered. The myoglobin is supposed to be denatured at 300 MPa.

Pressure-Denatured Structure of Myoglobin. In order to compare different structures with each other, it is inappropriate to use the Rg of one protein concentration since the apparent Rg is affected by interparticle interference [21]. In order to ex-

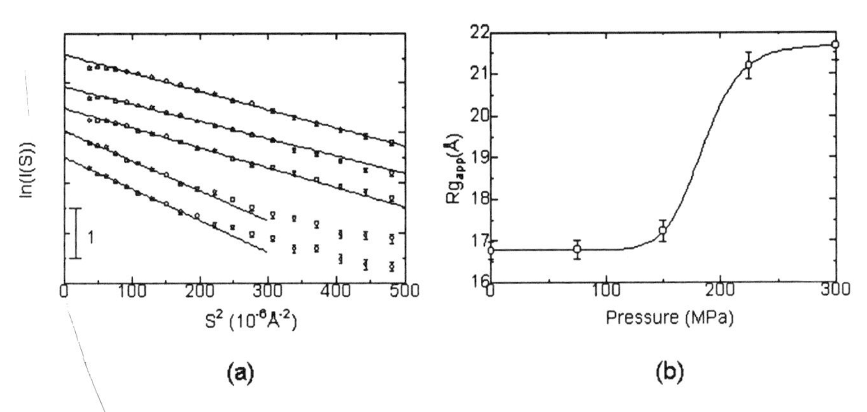

Fig. 6.2. (**a**) Guinier plot of metmyoglobin under various pressures. (**b**) Apparent Rg as a function of pressure. Protein concentration was 16.7 mg/mL. The buffer condition for both plots was 50 mM acetate buffer, pH 4.4. Temperature was maintained at 25℃. In (**a**) the pressures are from top to bottom: 0.1 MPa, 75 MPa, 150 MPa, 225 MPa, and 300 MPa. For clarity, each plot is vertically shifted

trapolate Rg to that for an infinitely dilute solution, a series of measurements on protein concentrations is necessary. This requirement is very difficult for high-pressure SAXS measurement: even if a series of experiments between sample and buffer for one concentration is passed, another series may differ from the previous series and be rejected. This could happen due to various factors, such as, for example, a sudden drop of incident intensity, poor reproducibility of cell setting and so forth. Figure 6.3 shows protein concentration dependency on Rg^2. Generally Rg^2 decreases with increasing protein concentration. This is mainly due to the exclusion volume effect, *i.e.*, a negative value of the second virial coefficient. At 300 MPa, however, the situation is reversed, the tendency for particle association is indicated, although myoglobin remains a monomer: *I(0)/C*, which is the scattering intensity at 0 degrees divided by the weight concentration *C* (mg/mL) of protein and is proportional to molecular weight, showed slight decrease by 5 %. The change in the interparticle interaction seems to be related to the dehydration effect under high pressure.

The effect of dehydration on Rg upon compression is conspicuous in the high-pressure SAXS experiment. The apparent Rg of native lysozyme decreases from 15.3 Å at ambient pressure to 14.8 Å at 400 MPa [3]. The decrease in Rg per 100 MPa is 1.0% as measured by SAXS but only 0.33% as measured by X-ray crystallography [30]. As discussed in the previous section, a contrast effect cannot explain the decrease. In order to compensate this difference two factors can be taken in account: (a) more compressibility of protein in solution than in crystal, and (b) the dehydration of protein under high pressure. The first contribution is limited without introducing a large structural change upon compression in the native protein itself, which was rejected by a number of reviews [31, 32]. The contribution of hydration to Rg is not small. The experimentally derived value of Rg from SAXS is 11 % larger than that from crystal coordinates, while Rg from neutron scattering in H_2O, where the contribution of the hydration layer is cancelled out, is of a value similar to that in the crystal [33]. The discrepancy between the value measured with SAXS and that in crystal is attributed to hydration. The dehydration effect therefore dominates the decrease in Rg of native protein upon compression. When dehydration proceeds under high pressure, water molecules become poor solvents for proteins in which biopolymers try to associate with each other.

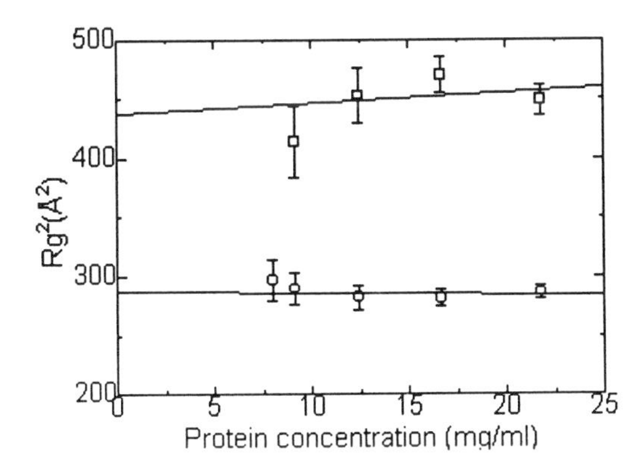

Fig. 6.3. Protein concentration dependence on Rg^2. Values are given for samples at 0.1 MPa (*open circles*) and at 300 MPa (*open squares*), respectively. Bars indicate one sigma.

Table 6.1. Values of radii of gyration (*Rg*) of myoglobin in various conformational states from SAXS

Conformational state	*Rg* (Å) (the increase in *Rg* from holo-native-state)	Reference
Holomyoglobin native state	17.0±0.4[a]	This work
	17.5±0.1[b]	[16]
Pressure-induced denatured state	20.9±0.9[b] (23%)	This work
Apomyoglobin native state	19.7±0.6[a] (13%)	[16]
TCA-induced molten globule	23.1±1.3[a] (32%)	[16]
Acid-unfolded state	30.2 ±2.5[a] (73%)	[16]
Urea-induced unfolded state	34.2±1.5[a] (95%)	[16]
GuanidineHCl unfolded state	35.8±1.0[a] (105%)	[16]

[a] Sperm whale myoglobin (pH 6.0), [b] Horse myoglobin (pH 4.4)

Table 6.1 summarizes the value of *Rg* for various conformational states of myoglobin. The extrapolated values of *Rg* were 17.0 ±0.4 Å and 20.9 ±0.9 Å for atmospheric pressure and 300 MPa, respectively. The *Rg* of the pressure-induced denatured state was much smaller than that of the denaturant-induced denatured state, and was much much smaller than that of the molten globule state. In the visible region of the spectrum, however, the spectrum of pressure-denatured myoglobin resembles that of the acid-denatured protein at atmospheric pressure [7]. The picture from SAXS is that at acidic pH pressure-denatured myoglobin is less disrupted than molten globule myoglobin. This unexpected small dimension of myoglobin leads to the possibility of an intermediate state. The assessment of secondary structure under pressure is now under progress by FTIR combined with resolution-enhancement techniques.

The myoglobin result is in sharp contrast with the pressure-induced denatured state of staphylococcal nuclease (SNase): pressure-denatured SNase is more compact than a random coil but its secondary structure is more disrupted than that of the molten globule state [15]. The apparent *Rg* of pressure-denatured wild type Snase was 36 Å, which is 112% larger than that of the native state and is almost comparable to the 33 Å of the urea-denatured state [34].

It is not known whether pressure-denatured myoglobin at neutral pH gives a similar *Rg* to that at acidic pH or not. Even if we admit the difficulty of generalizing the structure of various proteins in a pressure-denatured state, it is unlikely to assume a compact denatured structure at neutral pH under high pressure. If that is the case, describing the unfolding state quantitatively, not just the native and unfolded states, becomes very important in the contour map of pressure-temperature or pressure-pH curves.

6.3.2 Pressure Dissociation of LDH

Hysteresis Effect Observed by SAXS

***Rg* and *I(0)* Change During the Pressurization and Depressurization Cycle.** Figure 6.4 shows the *Rg* and *I(0)/C* change during the compression and decompression cycle of LDH. The overall structural change showed the hysteresis effect as

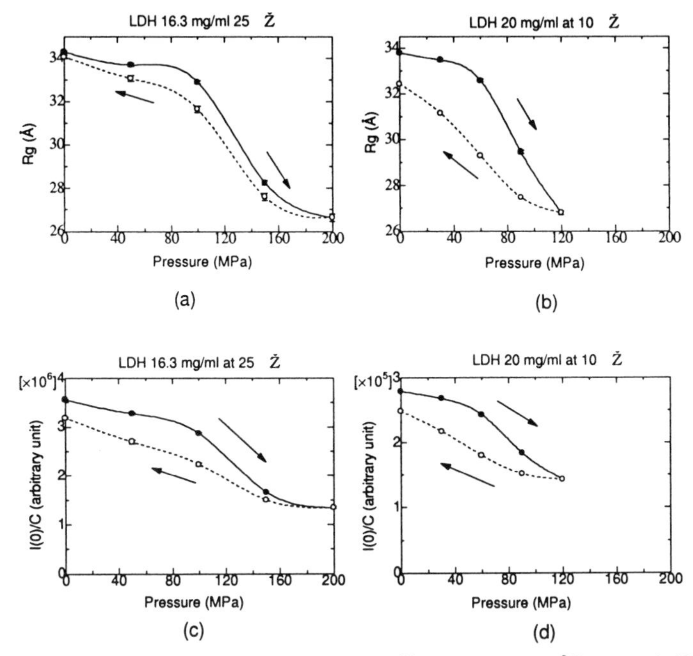

Fig. 6.4. *Rg* change of LDH with pressure at 25℃. (**a**) and at 10℃ (**b**). *I(0)/C* change with pressure at 25℃ (**c**) and at 10℃ (**d**). Protein concentrations are 16.3 mg/mL and 20 mg/ml for 25℃ and 10℃ respectively. Compression (*solid line*) and decompression (*dashed line*) are shown in each panel. Compression beyond 200 MPa and 120 MPa resulted in protein precipitation for 25℃ and 10℃, respectively

observed by the local structure. The dissociation mid-points were 120 MPa (Fig. 6.4a) and 78 MPa (Fig. 6.4b) for 25℃ and 10℃, respectively. These values are compatible with ~145 MPa at 25℃ and ~97 MPa at 5℃, as determined by fluorescence spectroscopy [17, 18]. Lowering the temperature shifted the mid-point to –42 MPa in terms of overall structure. The recovery of *Rg* after the compression–decompression cycle was incomplete at 10℃, while at 25℃ the recovery was finished. The delay of this recovery is more conspicuous in SAXS compared with spectroscopy [18]. In the case of Snase SAXS also shows a slower tail transition compared to FTIR [15].The most important finding from this study is shown in Fig.6.4c and d. *I(0)/C*, which is proportional to molecular weight, is a direct measure of dissociation. *I(0)/C* decreased to 38% and 51% of its initial state at 10℃ and 25℃, respectively. This shows that the dissociation proceeded not to a monomer but to a dimer. The application of more pressure resulted in irreversible aggregation. The result is completely different from many other observations made using fluorescence spectroscopy [17, 18, 35]. The difference may originate from higher protein concentrations made using measurement and/or addition of glycerol, which is known to induce preferential hydration of the protein-water interface [36].

Structure of Difted LDH. We were interested in the structural difference between before and after compression in terms of subunit displacement. Monod proposed that

Table 6.2. Values of *Rg* in different states of LDH

State of LDH	Apparent *Rg* (Å)
Atmospheric pressure	
Apo	34.3±0.1
	34.6±0.1[a]
Holo	33.8±0.2[a]
200 MPa	26.6±0.4
Reconstituted LDH at atmospheric pressure	34.1±0.1

[a] These are not apparent values, but for *Rg* in an infinitely dilute solution measured using a static cell

oligomeric protein exists in two main conformations, T and R, determined by the disposition of the subunits [37]. In crystal apo- and holo-states, LDH has a very similar *Rg* [38, 39], while clear compaction of the molecule was observed in solution at atmospheric pressure (Table 6.2). The displacement of a subunit should be detectable from *Rg*. The result was very small: 34.3 ±0.1 Å and 34.1 ±0.1 Å for native and drifted LDH, respectively. The change in the association property of each subunit did not affect the subunit arrangement.

Tertiary-Structure Change. The structural behavior in the compression-decompression cycle in more pronounced in the Kratky plot (*I(S)·S·S vs.S* plot) shown in Fig. 6.5. In the case of the denaturation of a small protein like Snase, the peak height of the first peak decreases as denaturation proceeds, keeping its peak position, which indicates that the scattering curve is a mixture of only native and denatured particles, *i.e.*, a two-state transition. The peak height, which shows the compactness of mole-

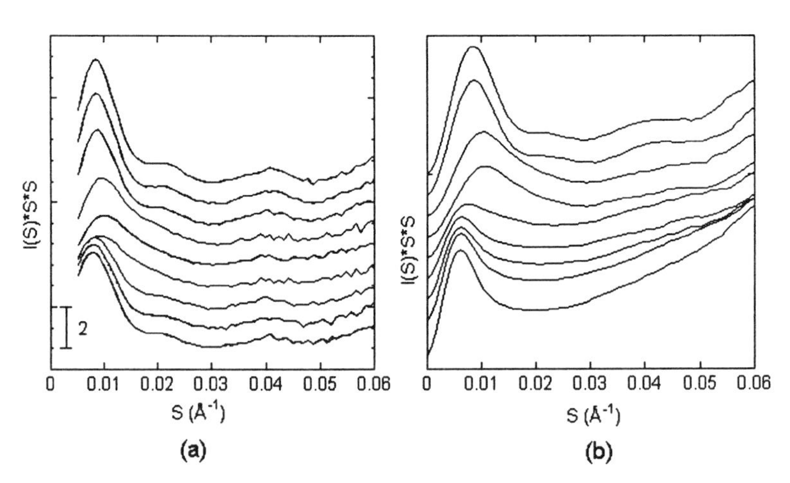

Fig. 6.5. (**a**) Kratky plot of LDH at 25℃. Protein concentration was 40 mg/mL. (**b**) Irreversible aggregation detected by Kratky plot 10℃. Protein concentration was 16.3 mg/mL. For both panels, the plots are arranged in the following order, from top to bottom, 0.1 MPa, 50 MPa, 100 MPa, 150 MPa, 200 MPa, 150 MPa, 100 MPa, 50 MPa, 0.1 MPa. Lines are shifted for clarity

cule, correlates to secondary-structure formation [16]. In Fig. 6.5a at ambient pressure the two peaks at 0.023 $Å^{-1}$ and 0.041 $Å^{-1}$, which cannot be seen in a small globular protein, come from the interference function of four centers of subunits [40]. As dissociation proceeds, the interference becomes less conspicuous. The first peak shifts to a wider *S* value as a result of the decrease in size of the particle but no first peak lies outside of the initial and final positions in Fig. 6.5a. The observation is compatible with a two-state model of a tetramer and a dimer both in dissociation and association. After the pressurization-depressurization cycle, the recovery of tertiary structure and its hysteresis effect were obvious in this plot.

Irreversible Aggregation. SAXS theory assumes that the particles are oriented randomly. In the case of precipitation, in the small *S* range, the scattering profile is strongly affected, so that the determination of *Rg* is impossible. In the higher *S* range, however, the contribution of the interparticle effect becomes small and no perturbations are expected unless the particles stack in a certain orientation [21]. If it exists, the peak corresponding to the correlation length would appear in SAXS profiles, which is not the case here. If we dared to make the same assumption as above, some speculation on pressure-induced irreversible aggregation or 'wrong aggregation' is possible. Figure 6.5b shows the formation process of wrong aggregation by applying too much pressure to LDH. Upto 150 MPa, LDH dissociates into a dimer as in Fig. 6.5a. At 200 MPa the first peak becomes flattened, which shows the collapse of the dimer resulting in an unfolded monomer. The association of the monomer and the collapse of its structure seem to occur simultaneously, because the position of the first peak has already shifted to a lower scattering angle at 200 MPa. During decompression, the height increased, the equilibrium shifted to a stable large aggregation where bulk structure is formed. The subunit structure of wrong aggregation is completely lost, because no subunit interference is seen.

LDH Structure at 200 MPa at 25℃

Subunit Orientation in an LDH Dimer. As shown in Fig. 6.4d LDH dissociated into a dimer at 200 MPa, 25℃. In the analysis of LDH at 200 MPa, we introduced the following assumptions: (1) The SAXS pattern at 200 MPa is only from the LDH dimer, although it should contain a small amount of tetramers. (2) Non-crystallographic symmetry exists even in the LDH dimer.

Assumption (1) could be justified indirectly from Fig. 6.5a, since no conspicuous peaks due to tetramers can be seen there. If assumption (2) is invalid, no interference can be seen in the medium *S* region, which is common for most oligomeric proteins without symmetry or distinct shape. In the analysis of the medium *S* region, it is possible to tell the difference between a given structure and a solution structure; however, it is very difficult to determine the structure without additional information. Since apo-LDH has four subunits with 422 non-crystallographic symmetries [39], there are only 3 combinations of subunit orientation for the dimer, as shown in Fig. 6.6. It is natural to classify the subunit orientation of the initial intermediate into these three candidates.

The *P(r)* function in Fig. 6.7a indicates that a solution structure at 200 MPa exists as a dimer and that the molecular dimension in solution does not deviate highly from model structures. It is, however, very difficult to select one orientation. It is because the contribution of the interference around a few nm, where the difference is sup-

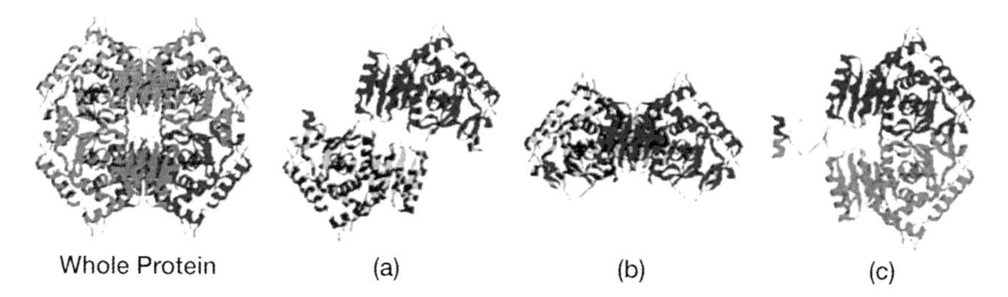

Fig. 6.6. Combination of subunit orientations in LDH. The whole protein, and three models (**a**, **b**, and **c**) are shown. Crystallographic coordinates are used from 6LDH (Protein Data Bank) [39]. Colors show the differences in chains. The figure was drawn by the RasMol2 program

posed to exist, is several hundredths of the forward scattering intensity. While in reciprocal space, S, as the form factor decreases rapidly as S increases, the interference originating from the subunit orientation becomes more evident than that in real space. In order to make conspicuous the difference between candidate structures and the solution structure in the medium S region, we took first derivative of $I(S)$. The procedure exaggerates peak positions which appear in the medium S range, but their height information is lost.

In Fig. 6.7b one can see that the scattering patterns of the three models are very different from each other. The experimental data has five peaks, which indicates important evidence that the dissociated dimer has a rigid tertiary structure. If not, fewer peaks are expected because of a collapse in the core structure, and, consequently, a loss of spatial interference as observed in the molten globule state of monomeric protein [16]. This is very different from the Arc repressor, where the dissociated monomer takes a molten globule structure [41].

The existence of five peaks is a very important clue to the subunit orientation, because the combination of subunit orientation satisfying this pattern is very limited. Among the three candidates shown in Fig. 6.6 neither model (b) nor (c) has any fine structures. Model (a) has the same number of peaks as the experimental data. The positions of the five peaks are different from the experiment data, which is not important in regard to orientation, because peak positions are dependent on the separation of two subunits. In this case experimental positions are shifted to larger S, which reflects the shortening of the separation in solution. This is consistent with the observation that Rg at 200 MPa is smaller than the Rg of model (a) (Table 6.3). The solution structure therefore keeps an orientation similar to model (a) with a shortened subunit separation. It is very interesting for LDH to choose model (a) subunit configuration rather than the other possibilities. More detailed discussion will be given in the following section.

Structural Pathway of Pressure-Induced Dissociation of LDH. When LDH dissociates into dimer molecules, the initial stage involves the choice of subunit orientation from three possibilities. In order to clarify the origin of this process, physical parameters, Rg, Voronoi volume and surface area of the three models are summarized in Table 6.3. It is obvious that none of these three models from crystallographic coordinates cannot be the real solution structure under pressure. This could

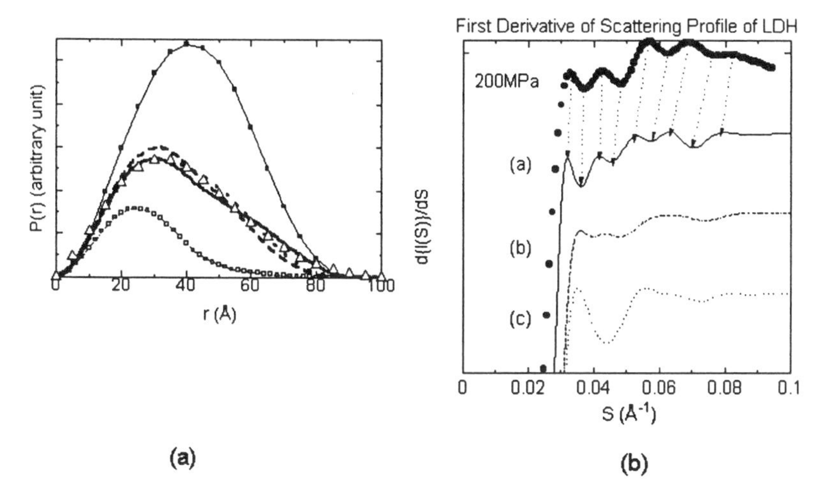

Fig. 6.7. Characterization of model structures by SAXS. (**a**) *P(r)* function of model structures and experimental data at 200 MPa. *P(r)* functions of model structures: LDH whole molecule (*solid squares with line*), model (**a**) (*solid line*), model (**b**) (*broken line*), model (**c**) (*dotted line*), LDH monomer (*open squares with line*), and experimental data at 200 MPa (*open triangles*). *P(r)* of experimental data was scaled so as to fit model *P(r)* values. (**b**). The first derivative of *I(S)*. From the top: experimental data at 200 MPa (*dotted line*), model (**a**), model (**b**), and model (**c**). Curves are shifted for clarity. Dashed array shows the correspondence of peaks and troughs to model data

be justified if *Le Chatlier*'s principle is considered: the final volume of a system should be decreased under pressure. The solution structure under pressure should need further conformational change after dissociating into two subunits, otherwise the total volume of a dissociated subunit with hydration shell would be larger than the initial volume of an oligomer particle. This speculation is also consistent with the fact that the radii of gyration of the models are all larger than that of the solution. In the process of reducing the total volume of a system by pressure it is very interesting that model (a), which is the most plausible orientation shown above, has neither the smallest volume nor the smallest surface area. It is model (b) that has the most compact structure. If the dissociation pathway always follows the path to minimize the system volume, the subunit orientation would be as in model (b). It is very difficult to answer why the dissociation did not take the orientation of model (b). If one looks at Fig. 6.6, model (a) has the most room in between two subunits among the three models, which consequently facilitates further compaction by shortening the distance between two subunits from the model (a) structure. The dissociation pathway picture seems to be as follows: with the application of pressure, LDH dissociates into dimers so as to keep subunit orientation as in model (a), then, holding the orientation constant, the distance between the two subunits shrinks by several angstroms without collapse of the structural core of each subunit.

Table 6.3. Structural parameters of the models

	Rg (Å)	Volume (Å^3)	Surface area (Å^2)
Whole protein	32.43	224,480	42,130
Model (a)	30.17	105,750	26,210
Model (b)	28.28	105,720	24,230
Model (c)	29.48	105,650	28,140
Monomer	22.51	48,050	11,150

6.4 Conclusion and Future Prospects

One of the major questions concerning the protein-folding problem is the relationship between tertiary-structure formation and secondary-structure formation, that is, whether bulk structure forms prior to secondary structure or not. In various denaturant-induced unfolding states at atmospheric pressure, a strong correlation between the compactness and secondary structure such as α-helix content was found [16]. Although, because of a strong tendency to keep the volume compact under high pressure, there could be some cases where the correlation does not hold, as suggested by pressure-denatured myoglobin at acidic pH: the size seems more compact than what it is supposed to be from secondary structure. The study of unfolded structure by high-pressure SAXS would contribute not only to an understanding with respect to the physical chemistry at high pressure but alsoto the general folding problem.

In studying weak interactions such as oligomer formation, high-pressure SAXS becomes very powerful: it can be used to study structures from a small monomeric protein to a large virus or protein complex. In many cases destroying a weak interaction by high pressure does not introduce a large structural change in each component, as found where crystallographic coordinates are available for analyzing the SAXS profile. The merit of this was demonstrated in the case of LDH. The quality of the output of SAXS for oligomeric proteins, however, is very dependent both on the reversibility of dissociation and on the initial shape of the object studied: the higher the symmetry, the more scattering peaks are expected; this becomes a very important clue in solving the structural puzzle of the SAXS profile. Since SAXS uses a higher protein concentration, aggregation occurs easily upon pressurization so that finding a good condition for reversibility is very important issue. Exchanging more information with biochemists is be necessary.

From the viewpoint of technological development, time-resolved experiments such as a high-pressure jump or a high-pressure stopped flow should be done using strong X-ray synchrotron sources, one of which has been demonstrated in the pressure-jump experiment on ribonuclease [15]. The rapid progress of synchrotron facilities will accelerate this field. RIKEN is constructing an SAXS beamline for structural biology at SPring-8, the third-generation synchrotron facility. The beam employs a rather short wavelength (about 1.0 Å) and a smaller size of the beam at the sample position. In combination with an efficient two-dimensional detector and an X-ray image intensifier with cooled CCD, a high-pressure stopped flow with a time slice of tens of milliseconds is planned [42]. It is expected that kinetics will be able to be studied under high pressure.

Acknowledgements. The encouragement and support of Professor T. Ueki (JASRI), Professor Y.Taniguchi (Ritsumeikan University), Professor T.Iizuka (Hosei University), and Professor Y.Inoue (RIKEN) are greatly appreciated. We are grateful to Dr. Y. Inoko (Osaka University) and Dr. K.Kobayashi (PF) for their help with maintaining the beamline. This work was in part supported by a Grant-in-Aid for Scientific Research from the Japanese Ministry of Education, Science and Culture. These SAXS measurements were performed under approvals from the Program Advisory Committee of the Photon Factory (proposal Nos. 93-G034, 93-G038, 95G086, and 95G092).

References

6.1 G. Cevc, D. Marsh: *Phospholipid Bilayers* (John Wiley & Sons Inc., NewYork, 1987)
6.2 R. Winter, M.Boettner: in *High Pressure Chemistry, Biochemistry and Materials Science*, ed. by R. Winter, J. Jonas (Kluwer Academic Publishers, Dordrecht 1993)
6.3 M.Kato, T.Fujisawa, J.Synchr. Rad., **5**, 1282 (1998)
6.4 K.Suzuki, Y.Miyosawa, C. Suzuki, Arch.Biochem.Biophys. **101**, 225 (1963)
6.5 J.F., Brandts, R.J. Oliveira, C. Westort, Biochemistry, **9**, 1038 (1970)
6.6 S.A. Hawley, Biochemistry, **10**, 2436 (1971)
6.7 A.Zipp, W.Kauzmann, Biochemistry, **12** 4217 (1973)
6.8 T.Taniguchi, K.Suzuki, J.Phys.Chem., **87**, 5185 (1983)
6.9 P.T.T.Wong, K.Heremans, Biochim.Biophys.Acta, **956**, 1 (1988)
6.10 S.D.Samarasinghe, D.M. Campbell, A.Jonas, J.Jonas, Biochemistry, **31** 7773 (1992)
6.11 M.Kataoka, Y.Goto, Folding & Design 1, R107 and references cited therin, 1996
6.12 M.Kato, T.Fujisawa, Y.Taniguchi, T.Ueki, Biophysics, **34** (Suppl.), 61 (1994)
6.13 R.Kleppinger, K.Goossens, K.Heremans, M.Lorenzen, in *High Pressure research in Bioscience and Biotechnology*, ed. by K.Heremans (Leuven University Press, Leuven 1997) pp.135
6.14 M. Lorenzen, S.Fiedler, ibid, pp.139 (1997)
6.15 G.Panick, R.Malessa, R.Winter, G.Rapp, K.J.Frye, C.A. Royer, J.Mol.Biol., **275**, 389 (1998)
6.16 M.Kataoka, I.Nishii, T.Fujisawa, T.Ueki, F.Tokunaga, Y.Goto, J.Mol.Biol., **249**, 215 (1995)
6.17 L.King, G. Weber, Biochemistry, **25**, 3632 (1986)
6.18 L.King, G.Weber, Biochemistry, **25**, 3637 (1986)
6.19 T.Fujisawa, M.Kato, Y.Inoko, Biochemistry, **38**, 6411 (1999)
6.20 L.A. Feigin, D.I. Svergun, *Structure Analysis by Small-Angle X-ray and Neutron Scattering* (Plenum Press, NewYork 1987)
6.21 A.Guinier, G.Fournet, *Small-Angle Scattering of X-rays* (John Wiley & Sons Inc., NewYork, 1955)
6.22 K. Gekko, Y. Hasegawa, Biochemistry, **25**, 6563 (1986)
6.23 T. Ueki, Y. Hiragi, M. Kataoka, Y. Inoko, Y. Amemiya, Y. Izumi, H. Tagawa, Y. Muroga, Biophys. Chem., **23**, 115 (1985)
6.24 T.Fujisawa, T.Uruga, Z.Yamaizumi, Y.Inoko, S.Nishimura, T.Ueki, J.Biochem., **115**, 875 (1994)
6.25 P.B.Moore, J.Appl.Cryst. **13** 168 (1980)
6.26 D.I. Svergun, A.V. Semenyuk, L.A.Feigin, Acta Cryst. Sect. A **44**, 244 (1988)

6.27 D.Svergun, C.Baberato, M.H.J.Koch, J.Appl.Cryst., **28**, 768 (1995)
6.28 M.Gerstein, J.Tsai, M.Levitt, J.Mol.Biol., **249**, 955 (1995)
6.29 Y. Harpaz, M.Gerstein, C. Cothia, Structure **2**, 641 (1994)
6.30 C.E.Kundrot, F.M.Richards, J.Mol.Biol., 193, 157 (1987)
6.31 M. Gross, R. Jearnicke, Eur.J.Biochem., **221**, 617 (1994)
6.32 V.V. Mozhaev, K.Heremans, J. Frank, P. Masson, C. Balny, Proteins **24**, 81 (1996)
6.33 D.I. Svergun, S. Richard, M.H. Koch, Z. Sayers, S. Kuprin, G. Zaccai, Proc. Natl. Acad. Sci. USA **95**, 2267 (1998)
6.34 J.F. Flanagan, M.Kataoka, T.Fujisawa, D.Engelman, Biochemistry **32**, 10359 (1993)
6.35 R. Jaenicke, Naturwissenschaften **70**, 332 (1983) and references cited therin
6.36 K.Gekko, S.N.Timasheff, Biochemistry, **20**, 4667 (1981)
6.37 J. Monod, J.Wyman, J.P.Changeux, J.Mol.Biol. **12**, 88 (1965)
6.38 U.M.Grau, M.G.Rossmann, Biochemistry **17**, 4 (1978)
6.39 C.Abad-Zapatero, J.P.Griffith, J.L.Sussman, M.G.Rossmann, J.Mol.Biol., **198**, 445 (1987)
6.40 T.Ueki, Y.Inoko, M.Kataoka, Y.Amemiya, Y.Hiragi, J.Biochem, **99**, 1127 (1986)
6.41 X.Peng, J.Jonas, J.Silvia, Proc. Natl. Acad. Sci. USA **90**, 1776 (1993)
6.42 T.Fujisawa, K.Inoue, T.Oka, H.Iwamoto, T.Uruga, T.Kumasaka, Y.Inoko, N.Yagi, J.Appl.Cryst. **33**, 797 (2000)

7 Accurate Calculations of Relative Melting Temperatures of Mutant Proteins by Molecular Dynamics/Free Energy Perturbation Methods

Minoru Saito

Faculty of Science and Technology, Hirosaki University, 3 Bunkyo-cho, Hirosaki, Aomori 036-8561, Japan
E-mail: msaito@si.hirosaki-u.ac.jp

Abstract. Melting-temperature shifts of mutant proteins were successfully calculated to high accuracy by improving the methods of molecular dynamics (MD) simulation and free energy perturbation calculations. First, MD simulations were performed by explicitly calculating long-range Coulomb interactions by the Particle–Particle and Particle–Cell (PPPC) method without truncating the interactions as in the conventional cutoff method. Second, free energy differences between the wild-type proteins and mutant proteins were estimated by the Acceptance Ratio Method (ARM) instead of the ordinary free energy perturbation method (FEPM). Melting-temperature shifts calculated for 14 mutant proteins of RNaseHI, human lysozyme, and the Myb R2 domain agreed well with their experimental values, although the calculation methodology does not include any adjustable parameters or experimental data for the mutants.

The conventional MD simulation/free energy calculation methodology is based on a cutoff and FEPM cannot successfully calculate the melting-temperature shifts for the following reasons: First, the truncation of the long-range Coulomb interactions by the cutoff method causes the protein structures for sampling conformations to be artificially deformed during the MD simulations. Second, the free energy calculations based on FEPM cannot give reliable values of changes in bond free energy due to the mutations because it utilizes conformations sampled from either state. Therefore, the conventional methodology has a large hysteresis error (mutation-path dependence) and statistical error.

In contrast, MD simulations explicitly including the long-range Coulomb interactions by the PPPC method maintained the protein structures near their X-ray structures and, furthermore, the thermal fluctuations around the equilibrium structures correlated well with both the experimental fluctuations deduced from X-ray temperature factors and with the order parameters of NMR spectroscopy. In addition, the free energy calculations based on ARM successfully gave reliable values for changes in bond free energy because it utilizes conformations sampled from both states. Therefore, the present MD/free energy calculation methodology based on PPPC and ARM suppressed the hysteresis error and statistical error.

7.1 Introduction

Proteins undergo phase transitions between their folded native states and unfolded denatured states at their intrinsic melting temperatures, T_m. At room temperature (below T_m), proteins are in their native states and have stable tertiary structures specific to their amino acid sequences. At high temperatures (above T_m), native proteins unfold into amino acid chains and spontaneously re-fold to their initial tertiary structures when temperatures fall below T_m.

The melting temperature of a protein is specific to its amino acid sequence and is sensitive to alterations in that sequence. Evolutional alterations to a protein's amino acid sequence can sometimes greatly affect its melting temperature, even if its tertiary structure is conserved. For example, *Thermus thermophilus* Ribonuclease H (RNaseH), which has a 52% amino acid sequence identity with *Escherichia coli (E. coli)* RNaseHI, has a melting temperature 34 °C higher than that of *E. coli* RNaseHI, although their tertiary structures are similar [1].

The thermal stability of proteins attracts the special attention of protein scientists for the following reasons: Studies of protein stability are expected to reveal the physicochemical mechanisms that cause the tertiary structures of proteins to be most stable at room temperature. In addition, such studies are expected to provide guiding principles in the creation of thermostable enzymes able to accelerate chemical reactions in bioreactors.

In this decade, many experimental studies have been carried out to clarify the stability mechanisms of various proteins, such as RNaseHI, human lysozyme, T4 lysozyme, barnase, etc. [2–7]. These experimental studies were based on the following strategy: First, an amino acid in a protein is substituted with other amino

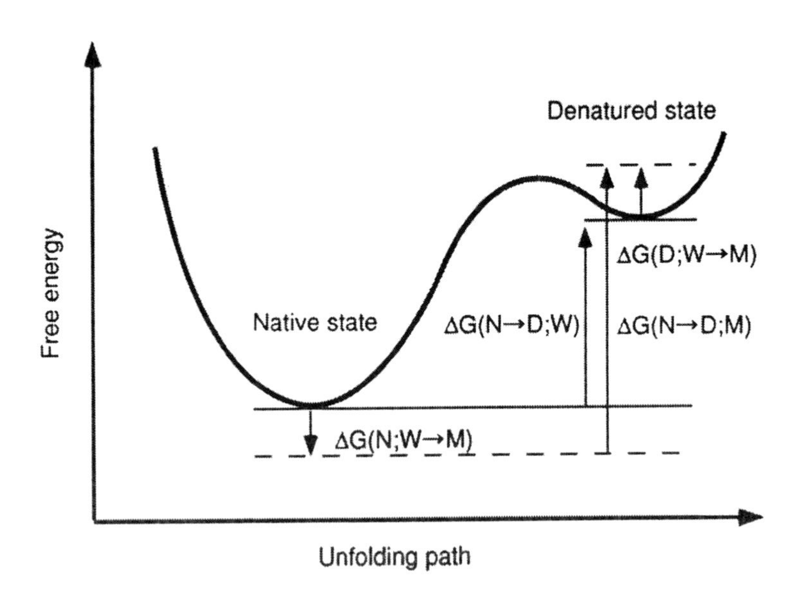

Fig. 7.1. Free energy profile for protein folding/unfolding. ΔG(N;W→M) and ΔG(D;W→M) denote the free energy changes due to mutation from the wild-type (W) to a mutant protein (M) for the native (N) and denatured states (D), respectively. ΔG(N→D;W) and ΔG(N→D;M) denote the unfolding free energies of the wild-type and mutant proteins, respectively. This figure shows an example of a mutation that stabilizes the protein. Source: [8]

acids with different physicochemical properties by site-directed mutagenesis. Second, the melting-temperature shifts, ΔT_m, of the mutant proteins are measured by circular dichroism (CD) and/or differential scanning calorimetry. Third, the kinds of physicochemical properties of the amino acids that stabilize (or destabilize) the proteins are inferred. For example, to clarify whether the negative charge of Glu in a wild-type protein is important or not, the Glu is substituted with the neutral amino acid Gln. If this mutation destabilizes the protein, we infer that the charge of Glu is a major factor stabilizing the wild-type protein. Much knowledge of the stabilization factors has been accumulated through these experimental studies of protein stability [2].

On the other hand, only a few theoretical approaches have been recently carried out to examine the mechanisms of protein stability [8–10]. These theoretical approaches have been based on the fact that protein stability is described by the same free energy profile (Fig. 7.1) as the phase transition phenomenon of many-body systems, such as between solid and liquid. The melting temperature, T_m, is determined by the unfolding free energy, ΔG(N→D). The unfolding free energy is defined as the difference in the free energy between the native state (N) and the denatured state (D) as follows:

$$\Delta G(N \rightarrow D) = G(D) - G(N). \tag{7.1}$$

Mutation from a wild-type protein (W) to a mutant protein (M) changes the free energies of the native and denatured states, G(N) and G(D), and thus the unfolding free energy, ΔG(N→D), as follows:

$$\Delta\Delta G(N \rightarrow D; W \rightarrow M) = \Delta G(D; W \rightarrow M) - \Delta G(N; W \rightarrow M). \tag{7.2}$$

This change in the unfolding free energy due to amino acid substitution, $\Delta\Delta G$(N→D; W→M), is related to the relative melting temperature (ΔT_m) through the simple equation (7.38), given in the Appendix. The theoretical approaches derive ΔT_m from $\Delta\Delta G$(N→D; W→M) by directly calculating the free energy changes due to the amino acid substitution, ΔG(D; W→M) and ΔG(N; W→M).

The free energy changes for the native and denatured states, ΔG(N; W→M) and ΔG(D; W→M), are calculated according to the following strategy : First, molecular dynamics (MD) simulations are carried out for a protein in both the native and denatured states (see Section 7.2). The protein in the native state is prepared by immersing the wild-type protein with known X-ray crystal structure into water. However, there is no experimental data on the precise three-dimensional structure of an unfolded protein in the denatured state. Therefore, as an assumption, the unfolded protein in the denatured state is represented by a short peptide segment centered at the mutation site. This segment, with extended conformation, is immersed into water. The validity of this assumption will be discussed in Section 7.9.

Second, during the MD simulations, an amino acid at the mutation site is substituted with a different amino acid by gradually altering the force field parameters of atoms in the amino acid side chain (Section 7.4). The free energy changes due to the amino acid substitution for the native and denatured states are calculated by the free energy perturbation method, a method based on statistical mechanics, from configurations generated from the MD simulations (Section 7.5).

Third, the major factors stabilizing (or destabilizing) a protein are revealed by finding leading terms in $\Delta\Delta G$. Based on the free energy component analysis (Section 7.6), we can determine: (1) which residue around the mutation site mainly contributes to the stabilization (or destabilization) of the protein; (2) what kinds of interaction energies (for example, van der Waals [vdW] and Coulomb) mainly stabilize (or destabilize) the protein; and (3) what role the solvent water plays in stabilizing (or destabilizing) the protein.

Experimental studies can determine the free energy difference between the native and denatured states of a protein but not the free energy difference between a wild type and its mutant in the same state. The computational approaches are able to calculate the free energy changes due to a mutation in either state. Therefore, the computational approaches can also determine which state (native or denatured) mainly contributes to the stabilization (or destabilization) of the protein, as shown in Fig. 7.1.

Since the theoretical approaches include no adjustable parameters in their strategy as mentioned above, they can quantitatively predict the melting-temperature shift, ΔT_m. The accuracy of this prediction strongly depends on the validity of the MD simulations and the free energy calculations [11]. In 1992, I improved the accuracy of the MD simulations by developing a new method, the PPPC (Particle–Particle and Particle–Cell) method [12, 13], and improved the accuracy of the free energy perturbation calculations by adopting the Acceptance Ratio Method (ARM) [14].

Based on these new methods, relative melting temperatures were calculated for 3 mutants of RNaseHI [8, 9], five mutants of human lysozyme [15], and recently 6 mutants of the Myb R2 domain [16]. The relative melting temperatures calculated for these 14 mutants (from −20 to 20 K) agreed well with experimental values. Furthermore, the stability mechanisms of the mutant proteins were clarified by the free energy component analyses. Details of the present methodology and its application to the two proteins RNaseHI and human lysozyme are described in this chapter.

7.2 Molecular Dynamics Simulation of Proteins

Over several decades, MD simulations have developed into a powerful tool for computer experiments on many-body systems such as liquids and solids. MD simulations for proteins are similar to those for liquids except that proteins contain more complicated interactions (bond, angle, and torsion) as well as vdW and Coulomb interactions, as follows [17]:

$$E_{total} = \sum_{bonds} K_d (d - d_0)^2 \tag{7.3}$$

$$+ \sum_{angles} K_\theta (\theta - \theta_0)^2 \tag{7.4}$$

$$+ \sum_{dihedrals} \frac{V_n}{2} [1 + \cos(n\phi - \gamma)] \tag{7.5}$$

$$+ \sum_{i<j} \left[\frac{A_{ij}}{r_{ij}^{12}} - \frac{B_{ij}}{r_{ij}^{6}} \right] \tag{7.6}$$

$$+ \sum_{i<j} \left[\frac{q_i q_j}{r_{ij}} \right] . \tag{7.7}$$

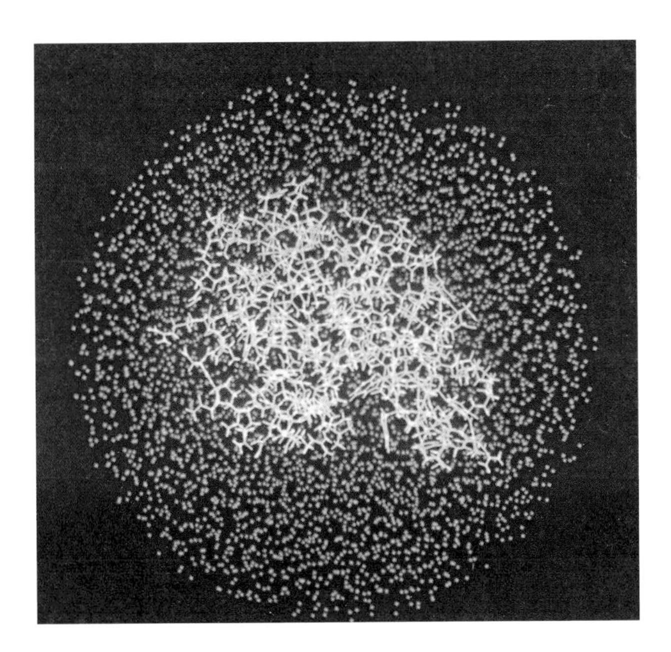

Fig. 7.2. A protein in water. Human lysozyme (130 amino acids) is immersed in a water sphere with a radius of 34 Å. This system consists of 4689 water molecules. Total number of atoms is 16108. Source: [12]

(a)

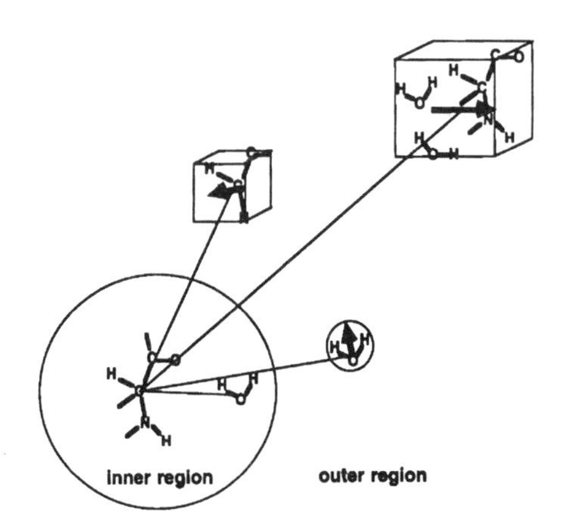

(b)

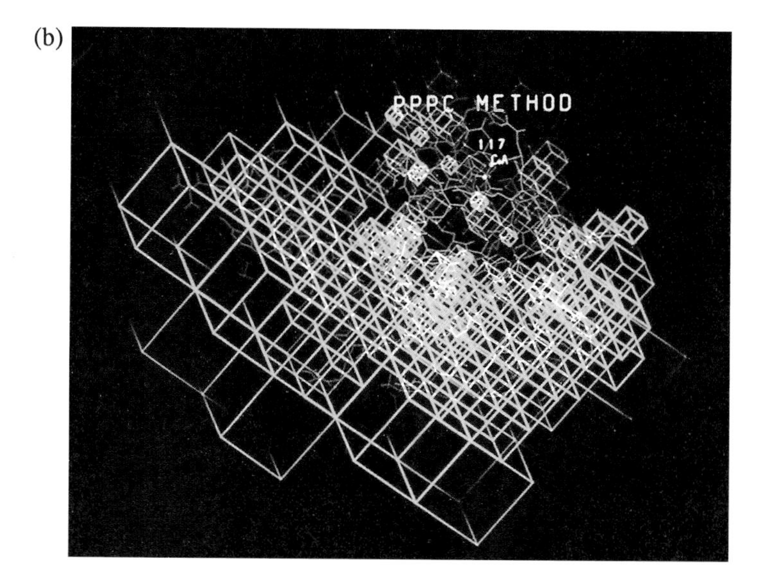

Fig. 7.3. The PPPC method. (**a**) A schematic description of the PPPC (Particle-Particle and Particle-Cell) method. Each atom interacts with nearby atoms through their charges (inner region) and with distant cells through a total charge and dipole moment described for the atoms contained in the cell (outer region). This method is based on the fact that Coulomb forces from distant atoms are not sensitive to either time development or the precise position of those atoms. Source [13]. (**b**) An example of space subdivision in the PPPC method. Coulomb forces acting on the 117th C_α atom are calculated. The sizes of the cells increase as a function of distance from the atom

Because of the limited speed of computers at the time (in 1977), the first MD simulation was carried out for a small protein, bovine pancreatic trypsin inhibitor (BPTI; 58 amino acids, about 900 atoms), in a vacuum [18]. With the appearance of super computers with vector processors (in about 1985), the speed of computers grew rapidly and MD simulations were applied to proteins in the presence of explicit water molecules instead of in a vacuum [11, 19]. Those systems consisted of thousands of water molecules and tens of thousands of atoms (for example, see Fig. 7.2).

Because of long-range force, the number of Coulomb interactions that need to be calculated increases rapidly (the square of the number of atoms) and becomes as high as 10^8 for such large systems as a protein in the presence of explicit water molecules. It is, therefore, practically impossible to evaluate such a large number of Coulomb interactions at every time step in a trajectory long enough to derive accurate free energies. Consequently, in those studies, Coulomb interactions between atoms, an important factor, were applied incorrectly – the interactions were truncated to distances shorter than the protein radii (the 10 Å cutoff method) [11, 19].

In 1992, I developed the PPPC method to efficiently calculate such a large number of long-range Coulomb interactions without sacrificing accuracy [12]. The PPPC method is based on the fact that Coulomb forces for distant atoms are not sensitive to either time development or the precise position of those atoms. In this method, distant atoms are grouped into cubic cells (Fig. 7.3a and b). Each atom interacts with near atoms as in the direct summation manner but with each distant cell through their total charge and dipole moment. The error in the Coulomb energy per atom of the PPPC method was significantly smaller than that of the 10 Å cutoff method (about 0.9 kcal/mol versus 18 kcal/mol, respectively; see Fig. 7.4). This method made it possible to simulate a protein in the presence of explicit water molecules with the long-range Coulomb interactions included and without requiring more computation time than the 10 Å cutoff method [12].

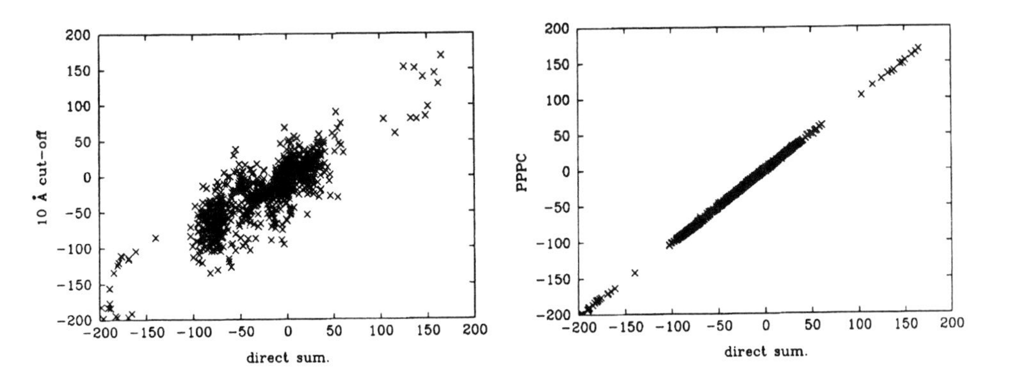

Fig. 7.4. Coulomb energy (kcal/mol) of each atom of human lysozyme and each water molecule. In both panels, the horizontal axes show the energy value calculated by the accurate direct summation method. The vertical axes show the energy values calculated by **(a)** the 10 Å cutoff method approximation and **(b)** the PPPC method approximation, respectively. Source: [12]

7.3 Equilibrium Structure and Thermal Fluctuation

MD simulations including the long-range Coulomb interactions determined by the PPPC method were performed for RNaseHI and human lysozyme in the presence of explicit water molecules with all degrees of freedom [8, 9, 12, 13, 20, 21]. These MD simulations were also performed by truncating the Coulomb interactions to 10 Å (the 10 Å cutoff method) to investigate artificial errors caused by the truncation. The reliability of MD simulations for proteins is usually checked by monitoring any deviations in structure from their initial X-ray structures and any fluctuations around the equilibrium structures. The deviations from their X-ray structures are measured by RMSD (root mean square deviation) of main-chain atoms:

$$RMSD = \sqrt{\frac{1}{n}\sum_{i=1}^{n}\left(r_i - r_i^0\right)^2}\,, \tag{7.8}$$

where n is the number of main-chain atoms and r_i^0 is the initial X-ray position of atom i.

RMSDs are plotted for lysozyme and RNaseHI in Fig. 7.5a and b, respectively. The simulations based on the PPPC method (no cutoff) maintained the protein structures within a RMSD of about 1.5 Å of the initial X-ray structures for each protein [12, 13, 20, 21]. These RMSDs were mainly caused by the relatively large deviations of a few tens of packing residues that interact with the neighboring proteins in the crystal structure through hydrogen bonds and salt bridges. Since there were no adjacent proteins in the MD simulations, the packing residues had no adjacent proteins and deviated to new equilibrium positions in water. The RMSDs of the PPPC simulations were thus caused by the differences between the environmental conditions of proteins in the X-ray crystallography and MD simulation studies.

On the other hand, the cutoff simulations gave significantly larger RMSDs than the PPPC simulations for each protein. As shown in Fig. 7.5b, the truncation of Coulomb interactions from 250 ps caused the lysozyme structure to deviate from the stable PPPC structure (about 1.5 Å RMSD) and shift to a new stable structure with a larger RMSD (about 2.2 Å). This large RMSD of the cutoff simulation was reproduced by another 10 Å cutoff simulation in which the Coulomb interactions were truncated from the beginning [13]. The structural deviations caused by the cutoff simulation were not restricted to the packing residues and were significantly larger for all charged residues than for neutral residues. I clarified the reason for the large deviations by analyzing the trajectories of the cutoff and PPPC simulations [13]. The truncation of the Coulomb interactions causes imbalances of the resultant forces acting on charged residues and thereby causes the charged residues to greatly deviate from their equilibrium positions. Therefore, the large RMSDs of the cutoff trajectories were artificial errors caused by the truncations of long-range Coulomb interactions.

(**a**)

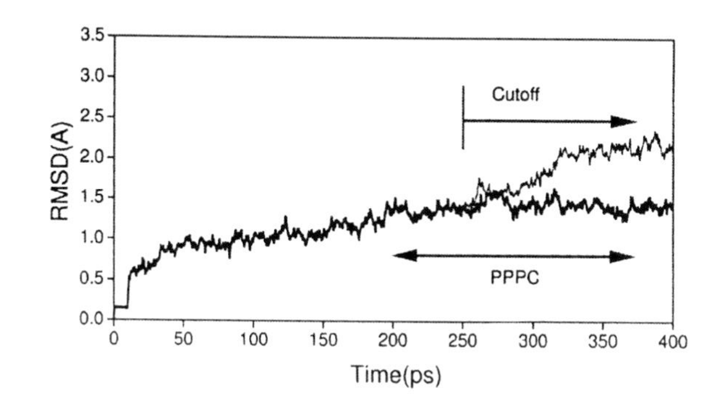

(**b**)

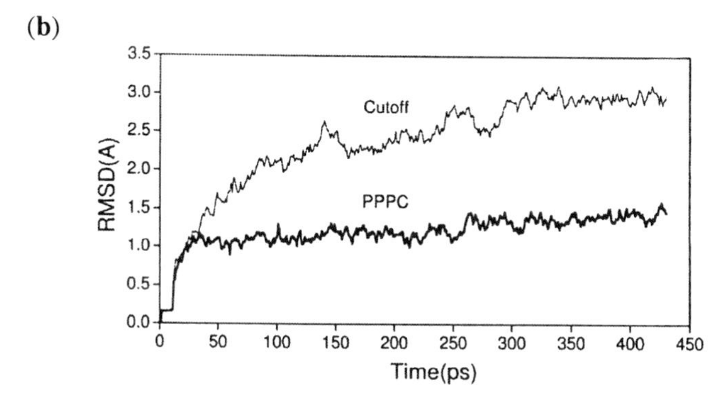

Fig. 7.5. RMS deviations of main-chain atoms of (**a**) human lysozyme and (**b**) RNaseHI from their initial X-ray structures. In each panel, the 2 lines show the deviations as a function of time for the simulations with the 10 Å cutoff method (*thin line*) and with the PPPC method (*thick line*). In (**a**), the calculation method for the Coulomb interactions was changed from the PPPC method to the 10 Å cut-off method at 250 ps. Source: [13].

Are MD simulations based on the PPPC method able to correctly describe the dynamics of proteins in water? To answer this question, RMSD fluctuations around the equilibrium structures were calculated and compared with the fluctuations deduced from X-ray *B*-factors [13, 20, 21]. The root mean square fluctuation (RMSF) values were calculated from the MD trajectories as follows:

$$RMSF = \sqrt{\left\langle \left(r_i(t) - \langle r_i \rangle \right)^2 \right\rangle}\,; \tag{7.9}$$

the experimental fluctuations were deduced from the X-ray *B*-factors (*B*) using the following equation:

$$\Delta r_{X-ray} = \sqrt{\frac{3B}{8\pi^2}} \,. \tag{7.10}$$

The RMSF and the X-ray fluctuation are plotted as a function of residue number for RNaseHI and lysozyme in Fig. 7.6a and b, respectively. The RMSF curves show the same residue dependence as the experimental fluctuation curves for both proteins. The RMSF values agree well with the experimental fluctuations for

(a)

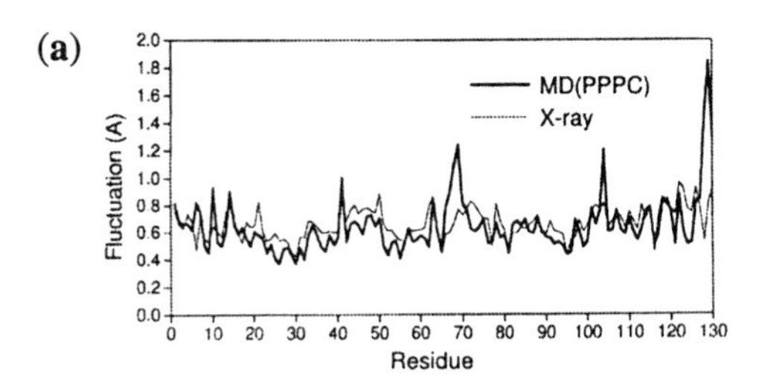

(b)

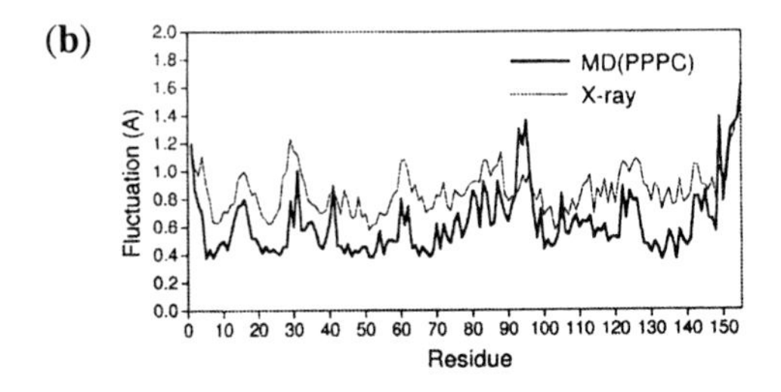

(c)

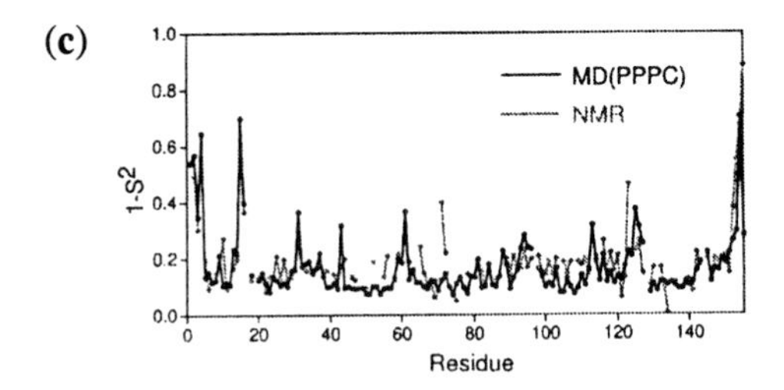

Fig. 7.6. Fluctuations of heavy atoms of (**a**) human lysozyme and (**b**) RNaseHI as a function of residue number. The thick lines denote root mean square fluctuation (RMSF) around their equilibrium positions obtained from the 200 to 250 ps MD trajectory for human lysozyme and from the last 60 ps (325 to 385 ps) of the MD trajectory for RNaseHI. The thin lines denote the fluctuation deduced from the temperature factor of the X-ray crystal. (**c**) Rotational fluctuations of backbone N–H bond vectors in RNaseHI. S^2 is the generalized order parameter. Experimental values were obtained by ^{15}N-NMR relaxation experiments. Source: [22]. Theoretical values were obtained by calculating the time correlation function of the N-H vectors from the MD trajectories. Source: [21].

lysozyme but are systematically smaller than the experimental fluctuations for RNaseHI, except for several packing residues. The systematically large experimental fluctuation is attributed to crystal disorder and to various small residual factors that preclude ultimate structure refinement. Several packing residues located in flexible loops (for example, the 94th to 97th residues in RNaseHI) have small fluctuations due to the interactions with the neighboring proteins in the crystal but large fluctuations in the MD simulations, where there is no adjacent protein. Rotational fluctuations of backbone N–H bond vectors in RNaseHI were observed by ^{15}N-NMR relaxation experiments through the generalized order parameter (S^2) [21, 22]. In the MD simulations, the order parameters were obtained from the MD trajectories for RNaseHI by evaluating the time correlation functions of the N–H vectors [21]. The ($1-S^2$) values calculated from MD and those observed from NMR were plotted for all residues (Fig. 7.6c). The orientation fluctuations calculated for the backbone N–H vectors agreed well with the experimental fluctuations. Therefore, the MD simulations with the long-range Coulomb interactions determined by the PPPC method correctly describe the thermal fluctuations as well as the equilibrium structures of the proteins.

7.4 Computational Mutagenesis

In experiments, an amino acid in the wild-type protein is substituted with other amino acids by site-directed mutagenesis. In MD simulations, the same amino acid substitutions are achieved by the following strategy [14, 23]: Since atoms in MD simulations are identified by their force fields of from (7.3) to (7.7), the atoms can be changed to different atoms by altering the force field parameters. Therefore, amino acids can be substituted with other amino acids by changing the force field parameters of atoms in the amino acid side chains (computational mutagenesis). The force field parameter change must be performed gradually through many intermediate stages because the free energy differences between intermediate stages must be small enough for the free energy perturbation approximations to be reliable. In practice, the wild-type proteins are gradually mutated to mutant proteins through 5 to 22 intermediate stages (Table 7.1).

In computational mutagenesis, the mutation path from the wild-type protein to a mutant protein can be controlled by changing the protocols for the force field parameter changes [8]. For example, protocols A and C in Table 7.1 are completely different from each other because the force field parameters are changed simultaneously in A but sequentially in C. The free energy difference, ΔG, between the wild-type protein and a mutant protein must be independent of the mutation path in principle because ΔG is a thermodynamic quantity. This feature is utilized to evaluate the reliability of the free energy calculations based on MD simulations [8].

The free energy differences between the wild-type protein and a mutant protein, $\Delta G(\mathrm{N\ or\ D};\ \mathrm{W}\rightarrow\mathrm{M})$, are obtained by summing the free energy differences between intermediate stages, $\Delta G_i(\mathrm{N\ or\ D};\ i\rightarrow i+1)$ ($i = 1$ to n):

$$\Delta G(N;W \to M) = \sum_i \Delta G(N;i \to i+1) \tag{7.11}$$

and

$$\Delta G(D;W \to M) = \sum_i \Delta G(D;i \to i+1). \tag{7.12}$$

From these free energy changes, the unfolding free energy change due to the mutation, $\Delta\Delta G$(N→D; W→M), is obtained from (7.2).

Table 7.1. Protocols used for MD simulations and free energy calculations

Protocol	Equilibration time(ps)	Stages (force field)[a]	Equilibration/ sampling times per stage (ps)	Total mutation time (ps)
RNaseHI; Ala95→Gly				
A	130	1~22 (charge, vdW, bond, angle)	5/10	330
B (reverse of A)		1~22 (charge, vdW, bond, angle)	5/10	330
C	130	1 (charge), 2~6 (vdW),	10/20	270
		7~8 (angle), 9~12 (bond)	5/10	
D	130	1~5 (charge, vdW, bond, angle)	10/20	150
RNaseHI; Val74→Ala				
E	100	1~20 (charge, vdW, bond, angle)	5/5	200
F (reverse of E)		1~20 (charge, vdW, bond, angle)	5/5	200
G	130	1~5 (charge, vdW, bond, angle)	10/20	150
H (reverse of G)		1~5 (charge, vdW, bond, angle)	10/20	150
RNaseHI; Val74→Ile				
I	100	1 (charge), 2~11 (vdW),	5/5	
		12 (angle), 13~20 (bond)	5/5	200
J (reverse of I)		1~8 (bond), 9 (angle),	5/5	
		10~19 (vdW), 20 (charge)	5/5	200
HLZM; Ile23→Val, Ile56→Val, Ile59→Val, Ile89→Val, Ile106→Val				
K	250	1~5 (charge, vdW, bond, angle)	10/20	150

[a] Force field parameters to be changed in the stages. In the protocols C, I, and J, only one force field parameter was changed linearly in each stage. In other protocols, all force field parameters were changed simultaneously.

7.5 Free Energy Perturbation Method

In general, it is difficult to calculate the absolute free energy for many-body systems except for simple model systems such as Einstein crystals and ideal gases. However, the relative free energy can be calculated for realistic many-body systems such as liquids, solids, or proteins by using the free energy perturbation method based on MD simulations.

Various perturbation schemes based on statistical mechanics have been proposed to calculate the free energy differences from configurations generated by MD simulations [24]. Until recently, the free energy calculations for proteins have been carried out by using the simplest perturbation scheme (the so-called FEPM). This method evaluates the free energy difference between 2 states, i and i+1, by the following equations:

$$\Delta G_{Forw}(i;i+1) = -\frac{1}{\beta}\log\left\langle e^{-\beta(U_{i+1}-U_i)}\right\rangle_i \tag{7.13}$$

and

$$\Delta G_{Back}(i;i+1) = \frac{1}{\beta}\log\left\langle e^{-\beta(U_i-U_{i+1})}\right\rangle_{i+1}, \tag{7.14}$$

where $< >_i$ and $< >_{i+1}$ denote sample averages over configurations of the stages i and i+1, respectively. U_i and U_{i+1} are potential energies of i and i+1, respectively. ΔG_{Forw} and ΔG_{Back} are usually different from each other because the sample set is not identical to the canonical ensemble. This scheme utilizes only configurations from either i or i+1 in spite of the free energy difference between the 2 states. Therefore, this scheme is not as efficient as other schemes.

In 1976, Bennett proposed a more efficient method that utilizes configurations from both states [25]. His method (ARM) evaluates ΔG according to the following equations:

$$\Delta G(i;i+1) = \frac{C}{\beta} - \frac{1}{\beta}\log\left(\frac{n_{i+1}}{n_i}\right) \tag{7.15}$$

and

$$n_i\left\langle\frac{1}{e^{\beta(U_{i+1}-U_i)-C}+1}\right\rangle_i = n_{i+1}\left\langle\frac{1}{e^{\beta(U_i-U_{i+1})+C}+1}\right\rangle_{i+1}, \tag{7.16}$$

where the value of C that satisfies (7.16) is determined numerically. n_i and n_{i+1} are the number of samples for i and i+1, respectively. I and a coworker have demonstrated that this method is more efficient than FEPM for calculating the hydration free energy difference between the alanine and glycine zwitterions [14]. Furthermore, I have confirmed that this scheme precisely determines the bond free energy difference between the wild-type protein and a mutant protein, something at which FEPM apparently fails (Section 7.10).

7.6 Free Energy Component Analysis

One of the most important purposes of theoretical studies on protein stability is to determine: (a) what kinds of physical interactions between the amino acid at the mutation site and the surrounding amino acids act to stabilize (or destabilize) a protein, and (b) which of the amino acids surrounding the mutation site contribute to the stabilization (or destabilization) of the protein through these physical interactions. These questions are answered by investigating the contributions of the physical interactions depicted with (7.3) to (7.7) and the contributions of the amino acids around the mutation site to $\Delta G(\mathrm{N; W}\rightarrow\mathrm{M})$, $\Delta G(\mathrm{D; W}\rightarrow\mathrm{M})$, and $\Delta\Delta G(\mathrm{N}\rightarrow\mathrm{D; W}\rightarrow\mathrm{M})$.

The contribution of a particular interaction energy to the free energy differences is obtained by evaluating U_{i+1} and U_i in (7.16) for only that interaction energy. Similarly, the contribution of a surrounding amino acid is obtained by evaluating U_{i+1} and U_i in (7.16) only between the amino acid at the mutation site and the surrounding amino acid. These contributions are simply called the free energy components of the interaction energy and the amino acid [9].

It should be noted that $\Delta\Delta G$ cannot be clearly divided into 2 different components, ΔG_α and ΔG_β, if the energy samples $\{U_{i+1}-U_i\}_\alpha$ and $\{U_{i+1}-U_i\}_\beta$ are strongly correlated [26]. For example, it is meaningless to divide a free energy change due to hydrogen bonding into the free energy components of vdW and Coulomb interactions, because the hydrogen bond is formed by both these interactions and $\{U_{i+1}-U_i\}_{\mathrm{vdW}}$ and $\{U_{i+1}-U_i\}_{\mathrm{Coulomb}}$ are strongly correlated. The free energy component analysis is meaningful only in cases where the correlation coefficient between $\{U_{i+1}-U_i\}_\alpha$ and $\{U_{i+1}-U_i\}_\beta$ is quite small [9]. The details of the stabilization mechanism for the Val74Ile mutant were revealed by free energy component analysis, as described in Section 7.8.

7.7 Calculation Results of ΔT_{m} and $\Delta\Delta G$

Recently, the relative melting temperatures of mutant proteins were observed for RNaseHI by Kimura et al. [3] and for human lysozyme (HLZM) by Takano et al. [5]. Kimura et al. substituted the 95th amino acid (Lys) with Ala and Gly and found that the mutant Lys95Gly had a larger melting-temperature shift ($\Delta T_{\mathrm{m}} = 6.8$ K) than Lys95Ala ($\Delta T_{\mathrm{m}} = 0.4$ K). In other words, the mutation of Ala95 to Gly caused a large melting temperature shift ($\Delta T_{\mathrm{m}} = 6.4$ K) in spite of the small difference in the side chain; CH_3 in Ala is changed to H in Gly. They also substituted the 74th Val with Ile and Ala and found that the mutation Val74 to Ile stabilized the protein ($\Delta T_{\mathrm{m}} = 2.1$ K) while the mutation Val74 to Ala largely destabilized the protein ($\Delta T_{\mathrm{m}} = -12.7$ K). I and a coworker decided to calculate the ΔT_{m} values of these mutants because the small changes in the side chain caused large shifts in melting temperatures [8, 9]. Takano et al. substituted all isoleucines of HLZM with valines and found that, despite the mutations all being of the same type (Ile→Val), ΔT_m depended on the mutation site. I calculated ΔT_m for these mutants also [15].

In order to calculate ΔT_m for the mutants, MD simulations and computational mutagenesis were performed according to the protocols in Table 7.1. From the configurations in the trajectories, the free energy changes of the native and denatured states, ΔG(N; W→M) and ΔG(D; W→M), were evaluated by ARM. The unfolding free energy changes $\Delta\Delta G$(N→D; W→M) obtained from ΔG(N; W→M) and ΔG(D; W→M) were translated to ΔT_m through (7.38) in the Appendix whose right-hand side includes experimental ΔH and ΔC_p values observed for the wild-type protein but does not include any experimental data for the mutant proteins. The ΔT_m values calculated and observed for 8 mutants of RNaseHI and HLZM are plotted in Fig. 7.7. The calculated ΔT_m values agreed well with the respective experimental values [8]. Since the calculations do not contain any adjustable parameters or any experimental data of the mutant proteins, it is possible to quantitatively predict the relative melting temperatures of mutant proteins.

The free energy changes of the native and denatured states due to the mutations are listed for the 8 mutants in Table 7.2. Since the mutation Ala95→Gly greatly decreased the free energy of the native state (–4.5 ± 0.7 kcal/mol) and slightly decreased the free energy of the denatured state (–2.1 ± 0.2 kcal/mol), the mutation Ala95→Gly stabilizes the protein. On the other hand, the mutation Val74→Ala caused a slight decrease in the native-state free energy

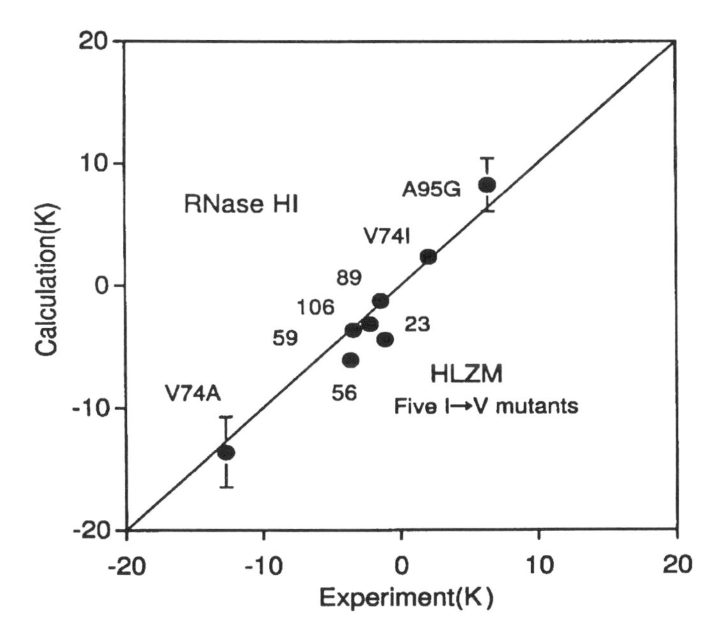

Fig. 7.7. Correlation between relative melting temperatures observed and calculated for 3 RNaseHI mutants and 5 Ile→Val lysozyme mutants. The error bars denote the hysteresis errors listed in Table 7.2. The experimental values were observed by Kanaya et al. for RNaseHI and Takano et al. for HLZM. Sources: 3, 5, 8.

Table 7.2. Changes in the free energies of native and denatured states, $\Delta G(\mathrm{N;W\rightarrow M})$ and $\Delta G(\mathrm{D;W\rightarrow M})$, unfolding free energy $\Delta\Delta G$, and melting temperature ΔT_m caused by mutations

Mutation	$\Delta G(\mathrm{N;W\rightarrow M})$ (kcal/mol)	$\Delta G(\mathrm{D;W\rightarrow M})$ (kcal/mol)		$\Delta\Delta G$ (kcal/mol)	ΔT_m (K)
RNaseHI					
Ala95→Gly	−4.5 ± 0.7	−2.1 ± 0.2		2.4 ± 0.6	8.2 ± 1.9
			Exp.	1.8[a]	6.4[a]
Val74→Ala	−1.7 ± 0.7	−4.9 ± 0.5		−3.2 ± 0.8	−13.8 ± 3.9
			Exp.	−3.4	−12.7
Val74→Ile	1.6 ± 0.2	2.3 ± 0.1		0.7 ± 0.2	2.5 ± 0.7
			Exp.	0.6	2.1
HLZM					
Ile23→Val	−2.2	−0.7		−1.4	−4.4
			Exp.	−0.4	−1.1
Ile56→Val	−1.7	0.3		−2.0	−6.1
			Exp.	−1.2	−3.6
Ile59→Val	−1.0	0.2		−1.2	−3.6
			Exp.	−1.1	−3.4
Ile89→Val	−1.2	−0.8		−0.4	−1.2
			Exp.	−0.5	−1.4
Ile106→Val	−1.4	−0.3		−1.0	−3.1
			Exp.	−0.7	−2.2

[a] The experimental values of $\Delta\Delta G$ and ΔT_m for Ala95→Gly were determined from the values observed for Lys95→Ala and Lys95→Gly, i.e., $\Delta\Delta G$ = 0.1 and 1.9 kcal/mol and ΔT_m = 0.4 and 6.8 K, respectively. Source: [3]

(−1.7 ± 0.7 kcal/mol) but a large decrease in the denatured-state free energy (−4.9 ± 0.5 kcal/mol). The relatively large free energy decrease of the denatured state makes the Val74Ala mutant less stable than the wild type.

In contrast to the mutations Ala95→Gly and Val74→Ala, the Val74→Ile mutation increased the free energies of both the native and denatured states by 1.6 ± 0.2 kcal/mol and 2.3 ± 0.1 kcal/mol, respectively. Since the increase is slightly larger (0.7 kcal/mol) for the denatured state than for the native state, the native state of the Val74Ile mutant is more stable than the wild type. Therefore, the 2 thermostable mutants, Ala95Gly and Val74Ile, were found to be stabilized by different mechanisms. Whether a mutation stabilizes the protein or not is determined by the delicate balance between the free energy changes in the native and denatured states, $\Delta G(\mathrm{N;\ W\rightarrow M})$ and $\Delta G(\mathrm{D;\ W\rightarrow M})$.

7.8 Stability Mechanism of Val74Ile RNaseHI Mutant

In an early stage of the study of Val74Ile mutant, its stability mechanism was intuitively explained on the basis of the X-ray structure of the mutant protein [4].

The extra methyl group of the 74Ile fills the cavity around the 74th site and was thought to improve packing interactions inside the native protein. The improved packing interactions were expected to strengthen the native structure and increase the melting temperature. In order to check whether this intuitive explanation was valid or not, a free energy component analysis was performed for this mutant [9].

The free energy changes ΔG(N; W→M), ΔG(D; W→M), and $\Delta\Delta G$(N→D; W→M) were broken down into vdW, electrostatic, and internal energy contributions, as shown in Table 7.3. The component of vdW interaction was small (0.25 kcal/mol) for ΔG(N; W→M) and relatively large (1.10 kcal/mol) for ΔG(D; W→M). Thus, the vdW component in $\Delta\Delta G$ was relatively large (0.85 kcal/mol). In contrast, the components of electrostatic interaction in ΔG(N; W→M) and ΔG(D; W→M) were nearly zero because the changes in the atomic charges due to the mutation from Val to Ile were quite small. Thus, the component of electrostatic interaction in $\Delta\Delta G$ was negligible (–0.02 kcal/mol). The internal energy components were large with similar values 1.32 and 1.13 kcal/mol for ΔG(N; W→M) and ΔG(D; W→M), respectively. Since they almost canceled each other out, the internal energy component in $\Delta\Delta G$ was small (–0.18 kcal/mol). Therefore, the stabilization of the V74I mutant was found to be caused mainly by the unfavorable vdW interactions in the denatured state.

Table 7.3. Free energy changes of the native (N) and denatured (D) states due to the Val→Ile mutation, ΔG(N;W→M) and ΔG(D;W→M), and the unfolding free energy difference between the wild type (W) and mutant (M), $\Delta\Delta G$(N→D;W→M), in kcal/mol

	ΔG(N;W→M)	ΔG(D;W→M)	$\Delta\Delta G$(N→D;W→M)
Total	1.60 ± 0.19	2.25 ± 0.00	0.66 ± 0.19
		Exp.	0.6[a]
Contribution			
vdW[b]	0.25 ± 0.20	1.10 ± 0.03	0.85 ± 0.21
Electrostatic	0.03 ± 0.02	0.01 ± 0.06	–0.02 ± 0.06
Internal[c]	1.32 ± 0.00	1.13 ± 0.10	–0.18 ± 0.10

[a] Source: [3] [b] The correlation coefficient between the vdW and internal components is smaller than 0.1. [c] Contributions from the bond, angle, 1–4 nonbond, and 1–5 nonbond in the side chain atoms of the residue 74 except for the C_β methine group

Furthermore, contributions from the solute and solvent to the vdW components in ΔG(N; W→M), ΔG(D; W→M), and $\Delta\Delta G$ (abbreviated as ΔG_{vdW} (N; W→M), ΔG_{vdW}(D; W→M), and $\Delta\Delta G_{vdW}$, respectively) were estimated. The results are shown in Table 7.4. The solvent contribution was –0.07 for ΔG_{vdW}(N; W→M), 0.73 for ΔG_{vdW}(D; W→M), and thus 0.80 for $\Delta\Delta G_{vdW}$. This small contribution of the solvent to ΔG_{vdW}(N; W→M) is consistent with the fact that the side chain atoms of Val (and Ile) 74 are completely buried inside the native protein. The solvent contribution to ΔG_{vdW}(D; W→M) was relatively large (0.73), because the side chain atoms are exposed to the solvent in the denatured state. This contribution of the solvent in increasing the free energy in the denatured state can be explained as a loss of solvation energy caused by the Cδ methyl group being created in water. On the other hand, the solute contributions to ΔG_{vdW}(N; W→M) and ΔG_{vdW}(D; W→M) were almost the same (0.32 and 0.37) and cancelled each other out in $\Delta\Delta G_{vdW}$. The positive contribution to ΔG_{vdW}(N; W→M) (0.32) means that the vdW interactions of the Cδ methyl group inside the protein do not stabilize the cavity-filling mutant Val74Ile. Therefore, the stabilization of the Val74Ile mutant was found to be caused by the loss of hydration energy in the denatured state and not by the favorable packing interactions in the native protein.

In order to clarify why the Cδ methyl group of Ile74 does not interact favorably inside the protein, the contributions from the surrounding 6 residues in Fig. 7.8 are also listed in Table 7.4. All the residues, except Leu49, contribute negative values to ΔG_{vdW}(N; W→M) and contribute to the stabilization of the Val74Ile mutant. In contrast, Leu49 has a positive value (0.89) and contributes to the destabilization of the native state. These results suggest that the cavity size between the Val74 and Leu49 side chains is too small to insert a methyl group without unfavorable vdW repulsion. Since the unfavorable vdW interactions with the Leu49 offset the favorable vdW interactions with the other surrounding residues, insertion of the Cδ methyl group into the cavity cannot stabilize the mutant native structure.

The RMSD from the initial X-ray structure was plotted (Fig. 7.9) for residue 74 and the 6 surrounding residues (see Fig. 7.8) during the equilibration and the round-trip mutation from Ile to Val and back to Ile. The plot shows that, during the forward and reverse mutation paths, the structure surrounding residue 74 fluctuates around a stable structure close to the X-ray structure. The 3-dimensional structure around residue 74, averaged over 5 ps, was drawn for the time after the equilibration, forward mutation, and reverse mutation (Fig. 7.8). This figure clearly shows that the side chains have almost the same structure within the thermal fluctuations (about 0.5 Å, see Fig. 7.6b) during the simulations. Therefore, the present MD simulations, without the truncation of Coulomb interactions, suppress the hysteresis error and prevent artificial structural deformations of the protein, as mentioned in Section 7.3.

Table 7.4. Contributions of the solvent and solute to the vdW free energy components, ΔG_{vdW}(N;W→M), ΔG_{vdW}(D;W→M), and $\Delta\Delta G_{vdW}$(N→D;W→M) in Table 7.3

	ΔG_{vdW}(N;W→M)	ΔG_{vdW}(D;W→M)	$\Delta\Delta G_{vdW}$(N→D;W→M)
Solvent[a]	−0.07 ± 0.01	0.73 ± 0.06	0.80 ± 0.06
Solute	0.32 ± 0.20	0.37 ± 0.03	0.05 ± 0.20
Residue[b]			
Val(Ile)74	0.47 ± 0.14	0.54 ± 0.12	0.07 ± 0.17
Leu49[c]	0.89 ± 0.29	-	−0.89 ± 0.29
Ala52	−0.06 ± 0.01	-	0.06 ± 0.01
Tyr73	−0.19 ± 0.02	−0.09 ± 0.12	0.19 ± 0.12
Trp104	−0.20 ± 0.04	-	0.20 ± 0.04
Leu107	−0.12 ± 0.03	-	0.12 ± 0.03
Trp118	−0.06 ± 0.00	-	0.06 ± 0.00
Others	−0.41 ± 0.23	−0.08 ± 0.06	0.33 ± 0.24

[a] The correlation coefficient between the solvent and solute contributions is smaler than 0.1.
[b] Those whose contribution is larger than 0.05.
[c] The correlation coefficients between the contributions from Leu49 and other residues are smaller than 0.3.

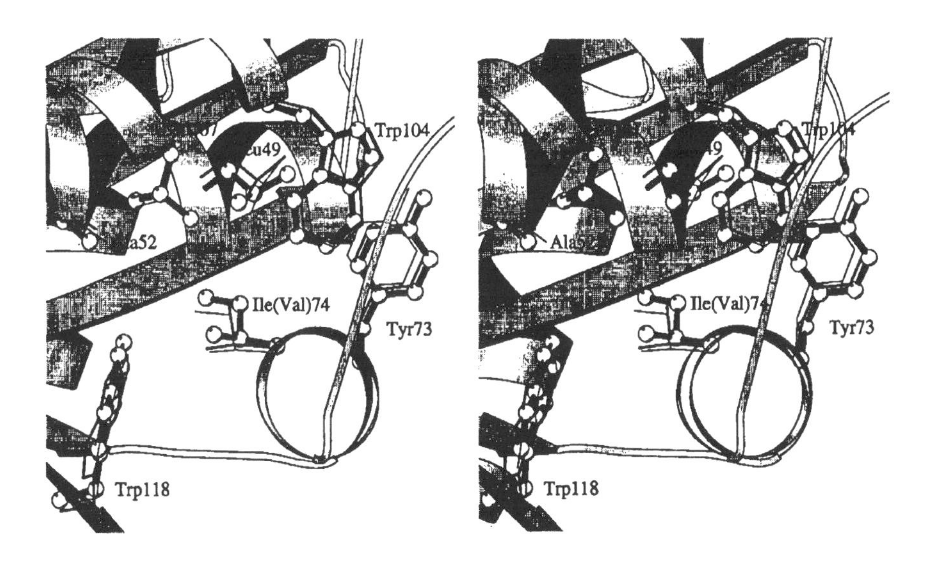

Fig. 7.8. Three-dimensional structure of residue 74 and the 6 residues surrounding it in RNaseHI. The ball-and-stick, thick line, and thin line denote the structures after the equilibration (A), forward mutation (B), and reverse mutation (C) in Fig. 7.9. These structures, A, B and C, were fitted by the least-square method for the heavy atoms of the 7 residues. RMS deviations between A and B, B and C, and C and A are 0.3, 0.4, and 0.5 Å, respectively. Source: [9]

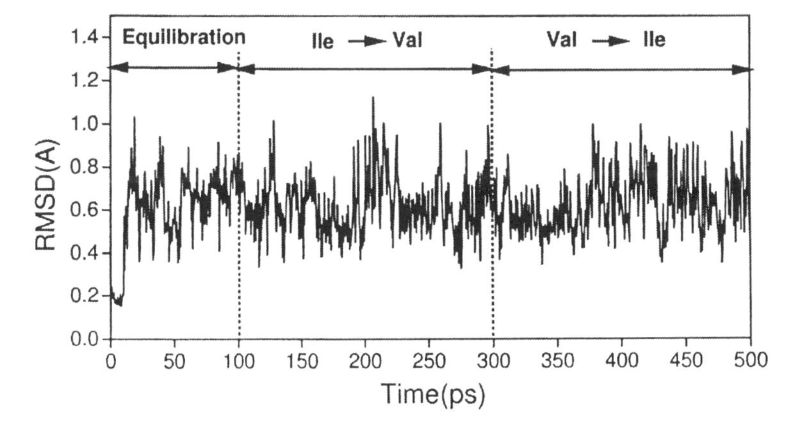

Fig. 7.9. Root mean square deviation (RMSD) from the X-ray structure for heavy atoms of the 7 residues in RNaseHI (see Fig. 7.8) during the equilibration (0 to 100 ps), forward mutation (100 to 300 ps), and reverse mutation (300 to 500 ps). Source: [9]

7.9 Stability Mechanism of Ile→Val Lysozyme Mutants

The 5 Ile→Val mutations of HLZM caused different melting temperature shifts (ΔT_m = −1 to −4 K), depending on the location of Ile in HLZM. In order to determine whether these site-dependent melting-temperature shifts were caused by the free energy change in the native or denatured state, The free energy changes of both the native and denatured states, ΔG(N; W→M) and ΔG(D; W→M), were plotted against the experimentally observed unfolding free energy change, $\Delta\Delta G$(N→D; W→M) in Fig. 7.10. The following features were found: The native-state free energy change ΔG(N; W→M) correlated well with the experimental unfolding free energy change $\Delta\Delta G$(N→D; W→M), as shown by squares in Fig. 7.10. However, the denatured-state free energy change ΔG(D; W→M) did not correlate with the experimental unfolding free energy changes, as shown by crosses in Fig. 7.10. These results are explained, as follows:

The native-state structure makes a more complicated environment for Ile than the denatured-state structure. That is, in the native state, each Ile is in contact with several amino acid residues but with a few solvent water molecules. However, in the denatured state, each Ile is completely exposed to the solvent water and is in contact with only the 2 neighboring amino acid residues. Therefore, the interactions of the Ile with the different surrounding residues in the native-state tertiary structure were expected to cause the different melting-temperature shifts ΔT_m (−1 to −4 K) of the Ile→Val mutants.

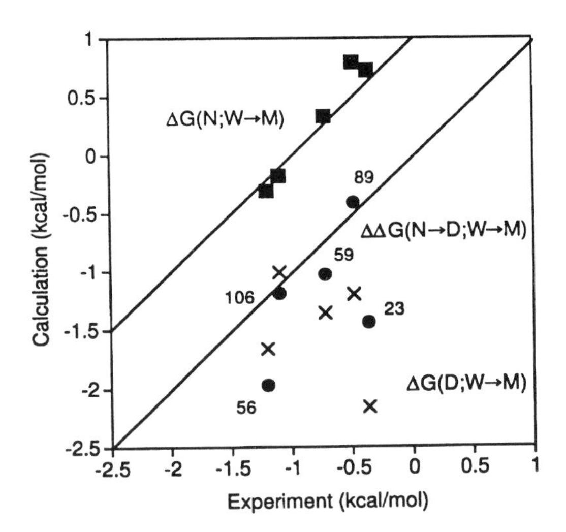

Fig. 7.10. Correlation between the calculated native-state free energy change ΔG(N;W→M) and the experimentally observed unfolding free energy change $\Delta\Delta G$(N→D;W→M). Against the observed $\Delta\Delta G$(N→D; W→M), the calculated ΔG(N;W→M) (squares), ΔG(D;W→M) (crosses), and $\Delta\Delta G$(N→D;W→M) (circles) are also plotted

The $\Delta\Delta G$ values calculated for two mutants of HLZM, Ile23Val and Ile56Val, deviate slightly from the correlation line with the experimental $\Delta\Delta G$ compared with the other 3 mutants, as shown by circles in Figs. 7.7 and 7.10. In addition, the denatured-state free energy change, ΔG(D; W→M), calculated for the Ile23Val and Ile56Val mutants was relatively smaller than those of the other 3 mutants (crosses in Fig. 7.10), although the native-state free energy change, ΔG(N; W→M), for the 5 mutants correlated well with the experimental values (squares in Fig. 7.10).

These results suggest that the discrepancies of $\Delta\Delta G$(N→D; W→M) for the Ile23Val and Ile56Val mutants were caused by estimates for the denatured-state free energy changes which were too low. Our methodology is based on the assumption that denatured-state proteins, whose structures are unknown, are represented by short peptide fragments centered at the mutation sites. Since this is the only assumption, its suitability should be doubted for the Ile23Val and Ile56Val mutant proteins.

7.10 Approximation Level Dependence

Previous MD simulations for the free energy calculations were carried out using some drastic simplifications [27, 28]. First, to decrease the total number of atoms, only a few hundred water molecules (CAP water) around the mutation site were included. Second, to decrease the degrees of freedom in the simulation, only protein atoms close to the site of interest were allowed to move. Third, to decrease the number of non-bonded interacting pairs that needed to be calculated, the long-range Coulomb interactions were truncated to distances shorter than typical protein radii. In addition, free energy calculations were performed for configurations obtained from the simplified MD simulations by using the most simple perturbation scheme (FEPM).

In contrast, the present methodology is based on more reliable MD simulations and a more efficient free energy perturbation scheme. The MD simulations are performed for proteins in water, i.e., fully solvated, with no artificial constraints applied, and with all long-range Coulomb interactions included by using the PPPC method. Furthermore, free energy changes are estimated from the MD simulation trajectories by ARM instead of FEPM. In order to check whether the previous simple methodology can successfully derive $\Delta\Delta G$(N→D; W→M), I re-calculated $\Delta\Delta G$(N→D; W→M) of the Val74Ile mutant using the 10 Å cutoff methodology instead of the present methodology.

The 10 Å cutoff simulations were carried out for 2 systems (RNaseHI in water and RNaseHI with CAP water) according to both mutation paths (Ile→Val and Ile←Val). The $\Delta\Delta G$(N→D; W→M) values calculated are listed in Table 7.5, together with the $\Delta\Delta G$(N→D; W→M) values calculated by the present methodology and the experimental value listed in Table 7.2. The present methodology gave $\Delta\Delta G$(N→D; W→M) values of 0.84 kcal/mol for the forward mutation path and 0.47 kcal/mol for the reverse mutation path, which were close to the experimental value (0.6 kcal/mol). In contrast, the 10 Å cutoff methodology for RNaseHI in water overestimated $\Delta\Delta G$(N→D; W→M) for the forward path

Table 7.5. Unfolding free energy change, $\Delta\Delta G$(N→D;Val→Ile) in kcal/mol, calculated for RNaseHI based on different approximation level MD simulations

Approximation level for MD simulation	Mutation path		
	Forward(Ile→Val)	Reverse (Val→Ile)	Average
Our methodology			
In water; no cutoff (PPPC)	0.84	0.47	0.65 ± 0.19
Conventional methodology			
In water; 10 Å cutoff	1.93	–0.16	0.89 ± 1.05
CAP water[a]; 10 Å cutoff	0.94	–1.02	–0.04 ± 0.98
		Exp.	0.6

[a] Only water molecules around the mutation site of the protein are included to reduce the total number of atoms to be simulated. Only protein atoms close to the mutation site are allowed to move in order to decrease the degrees of freedom in the simulations

Table 7.6. Reliability of two free energy perturbation methods, ARM and FEPM, in the calculation of bond free energy differences, ΔG_{bond}(N;Val→Ile) and ΔG_{bond}(D;Val→Ile) (in kcal/mol)

Mutation path	ARM	FEPM		
		Forward	Backward	Average
Native state				
Forward path	–0.33	–1.19	0.21	–0.49 ± 0.70
Reverse path	–0.34	0.91	–1.34	–0.22 ± 1.13
Denatured state				
Forward path	–0.14	–0.06	0.11	0.03 ± 0.09
Reverse path	–0.03	0.79	0.70	0.75 ± 0.05

(1.93 kcal/mol) and underestimated $\Delta\Delta G$(N→D; W→M) for the reverse mutation path (–0.16 kcal/mol). The hysteresis error (path dependence) was very large (±1.05), comparable to the $\Delta\Delta$G itself, although the average value (0.89 kcal/mol) was close to the experimental value. The 10 Å cutoff methodology for RNaseHI with the CAP water gave the wrong average value (–0.04) as well as a large hysteresis error (± 0.98). Therefore, the free energy calculations based on the simplified MD simulations fail to predict the stability of the mutant proteins.

In order to compare the reliability of FEPM with that of ARM, I calculated the bond free energy changes due to the Val74Ala mutation by using FEPM instead of ARM (Table 7.6). FEPM yields the 2 free energy values, forward (ΔG_{Forw}) and backward (ΔG_{Back}), according to (7.13) and (7.14). ΔG_{Forw} and ΔG_{Back} were calculated for both states (native and denatured) according to both mutation paths (forward Ile→Val and reverse Ile←Val). The bond free energy change calculated by ARM was almost independent of the mutation path (–0.33 vs –0.34 for the native state and –0.14 vs –0.03 for the denatured state). In contrast, the forward and backward free energy changes calculated by FEPM were significantly dependent on the mutation path (their hysteresis errors were larger than the free energy changes themselves). Furthermore, the forward free energy change of the native state was significantly different from the backward free energy change. These discrepancies between the forward and backward free energies and their large hysteresis errors make FEPM unreliable. Therefore, ARM is superior to FEPM for calculating free energy changes of proteins due to mutations with all degrees of freedom including the bond interactions.

7.11 Conclusion

The melting-temperature-shifts, ΔT_m, and unfolding free energy changes, $\Delta\Delta G$, of mutant proteins can be determined in principle by calculating free energy changes due to mutations for the native and denatured states, ΔG(N) and ΔG(D). However, the conventional methodologies used for MD/free energy calculations, cutoff and FEPM, do not give reliable free energy values for the following reasons: First, the cutoff of Coulomb interactions causes artificial deformations of protein structures during the simulations. Second, FEPM utilizes only the sampled conformations of either state in spite of the difference in free energy between the two states.

In order to overcome these difficulties, I developed a new method (PPPC method) to efficiently calculate long-range Coulomb interactions with no truncations. The MD simulations based on the PPPC method maintained the protein structures near their X-ray structures without any restraints. In addition, the thermal fluctuations around the equilibrium structures agreed well with the experimental fluctuations deduced from the X-ray temperature factors and NOE order parameters. Furthermore, to utilize the conformations sampled from both states, the author adopted ARM.

The MD/free energy calculation methodology based on PPPC and ARM made it possible to calculate more accurately the differences in free energy by suppressing the hysteresis error and statistical error. I and a coworker adopted this methodology for 8 mutants of RNaseHI and HLZM. The calculated ΔT_m values agreed well with the experimental values, although the methodology did not include any adjustable parameters or experimental data for the mutants.

The reasons the mutants were stabilized (or destabilized) were clarified by analyzing the results of the MD/free energy perturbation and then answering the following questions: First, which state, native or denatured, did the mutations mainly stabilize (or destabilize)? Second, what kind of physical interactions mainly stabilized (or destabilized) the mutant proteins? Third, what amino acids surrounding the mutation site stabilized (or destabilized) the mutants? Answers to these questions cannot be directly obtained from experiments.

Recently, I and coworkers applied the present methodology to calculate ΔT_m of 6 mutants of the Myb R2 domain. The calculated ΔT_m values agreed well with the respective experimental values, with correlation coefficients of 0.96 [16]. In this review, charge-creating (or annihilating) mutations (such as Glu to Gln) were not included. However, the present methodology is also applicable to such charge-creating mutations, as demonstrated by calculating the relative deprotonation free energy change from which the pK_a shift of Asp10 in RNaseHI was successfully derived [20]. A reliable MD simulation/free energy calculation methodology makes it possible to predict relative melting temperatures of mutant proteins because it needs no adjustable parameters or experimental data for the mutant proteins.

7.12 Appendix: Relationship Between ΔT_m and $\Delta\Delta G$

In this section, $\Delta\Delta G$(N→D;W→M), $\Delta\Delta G$(N→D;W), and $\Delta\Delta G$(N→D;M) at temperature T are denoted by $\Delta\Delta G(T)$, $\Delta G(T)$, and $\Delta G'(T)$, respectively. Then, (7.2) is represented as follows:

$$\Delta\Delta G(T) \equiv \Delta G'(T) - \Delta G(T). \qquad (7.17)$$

The unfolding free energy of the wild type protein at temperature T, $\Delta G(T)$, is defined by the following equation:

$$\Delta G(T) = \Delta H(T) - T\Delta S(T), \qquad (7.18)$$

where $\Delta H(T)$ and $\Delta S(T)$ are the unfolding enthalpy and entropy of the wild type protein at the temperature, respectively. $\Delta H(T)$ and $\Delta S(T)$ are represented by the unfolding enthalpy and entropy at the melting temperature T_m, $\Delta H(T_m)$ and $\Delta S(T_m)$, and the unfolding heat capacity change of the wild type protein, ΔC_p, as follows:

$$\Delta H(T) = \Delta H(T_m) + \int_{T_m}^{T} \Delta C_p dT \qquad (7.19)$$

$$= \Delta H(T_m) + \Delta C_p (T - T_m) \qquad (7.20)$$

and

$$\Delta S(T) = \Delta S(T_m) + \int_{T_m}^{T} \frac{\Delta C_p}{T} dT \qquad (7.21)$$

$$= \Delta S(T_m) + \Delta C_p \log \frac{T}{T_m} \qquad (7.22)$$

$$= \frac{1}{T_m} \Delta H(T_m) + \Delta C_p \log \frac{T}{T_m}. \qquad (7.23)$$

In (7.19) and (7.21), the following equations were used:

$$\frac{\partial \Delta H}{\partial T} = \Delta C_p \tag{7.24}$$

and

$$\frac{\partial \Delta S}{\partial T} = \frac{\Delta C_p}{T}. \tag{7.25}$$

In (7.20) and (7.22), ΔC_p was assumed not to be a function of T. In (7.23), the fact that $\Delta G(T)$ equals zero at $T = T_m$ was used, i.e.,

$$\Delta G(T_m) = \Delta H(T_m) - T_m \Delta S(T_m) = 0 \tag{7.26}$$

and

$$\Delta H(T_m) = T_m \Delta S(T_m). \tag{7.27}$$

By substituting (7.20) and (7.23) into (7.18), the following equation for the wild type ΔG was obtained:

$$\Delta G(T) \cong \left(1 - \frac{T}{T_m}\right)\Delta H(T_m) + \Delta C_p T\left(1 - \frac{T_m}{T} - \log\frac{T}{T_m}\right). \tag{7.28}$$

Similary, the following equation of a mutant $\Delta G'$ was obtained:

$$\Delta G'(T) \cong \left(1 - \frac{T}{T'_m}\right)\Delta H'(T'_m) + \Delta C_p T\left(1 - \frac{T'_m}{T} - \log\frac{T}{T'_m}\right), \tag{7.29}$$

where the heat capacity change of a mutant protein, $\Delta C'_p$, was approximated by that of the wild type protein, ΔC_p. The mutant $\Delta G'(T)$ can be expanded to a Taylor series of $\Delta T_m/T_m$ ($\Delta T_m = T'_m - T_m$), as follows:

$$\Delta G'(T) = \left[1 - \frac{T}{T_m} + \frac{T}{T_m}\left(\frac{\Delta T_m}{T_m}\right) - \frac{T}{T_m}\left(\frac{\Delta T_m}{T_m}\right)^2\right] \times \left[\Delta H'(T_m) + \Delta C_p T_m\left(\frac{\Delta T_m}{T_m}\right)\right]$$

$$+ \Delta C_p T\left[1 - \frac{T_m}{T} - \frac{\Delta T_m}{T} - \log\frac{T}{T_m} + \frac{\Delta T_m}{T_m} - \frac{1}{2}\left(\frac{\Delta T_m}{T_m}\right)^2\right]. \tag{7.30}$$

By substituting (7.30) and (7.28) into (7.17), the following expression is obtained:

$$\Delta\Delta G(T) \cong \left(1 - \frac{T}{T_m}\right)\left[\Delta H'(T_m) - \Delta H(T_m)\right] + \frac{T}{T_m}\Delta H'(T_m)\left(\frac{\Delta T_m}{T_m}\right)$$
$$+\frac{1}{2}\Delta C_p T\left(\frac{\Delta T_m}{T_m}\right)^2 - \frac{T}{T_m}\Delta H'(T_m)\left(\frac{\Delta T_m}{T_m}\right)^2. \tag{7.31}$$

Furthermore, this equation can be simplified under the following conditions:

Approximation 1.

$$\Delta H'(T_m) = \Delta H(T_m), \tag{7.32}$$

$$\Delta\Delta G(T) \cong \frac{T}{T_m}\Delta H(T_m)\left(\frac{\Delta T_m}{T_m}\right) + \frac{1}{2}\Delta C_p T\left(\frac{\Delta T_m}{T_m}\right)^2 - \frac{T}{T_m}\Delta H(T_m)\left(\frac{\Delta T_m}{T_m}\right)^2. \tag{7.33}$$

Approximation 2.

$$T = T_m, \tag{7.34}$$

$$\Delta\Delta G(T) \cong \Delta H'(T_m)\left(\frac{\Delta T_m}{T_m}\right) + \frac{1}{2}\Delta C_p \frac{(\Delta T_m)^2}{T_m} - \Delta H'(T_m)\left(\frac{\Delta T_m}{T_m}\right)^2. \tag{7.35}$$

Approximation 3.

$$T = T_m \quad \text{and} \quad \Delta H'(T_m) = \Delta H(T_m), \tag{7.36}$$

$$\Delta\Delta G(T) \cong \Delta H(T_m)\left(\frac{\Delta T_m}{T_m}\right) + \frac{1}{2}\Delta C_p \frac{(\Delta T_m)^2}{T_m} - \Delta H(T_m)\left(\frac{\Delta T_m}{T_m}\right)^2. \tag{7.37}$$

Since the last term is much smaller than the second term, it can be neglected to a first approximation. Then, the most simple relationship between ΔT_m and $\Delta\Delta G$ is obtained:

$$\Delta\Delta G(T) \cong \Delta H(T_m)\left(\frac{\Delta T_m}{T_m}\right) + \frac{1}{2}\Delta C_p \frac{(\Delta T_m)^2}{T_m}. \tag{7.38}$$

I used this equation to translate $\Delta\Delta G$ to ΔT_m in this chapter.

References

7.1 K. Ishikawa, M. Okumura, K. Katayanagi, S. Kimura, S. Kanaya, H. Nakamura, K. Morikawa: J. Mol. Biol. **230**, 529 (1993)
7.2 B.W. Matthews: Curr. Opin. Struct. Biol. **3**, 589 (1993)
7.3 S. Kimura, S. Kanaya, H. Nakamura: J. Biol. Chem. **267**, 22014 (1992)
7.4 K. Ishikawa, H. Nakamura, K. Morikawa, S. Kanaya: Biochemistry **32**, 6171 (1993)
7.5 K. Takano, K. Ogasahara, H. Kaneda, Y. Yamagata, S. Fuji, E. Kanaya, M. Kikuchi, M. Oobatake, K. Yutani: J. Mol. Biol. **254**, 62 (1995)
7.6 A.E. Eriksson, W.A. Baase, B.W. Matthews: J. Mol. Biol. **229**, 747 (1993)
7.7 A. Horovitz, L. Serrano, B. Avron, M. Bycroft, A.R. Fersht: J. Mol. Biol. **216**, 10311 (1990)
7.8 M. Saito, R. Tanimura: Chem. Phys. Lett. **236**, 156 (1995)
7.9 R. Tanimura, M. Saito: Mol. Simul. **16**, 75 (1996)
7.10 D.L. Veenstra, P.A. Kollman: Protein Eng. **10**, 789 (1997)
7.11 W.F. van Gunsteren, A.E. Mark: Eur. J. Biochem. **204**, 947 (1992)
7.12 M. Saito: Mol. Simul. **8**, 321 (1992)
7.13 M. Saito: J. Chem. Phys. **101**, 4055 (1994)
7.14 M. Saito, H. Nakamura: J. Comp. Chem. **11**, 76 (1990)
7.15 M. Satio, to be submitted.
7.16 M. Saito, H. Kono, H. Morii, H. Uedaira, T.H. Tahirov, K. Ogata, A. Sarai, J. Phys. Chem. **104,** 3705 (2000)
7.17 C.L. Brooks III, M. Karplus, B.M. Pettitt: *Proteins: A theoretical perspective of dynamics, structure, and thermodynamics*. (John Wiley & Sons, New York, 1988) p.29
7.18 J.A. McCammon, B.R. Gelin, M. Karplus: Nature **267**, 585 (1977)
7.19 M. Levitt, R. Sharon: Proc. Natl. Acad. Sci. USA **85**, 7557 (1988)
7.20 M. Saito: J. Phys. Chem. **99**, 17043 (1995)
7.21 K. Yamasaki, M. Saito, M. Oobatake, S. Kanaya: Biochemistry **34**, 6587 (1995)
7.22 A.M. Mandel, M. Akke, A.G. Palmer III: J. Mol. Biol. **246**, 144 (1995)
7.23 P.A. Bash, U.C. Singh, R. Langridge, P.A. Kollman: Science **236**, 564 (1987)
7.24 D. Frenkel: in *Molecular-Dynamics Simulation of Statistical-Mechanical Systems*, ed. by G. Ciccotti, W. G. Hoover, (North-Holland, Amsterdam, 1986) p.151
7.25 C.H. Bennett: J. Comp. Phys. **22**, 245 (1976)
7.26 P.E. Smith, W.F. van Gunsteren: J. Phys. Chem. **98**, 13735 (1994)
7.27 L.X. Dang, K.M. Merz Jr., P.A. Kollman: J. Am. Chem. Soc. **111**, 8505 (1989)
7.28 S.Yun-yu, A.E. Mark, W. Cun-xin, H. Fuhua, H.J.C. Berendsen, W.F. van Gunsteren: Protein Eng. **6**, 289 (1993)

8 Enzyme Kinetics: Stopped-Flow Under Extreme Conditions

Claude Balny

INSERM, Unité 128, IFR 24, CNRS, 1919, route de Mende, 34293 Montpellier Cedex 5, France
E-mail: balny@crbm.cnrs-mop.fr

Abstract. Temperature and pressure (extreme conditions), as perturbing variables, are two powerful tools to investigate the thermodynamic parameters of chemical or biochemical reactions and to study the mechanism of enzyme-catalyzed reactions, via the determination of rate constants and the use of the transition-state theory. However, one difficulty in measuring kinetics using spectroscopic detection is the relative long dead-time of the low-temperature and the high-pressure techniques. Different devices have been proposed to eliminate sources of errors in time, temperature and pressure. These techniques include *in situ* initiation of the reaction after temperature and pressure equilibrium. Using different examples of reactions, the interdependence of these two major variables, namely temperature and pressure, is presented in this chapter, in which the role of the reaction media is considered as a third variable.

8.1 Introduction

Kinetic methods are a means of elucidating the number of intermediates on an enzyme reaction pathway. A complete study of catalytic reactions involving a succession of very rapid different steps consists of the exploration of the properties of these steps, including thermodynamic parameters obtained by the action of temperature and pressure. They can give information on the intermediates such as, for example, those involved in ligand- or substrate-induced conformational changes.

For a given reaction, ideally one should obtain the structure and the thermodynamic properties of each intermediate in a reaction pathway, which may mean that different conditions have to be found to allow for accumulation of each intermediate in turn. It is usually difficult to obtain all this information with the native enzyme under normal experimental conditions. To aid we must perturb and then observe. For an enzyme reaction, we can perturb using genetic variants or chemically modified enzymes or by using substrate analogues. We can also perturb by changing experimental conditions using non-aqueous media at different temperatures (including sub-zero) and/or at elevated pressures, i.e., using extreme conditions.

The stopped-flow apparatus operating at room temperature and atmospheric pressure is now a routine technique which was described many years ago. The first rapid-reaction apparatus was a continuous-flow system and is the ancestor of the present stopped-flow. Experiments under extreme conditions need both special requirements and special apparatus. It is the purpose of this chapter to present a description of methods, techniques and instrumentation which may be used for measuring rates of reactions occurring in extreme conditions. To illustrate this approach, using some published results on a broad range of biochemical reactions, I intend to indicate how kinetics results and thermodynamic parameters derived from them can be used for mechanistic and/or structural interpretations.

8.2 Basic Principles

8.2.1 Cryo-Baro-Enzymology

In carrying out their biological functions, enzymes go through a number of subtle conformational changes that are related to their dynamic structural flexibility. These processes are usually very rapid and therefore difficult to study. One way of reducing this rapidity is to carried out experiments at low temperatures using the cryoenzymologic method [1,2]. This artifice, according to the Arrhenius expression, slows down the transformations of the various intermediates permitting one to identify and determine their disposition in a reaction pathway. If the temperature is low enough, each intermediate can be 'frozen' in time and thus analyzed as a quasi-stable state. This allows one to obtain in favorable cases a series of 'stopped-action' pictures of the most highly populated complexes along the reaction pathway. However, such separate 'frame' studies yield little dynamic information. To obtain this, other techniques are needed to fill in the missing 'unfrozen' frames. One way of obtaining dynamic information about enzyme systems is by kinetic studies, especially at low temperatures; these can lead to data on the rate constants (k_+ and k_-) describing the interconversions of successive intermediates. Currently, it is generally postulated that between such intermediates there is an activated complex whose characteristics govern k_+ and k_- and thus the dynamics of the structural interconversion. By its very nature the activated state is not accessible to measurement and its properties can only be inferred from the thermodynamic parameters associated with k_+ and k_-.

For a rigorous quantitation of these parameters, when applied to complex enzyme systems, there are two main requirements: (a) if possible, a simple rate constant must be measured – measurement of a composite rate constant, such as k_{cat} can lead to ambiguous results; (b) one must obtain a maximum of the thermodynamic values associated with the kinetic constants of interest, and, for this, one must exploit as many intensive parameters as possible.

For good accuracy, this means using rapid-reaction techniques; to further increase the time resolution, experiments can be carried out at low temperatures. The range over which the particular intensive parameter is varied must be as wide as possible. With enzyme systems, the intensive parameters most exploited are temperature and, to a lesser extent, pressure; by varying these, one can obtain estimates of the thermodynamic quantities relating to the activated complexes, namely $\Delta H^{\ddagger}$, $\Delta S^{\ddagger}$ and $\Delta V^{\ddagger}$ (variation of enthalpy, entropy, and volume respectively), such data permitting the construction of volume profiles of reactions. As we will see below, using volume profiles of reactions, mechanistic features can be revealed via molecular interactions. However, when the reaction is multistep or has concurrent reactions then analysis of measurements into the individual contributions is more complicated, and the success of the kinetic analysis must always be tempered with the knowledge that transition-state theory contains several assumptions.

A large temperature range can be obtained under cryoenzymic conditions, which requires the addition of an antifreeze, usually an organic solvent which can play the role of perturbant, making enzymology in non-pure-aqueous media another very useful extreme condition [1,2]. This approach has been exploited, using mainly the fact that these organic solvents may reversibly and often selectively alter the equilibrium and rate constants of the elementary steps in such a way as to lead to the temporal resolution of kinetically significant steps. However, whereas thermodynamic entities in themselves are rather uninformative, their variation of the perturbation of the system under study can lead to useful information on the dynamics of the structural changes implicated in the formation of the activation complex [3, 4].

8.2.2 Exploitation of Data

The application of the transition-state theory of Eyring to enzyme reaction has been described in several reviews [5-8]. I will not extend this theory in the present paper but I would like to make some remarks:

(a) Certain experiments carried out with a model system [9] show that the transition-state theory is an oversimplified interpretation. A treatment which may approach the real situation more closely is provided by Kramers' equations [10]. In this theory the classical formalism of the Arrhenius law is retained

$$K = A \cdot \exp\left(-\Delta H^{\ddagger}/RT\right)^{\ddagger} \qquad (8.1)$$

However, there is a pre-exponential term, A, which is taken to be a function of the viscosity of the medium and of the electronic structural effects involving the protein. From this it is clear that the viscosity of the solvent environment could be an important factor in reducing the reaction rate with $A = A_0/\eta$, where η is the medium viscosity. From an experimental point of view, the consequence of the viscosity variation as a function of T is, *a priori*, a non-linearity of Arrhenius plots. That is the case when a very large temperature range is explored and when the medium becomes glass-like [9].

Under most of our experimental conditions, where the medium is maintained fluid at sub-zero temperatures (cryosolvents of low viscosity within the temperature and pressure ranges used), it seems unlikely that the viscosity factor is important, and our results can be interpreted simply in terms of the transition-state theory. However, with a highly viscous solvent such as 80% glycerol, the viscosity factor could become so important as to dramatically influence interconversions involving triggered conformational change [11].

According to this theory, the variation of viscosity with pressure must introduce curvature in ln k as a function of pressure. But, for experiments performed in the same conditions as previously described, these effects remain negligible, as analyzed by Butz [12].

(b) To confirm the validity of the calculated thermodynamic parameters the quantities $\Delta H^{\ddagger}$, $\Delta S^{\ddagger}$ and $\Delta V^{\ddagger}$ for a given reaction as a function of T and P respectively, must be related to the Maxwell relationships:

$$-(\delta \Delta V^{\ddagger}/\delta T)_P = (\delta \Delta S^{\ddagger}/\delta P)_T \text{ and } (\delta \Delta H^{\ddagger}/\delta P)_T = \Delta V^{\ddagger} - T(\delta \Delta V^{\ddagger}/\delta T)_P \tag{8.2}$$

$(\delta \Delta V^{\ddagger}/\delta T)_P$ is the temperature coefficient for the activation volume; it is the expansibility term. $(\delta \Delta S^{\ddagger}/\delta P)_T$ and $(\delta \Delta H^{\ddagger}/\delta P)_T$ are the pressure coefficients for the activation entropy and enthalpy, respectively.

However, the verification of these equations does not mean that the transition-state theory is verified, it is only a mathematical consequence of the definition of the thermodynamical quantities. The verification of the Maxwell relationships implies only that the analyzed results are internally consistent [13].

For many examples, the overall enzyme reactions can be simply described by the classical Michaelis-Menten mechanism. Under some experimental conditions, the binding of substrate to enzyme can be analyzed using the induced-fit theory, which implies that the binding of the substrate to the enzyme is a two-step process: formation of a collision complex is a rapid equilibrium related to the association constant K_1; this is followed by an isomerization step described by the rate constants k_1 and k_2 [14].

8.3 The High-Pressure, Variable-Temperature, Stopped-Flow Technique (HP-VT-SF)

8.3.1 General Design

The general design of a stopped-flow apparatus classically consists of a driving mechanism, two vertical syringes, a mixing chamber, an observation chamber and a waste syringe. Apparatus working at sub-zero temperatures and at atmospheric pressure was described by my coworkers and me many years ago [15].

To perform enzymatic experiments at elevated pressures, the relatively long dead-time of high-pressure techniques using spectroscopic detection is the first important limitation to exploiting pressure enzyme kinetics. If the system under study can be characterized via optical detection (absorbance, fluorescence), the stopped-flow method is easily used. The second limitation deals with the temperature control, which must be efficient to compensate for the heat of compression. To solve these problems, devices were designed to mix samples under high pressure, at controlled temperatures. The first apparatus was described by Grieger and Eckert [16] and modified by Sasaki et al. [17]. Both systems were thermostated and used the breakage of a foil diaphragm to force the mixing of two components. After different improvements [18, 19], the Heremans' group described a stopped-flow apparatus designed for spectroscopic detections of fast reactions at pressures up to 120 MPa by means of immersing a stopped-flow unit in a high-pressure bomb [20].

To further reduce the dead-time with respect to the reactions studied, we have developed an apparatus which permits rapid mixing both at sub-zero temperatures and at high pressure (see Fig. 8.1). The principle of our instrument incorporates certain features of previous stopped-flow systems described for cryoenzymological studies and for investigations under high pressure. The principle of this apparatus, already published [21, 22], consists of a powerful pneumatic system driving the syringe

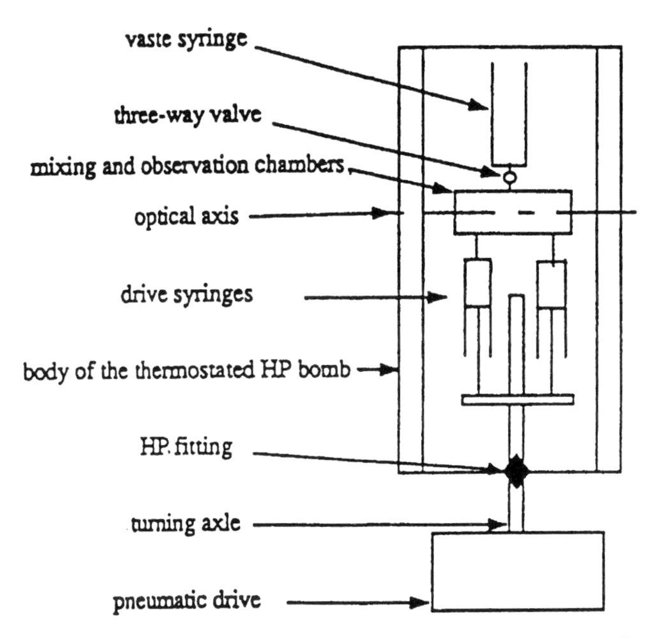

Fig. 8.1. Schematic of the high-pressure (HP), variable-temperature stopped-flow apparatus

mechanism, two vertical drive syringes containing the samples to be mixed, a mixing chamber, an observation chamber with quartz windows, and a waste syringe. Both temperature and pressure homogeneities are maintained by housing the whole apparatus in a high-pressure thermostated bomb. The stopped-flow apparatus can operate in absorbance or fluorescence mode over temperature and pressure ranges of +40 to −35°C and of 1 to 300 MPa, respectively. The system is mounted either on an Aminco DW2 spectrophotometer permitting detection in a dual wavelength mode or on a spectrofluorometer especially designed in the laboratory (wavelength limits: 230–650 nm). The dead-time, nearly independent of pressure, is less than 5 ms when measuring in aqueous solutions at room temperature. In the presence of organic solvent, this value increases to reach, for example, a value of 50 ms in a solvent system containing 40% ethylene glycol (v/v) at −15°C.

To be operative, the desiderata of a stopped-flow apparatus for cryo-baro-enzymological studies include:

- efficient rapid mixing;
- good thermal and pressure equilibrium of the samples, the mixing chamber and the connecting parts;
- a recording system that can be used to follow rapid and very slow reactions with the same sample.

Neglect in any of these area can lead to experimental artifacts or erroneous interpretation of the data, as discussed below.

8.3.2 Source of Artifacts in Stopped-Flow Operating Under Extreme Conditions

In our experience, spurious kinetic results can arise from: (a) incomplete mixing, (b) incomplete thermal equilibrium, mainly after compression, and (c) turbidity of samples. These problems may be difficult to diagnose because each depends on the enzyme concentration. The 'effect' frequently disappears when enzyme is omitted and can therefore appear 'real'. More than 15 years ago, using as a model system the hydrolysis reaction of several anilide substrates by bovine and porcine trypsin and porcine elastase between −30 and +20°C, we showed, at atmospheric pressure, that confusing spectral changes can arise from incomplete mixing, thermal gradients or heterogeneity of the substrate. Several of the solutions in mixed organic/aqueous solvents became turbid as the samples were cooled at sub-zero temperatures. Mixing of these solutions by the stopped-flow apparatus gave rise to time-dependent optical changes of a spurious nature.

In conclusion, stopped-flow, low temperatures and high-pressure techniques are powerful tools for mechanistic studies of enzyme-catalyzed reactions, but in order to use them effectively, one must recognize their limitations and potential problems. It appears that extreme care must be exercised to avoid artifacts in optical spectroscopic measurements, particularly at low temperatures.

8.3.3 Recent Progress

Whatever the design, the general principle of a stopped-flow apparatus working under extreme conditions is the same: the mixing of two compounds contained in two syringes using a pneumatic or electric driving system, an observation chamber and a waste syringe.

Some year ago, the van Eldik's group described a home-built, stopped-flow unit working up to 130 MPa. They recently used it to study the carbonic anhydrase catalysis to obtain the volume profile of the reaction [23].

Another instrument working at atmospheric pressure but at sub-zero temperatures with a dead-time below 10 ms has been constructed and used to study the quaternary structural change of *Escherichia coli* aspartate transcarbamylase by time-resolved X-ray solution scattering (in the presence of 30% ethylene glycol at −11°C) [24]. In this design, reservoirs and mixing blocks are thermo-regulated by a Peltier device and the mixing efficiency is improved by employing two-step mixing with two-jet and four-jet mixers. The volume ratio of the two mixed solutions could be changed from 1 : 1 to 1 : 3.5, depending on the experimental conditions [25].

A new generation of stopped-flow apparatus has recently been described by P. Bugnon et al.; it consists of one autoclave, including the static and dynamic components of the stopped-flow circuit, the drive mechanism and the high-pressure control unit (Fig. 8.2). Measurements in absorbance or fluorescence mode can be performed at pressures up to 200 MPa, in a temperature range between −40 and +100°C. The sample circuit is chemically inert and there are no metallic parts in contact with sample solutions. The dead-time of the instrument (i.e., the time needed for the reaction mixture to flow from the mixing chamber to the observation cell) is less than 2 ms in aqueous solution at room temperature up to 200 MPa [26]. However, no

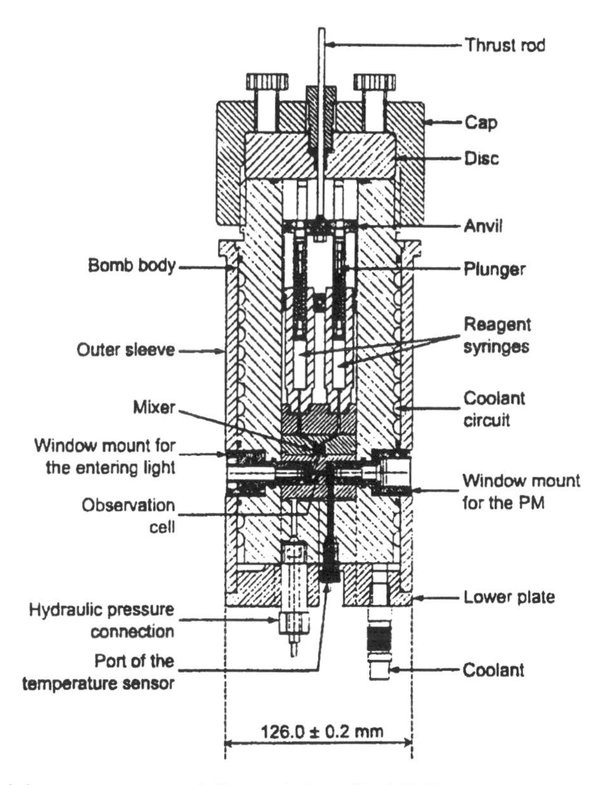

Fig. 8.2. The high pressure, stopped-flow unit described [26]

information is given concerning the performance of the apparatus at sub-zero temperatures in viscous media. This apparatus was used to study the kinetics and mechanism of cation-binding to Na^+, K^+-ATPase [27].

Other high-pressure, stopped-flow instruments have been reported, and a commercial unit is availble from Hi-Tech Scientific, Salisbury, England.

8.4 Examples of Application

HP-VT-SF has been used for different purposes, which will be analysed and discussed below. The examples given are not exhaustive but show the potential of the method.

8.4.1 Steady-State Kinetics of Enzymes of Monomeric or Polymeric Quaternary Structure

Thermolysin Reaction

Thermolysin is a thermostable monomeric microbial neutral protease containing zinc as a cofactor, where the activity shows a considerable activation by pressure for di- or tripeptide substrates (an increase of 45-fold at 220 MPa). S. Kunugi et al. have used HP-VT-SF: (a) to mix the enzyme and substrate solutions under the indicated pressure and temperature, and (b) for the measurement of the temperature-dependent and pressure-dependent deactivation process of the enzyme. For the latter, the container syringe of the enzyme solution of the apparatus was utilized as an incubation vessel, and the reaction upon mixing with the substrate solution, stored in the other syringe, was recorded after various time intervals during the elevation of pressure. Thermolysin showed distinct pressure-induced activation, with a maximum observed at 200−250 MPa for a dipeptide amide substrate and at 100−120 MPa for a heptapeptide substrate. By examining the pressure dependence of the hydrolytic rate for the former substrate, the activation volume of the reaction was –71 ml mol^{-1} and −95 ml mol^{-1} at 25 and 45°C respectively (negative activation expandability). A prolonged incubation of thermolysin under high pressure, however, caused a time-dependent deactivation. These change have been explained by a simple two-state transition model accompanied by a large negative change in the volume of reaction [28].

β-Galactosidase

Another example of the use of HP-VT-SF is given by P. Lemay et al. in an investigation of β-galactosidase, a tetrameric sucrase containing one active site per monomer. Using *o*-nitrophenyl-β-D-galactopyranoside (ONPG) as a substrate, it was observed at the level of V_0 in the reaction $\{V_0 = V_{max}[ONPG]/(K_m + [ONPG]\}$, an activation volume strongly dependent on temperature.$\Delta V^{\ddagger}$ = −14, 0 and +27 ml mol^{-1} at 4°, 25° and 35°C, respectively. The calculation of the Maxwell relationships and the extrapolation of the pressure dependence of the $\ln V_0$ versus pressure curves show that the system is thermodynamically consistent and that the pressure up to 120 MPa does not lead to dissociation and subsequent inactivation of the tetrameric enzyme [29]. These initial experiments allowed the development of an immunoaffinity fusion protein expression vector which can be subjected to high - pressure treatment for separation processes [30].

Enolase

The previous example showed that pressure did not affected the quaternary structure of β-galactosidase. This was not the case for the yeast enolase reaction for which M. J. Kornblatt et al. used pressure as a gentle and reversible perturbant [31]. Enolase catalyzes the interconversion of 2-phosphoglyceric acid and phospho-enolpyruvate. It is a dimer which contains two active sites and where each monomer consists of two domains. The goals of experiments using HP-VT-SF (up to 150 MPa) associated with fourth derivative spectroscopy [32] were to produce monomers in the presence of Mg (bound at the active site) and then (a) determine whether or not the monomers had catalytic activity and (b) compare dissociation produced by hydrostatic pressure with that produced by EDTA and KCl. The monomers produced by both methods are inactive. Keq, ΔV, $\Delta V^{\ddagger}$ for the dissociation/inactivation produced by hydrostatic pressure have been determined under various conditions. Removing the Mg(II) from enolase displaces the equilibrium towards monomers and decreases both ΔV and $\Delta V^{\ddagger}$. Loss of Mg(II) contributes to the negative ΔV for dissociation occurring in the transition state for dissociation. On the basis of CD spectroscopy, it was proposed that pressure changes the conformations of the mobile loops of enolase (there are three loops which change position significantly when substrate is bound).

Butyrylcholinesterase

P. Masson et al. have undertaken a series of experiments on the temperature and pressure dependence of the kinetics of the hydrolysis of various substrates by human plasma tetrameric cholinesterase (BuChE). Initial results were obtained using *o*-nitrophenylbutyrate as the substrate. HP-VT-SF experiments performed at various temperatures and pressures ranging from 0 to 40°C and from 1 bar to 200 MPa, have shown a break in the Arrhenius plot around 21°C and discontinuities in the thermodynamic quantities obtained from temperature and pressure dependence. Even though my coworkers and I know that the overall constants K_m and V_{max} measured under pressure lead to ambiguous results, the discontinuities in the thermodynamic quantities have given information concerning the cryptic conformational change of the enzyme modulated by temperature and pressure [33]. Using soman (an organophosphorous compound) as the inhibitor, the break observed before at 21°C is not seen. Moreover, the pressure dependence of the substrate hydrolysis also revealed differences between the native enzyme and the enzyme in the presence of soman: the sign and magnitude of the apparent activation volume ($\Delta V^{\ddagger}$) were different for the two reactions. Beyond 30 MPa, in the presence of soman a plateau ($\Delta V^{\ddagger}$ close to 0) was observed over a large pressure range, depending on temperature. Such behavior with respect to temperature and pressure can reflect a soman-induced enzyme conformation state [34].

Thus, temperature (from 4 to 30°C) and pressure (up to 160 MPa) perturbations of the kinetics measured using HP-VT-SF allow one to complete the inhibition scheme of BuChE by soman. Data suggested that upon soman binding, the enzyme undergoes a long-lived soman-induced-fit conformational change preceding the phosphonylation step. To determine whether this effect depends on soman itself or is dependent on the presence and nature of the substrate or ligand, my coworkers and I have examined the effect of amiloride, a reversible cholinesterase effector, upon the BuChE-catalyzed hydrolysis of nitrophenyl esters, using HP-VT-SF up to 160 MPa [35]. Results have shown that the effect of reversible soman binding on BuChE activity in the presence of amiloride depends on the position of the substrate nitro group and amiloride concentrations. On the other hand, reversible binding of amiloride and/or soman induces a new active conformational state that may be either a binary (or ternary) enzyme-ligand complex or a new free enzyme conformation resulting from a long-lived ligand-induced enzyme conformational change when *o*-nitrophenylbutyrate is the substrate. These ligand-induced states are stabilized by high pressure.

Another interesting experiment has been performed with regard to on the pressure effects on the single-turnover kinetics of the carbamylation of BuChE, the rate of which being monitored as the accumulation of N-methyl-7-hydroxyquinolinium using HP-VT-SF in its fluorescence version (excitation and emission wavelengths: 410 and 510 nm respectively). Elevated pressure favored formation of the enzyme-substrate complex but inhibited carbamylation of the enzyme. Because a single reaction step was recorded, it has been possible to interpret the data obtained under high pressure in the form of Michaelis—Menten equations. A substrate-induced, pressure-sensitive, enzyme conformation state was desmonstrated [36]. Hydration changes of BuChe have also been investigated using the same HP-VT-SF fluorometer, and experiments were carried out in different media: buffered water, water containing 0.1 M lithium chloride and deuterium oxide as solvents [37]. Kinetic data have shown that the binding of substrate to the enzyme leads to a pressure-sensitive enzyme conformational state which cannot perform the catalytic function. Large solvent effects indicate that enzyme sensitivity to pressure depends on the solvent structure, suggesting that the substrate-dependent pressure effect is modulated by the solvation state of the enzyme.

8.4.2 Structure-Function Relations: Case of Muscle Contraction

More than 15 years ago, it was shown that perturbations of the interaction between actin and myosin by hydrostatic pressure yielded important information on the mechanism of energy transduction by actomyosin ATPase. Increased pressure reduced the tension of isometrically contracting skinned muscle fibres. An isomerization of the actomyosin complex is known to be pressure sensitive, but the pressure sensitivity of other steps in the ATPase pathway had not been characterized. M. Geeves et al. have reported the effect of pressure on the ATP hydrolysis step of subfragment 1 ATPase, ADP binding to actomyosin subfragment 1 [acto.S1] and the rate of ATP-induced dissociation of [acto.S1]. For experiments on both the dissociation of [acto.S1] by ATP and the affinity of ADP for [acto.S1], the kinetics were recorded using HP-VT-SF in fluorescence mode (excitation and emission at 366 and 385 nm respectively) at various temperatures and pressures (up to 100 MPa). For the hydrolysis experiments, HP-VT-SF was used in its absorbance mode. The steps in the actomyosin ATPase pathway that were tested are unlikely to cause the 10% decrease in isometric tension in skinned fibres at 10 MPa. This supported the argument that the tension drop is due to an effect on the specific isomerization of the actomyosin complex. Assignment of this effect to a single step allows interpretation of transients observed in skinned fibres on pressure release [38].

8.4.3 Micellar Enzymology

Catalysis by enzymes solubilized in reversed micelles of amphiphilic (sodium bis-(2-ethylhexyl) sulfosuccinate) (aerosol OT : AOT) in octane is of interest for many research groups (for review see, for example, [39,40]). α-Chymotrypsin catalysis in reversed micelles has been described in detail [39]. Until recently, few attempts had

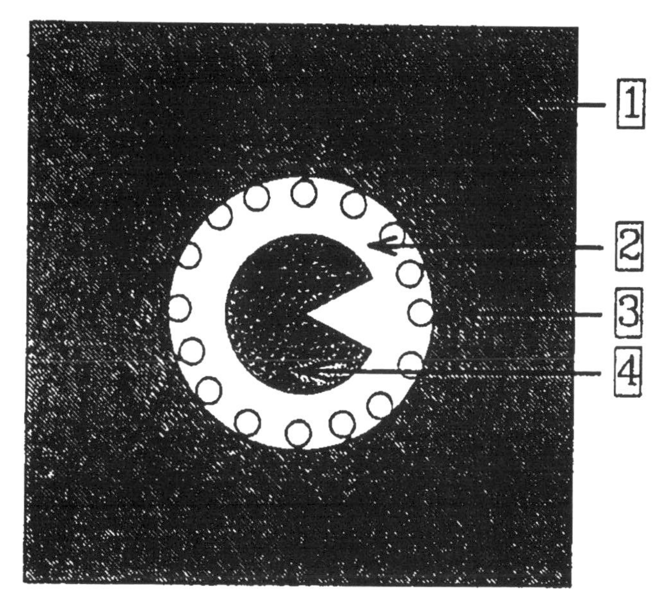

Fig. 8.3. Schematic of enzyme-containing reversed micelles formed by surfactant (Aerosol OT) in a non-polar solvent. 1: non-polar solvent; 2: water ; 3: Aerosol OT ; 4: enzyme

been made to study the effect of pressure on enzymes in mainly organic media. However, the combined effect of high pressures and organic solvents may provide a new picture of enzyme mechanisms and open new possibilities for modeling molecular structure and the catalytic behavior of enzymes. A micellar system includes three microphases (Fig. 8.3.): an organic apolar solvent, surfactant-containing interphase, and inner aqueous cavity, each phase having a different compressibility. The complexity is increased by the presence of enzyme in the water cavity, localization and structure of which can also be changed by pressure. Using the HP-VT-SF developed at Montpellier, a group from the Moscow State University has studied - for the first time - α-chymotrypsin-catalyzed hydrolysis of N-carbobenzoxy-L-tyrosine p-nitrophenyl ester (SPNA) and N-succinyl-L-phenylalanine p-nitroanilide. Coworkers and I have found that the enzyme retains high activity in these water-in-oil microemulsions up to a pressure of 200 MPa. The value of the activation volume for the enzyme reactions shows a dependence on water content in the system. When the size of the micellar aqueous inner cavity (defined as $w_0 = [H_2O]/[AOT]$) approaches the molecular size of α-chymotrypsin (CT), the activation volume becomes significantly lower that the value in aqueous solution (-13 and -27 ml mol^{-1} for water and micelles of $w_0 = 7.5$ respectively, for the reaction with SPNA as a substrate). This shows the possibility of regulating the enzyme activity by pressure in systems with low water content [41].

Similar investigations were performed to determine whether the conformational plasticity of human butyrylcholinesterase is altered by entrapment in reversed micelles [42]. The presence of soman, an irreversible inhibitor of cholinesterase, was used to bring to the fore a possible modification of the enzyme behavior in this system under pressure. Using HP-VT-SF we have shown differences between the enzyme in a conventional medium and in reversed micelles regarding the mechanism of enzyme-catalyzed hydrolysis of acetylcholine. In both systems, the enzyme displays a non-Michaelian behavior with this substrate. In a conventional medium the kinetics is multiphasic, with an activation phase followed by an inhibition phase at high concentration. In reversed micelles there is inhibition by excess substrate but the activation phase is missing. This behavior may be the result of a change of the enzyme conformational plasticity when is entrapped in reversed micelles.

However, some enzymes solubilized in reversed micelles undergo thermal inactivation and their stability decreases significantly when temperature increases (25 to 40°C). The systematic study of thermostability of α-chymotrypsin in the system of reversed micelles has been started recently [43]. The half-time of α-chymotrypsin in micelles reveals a bell-shaped dependence on the degree of hydration of AOT (w_0) analogous to the previously obtained dependence on w_0 for the enzyme activity. The optima of the catalytic activity and thermostability have been observed under the conditions where the diameter of the inner aqueous cavity of micelle is close to the size of the enzyme molecule ($w_0 = 10$).

As a hypothesis, a decisive factor, affecting the α-chymotrypsin stability in reversed micelles, is enhanced mobility and/or flexibility of the components of the micellar matrix which destabilize the protein structure when temperature increases. For this reason my coworkers and I had used high pressure as a factor which is capable of modulating the enzyme catalytic activity both in micelle free solutions and in micellar system. It was found out that the α-chymotrypsin stability increased significantly when a hydrostatic pressure (up to 150 MPa) was applied to the protein-containing micellar system [43]. It is known that under such pressure, micelles retain their integrity and the structure of enzymes entrapped in reversed micelles is conserved. In our opinion, application of high pressure increases the structural order of surfactant aggregates (micelles), that is, the factor leads to α-chymotrypsin stabilization.

It is well known that the use of organic cosolvents instead of water in the inner cavities of micelles leads to an increases in the time one molecule of surfactant exists in one micelle. Thus, one could expect that the effect caused by glycerol on the α-chymotrypsin stability in reversed micelles would be similar to that caused by pressure. For this purpose, work has been undertaken to study the joint influence of glycerol and pressure on the α-chymotrypsin stability in reversed micelles of AOT in octane [44]. Thermostability of α-chymotrypsin at normal pressure in reversed micelles depends on both an effective surfactant solvation degree and the glycerol content of the system. Difference in α-chymotrypsin stability in reversed micelles at various glycerol concentrations has been found to be more pronounced at high degrees of surfactant solvation. It has been found that after 1 h incubation in 'aqueous' reversed micelles (in the absence of glycerol) α-chymotrypsin retained only 1% of its initial catalytic activity, and 59% of its residual activity in glycerol-solvated micelles with 50% (v/v) glycerol, for example. The explanation of the effects observed is given in terms of the micellar matrix rigidity increasing in the presence of glycerol as a water-miscible cosolvent. It is possible that increasing the rigidity of the matrix results in a decrease in the mobility of the α-chymotrypsin molecule, and, thus, an increase in its stability. It has been shown that glycerol and hydrostatic pressure work additively. It has been found that α-chymotrypsin solubilized in reversed micelles, containing 50% (v/v) glycerol, was more stable under pressure of 50 MPa at all solvation degrees than in 'aqueous' micelles under higher pressure (150 MPa).

8.4.4 Transient Enzyme Kinetics

One of the first applications of the HP-VT-SF apparatus developed in our laboratory was the cryo-baro-enzymologic study of the creatine kinase-ADP-Mg-nitrate-creatine complex [4]. In the presence of creatine, Mg, ADP, and nitrate, creatine kinase forms the presumed transient state analogue complex E-ADP-NO_3-creatine where NO_3 mimics the P_i in the transient state complex E-ADP-P_i-creatine. We have used the formation of the analogue complex as a model for an elementary step, i.e., the conformational change of an enzyme intermediate determined by its kinetic constant k. We measured k in water, in 40% ethylene glycol in the temperaure range 1 to −15°C and at pressures up to 120 MPa. This permitted the determination of $\Delta H^{\ddagger}$, $\Delta S^{\ddagger}$ and $\Delta V^{\ddagger}$ and their interdependency as a function of both temperature and pressure.

Using a similar approach, my coworkers and I have studied, in collaboration with J. Frank, the kinetics of the reduction of the quinoprotein glucose dehydrogenase by substrate as a function of 3 parameters: pressure (0.1 to 100 MPa), temperature (down to −25°C) and solvent (water and 40% dimethyl sulfoxide, DMSO). A two-step formation of the reduced enzyme by its substrate (xylose) was observed. A rapid equilibrium described by the constant K_1 was followed by a slower process described by the constants k_2 and k_{-2}. By using the transition-state theory, the thermodynamic quantities $\Delta V^{\ddagger}$ were determined for these various kinetic constants under different experimental conditions [45]. We were able to determine the effect on K_1 and k_2 only at sub-zero temperature (−14°C in 40% DMSO). It was observed that the absolute value of the activation volume observed for step 2 was rather large compared to that of step 1: +21 and −7 ml mol^{-1} respectively. This observation was compared with experiments performed for the formation of horse radish peroxidase (HRP) compound I where the use of sub-zero temperatures (−30°C in 60% DMSO or −38°C in 50% methanol) permitted the measurement of K_1, too fast to be determined under 'classical' experimental conditions (i.e., water in the positive temperature range) [46]. In the latter case the second step of the reaction was related to an isomerization without specifying which portion of the peroxidase molecule was involved, and may not be related to a substrate-induced conformational change, since the spectral changes observed upon binding of peroxide could have originated from some movements within the peroxide molecule. For the quinoprotein glucose dehydrogenase reaction, it was postulated that the large $\Delta V^{\ddagger}$ related to step 2 was associated with a substrate-induced conformational change in the enzyme.

Another example of the potentiality of the pressure variable used together with water-miscible solvents is given in the case of reactions of the multiheme hydroxylamine oxidoreductase (HAO) followed using HP-VT-SF. This enzyme from *Nitrosomonas europaea* catalyzes the oxidation of NH_2OH. The kinetics of reduction of the enzyme by NH_2OH were studied in various media, in a temperature range of +20 to −15°C, and at pressures up to 100 MPa. Changing the solvent from water to 40% ethylene glycol resulted in an increase in activation volume from −3.6 to 57 ml mol^{-1} and changes in other thermodynamic parameters. The Arrhenius plot at atmospheric pressure has a downward inflection at about 0°C, which is enhanced at high pressure. For example, at 80 MPa, $\Delta H^{\ddagger}$ is 51 kJ mol^{-1} at sub-zero temperatures, whereas above 0°C the activation energy is −18 kJ mol^{-1}. This was interpreted as a conformational change of the enzyme with an unusual consequence: at high pressure, a decrease in temperature induced an increase in the velocity of the reaction. However, the Maxwell relationships were well verified [47].

Similar phenomena were observed for the binding reaction of ATP to the subfragment of myosin in 40% ethylene glycol. The binding reaction occurred in two steps, according to the description given above for the quinoprotein glucose dehydrogenase reaction. When ATP interacts with myosin, some tryptophan residues are perturbed (increase in absorption at 290 nm) and kinetics give k_2. This parameter is very sensitive to experimental conditions, a phenomenon interpreted to be related large solvation variation. We note also that for the reaction performed at 90 MPa, $\Delta H^{\ddagger}$ is negative (-12 kJ mol^{-1}) even though at atmospheric pressure this value is positive (+126 kJ mol^{-1}). This is another example where at high pressure the velocity of the reaction increases as a function of the decrease in temperature [48].

8.4.5 Carbon Monoxide (CO) Binding

We have investigated the CO binding to various reduced hemoproteins by stopped-flow mixing as a function of pressure (from 0.1 to 200 MPa) and temperature, from 4 to 35°C, and from -20 to +35°C, in water and in an hydro-organic medium, respectively. The first series of experiments was achieved with the records of the kinetics at 423 nm of the binding of CO to ferrous HRP in 40% ethylene glycol (EGOH) and in 50% methanol, in which we were able to extend the temperature range to sub-zero conditions [5] and to perturb the system [49]. The thermodynamic quantities $\Delta H^{\ddagger}$, $\Delta S^{\ddagger}$ and $\Delta V^{\ddagger}$ were determined under different experimental conditions and were found to be greatly modulated by the physico-chemical parameters of the media. The results suggested that the macroscopic thermodynamic response was mainly controlled by the solvent. By adjusting two of the variables among temperature, pressure and solvent, it was possible either to amplify or to cancel out the effect of the third. Typical kinetic traces are shown in Fig. 8.4. Similar experiments have been performed, in aqueous medium, with several varieties of cytochrome P-450, as well as chloroperoxidase, and my coworkers and I have compared the results to data reported for other proteins. The CO binding activation enthalpy and entropy varied greatly between the proteins, correlated through a compensation effect. There was no apparent relation to structural features. The pressure effect depended on the nature of the proximal axial heme ligand: the activation volume was very smell for cysteine ligand hemoproteins, and markedly negative for histidine ligand hemoproteins. Furthermore, the transition-state volume of the histidine ligand class enzymes, but not that of the cysteine ligand enzymes, depended on the solvent composition. These results suggested that the CO-binding transition state of the S-ligand class has a molecular conformation similar to the ground state. In the histidine class, however, the transition state appears to involve protein conformational changes and/or solvation processes.

8.4.6 Electron-Transfer Reactions

In biochemistry, the field of pressure dependence of electron transfer is limited to some pioneering works, although a more recent one, devoted to the pressure dependence of inorganic electron-transfer reactions in solution, has demonstrated that

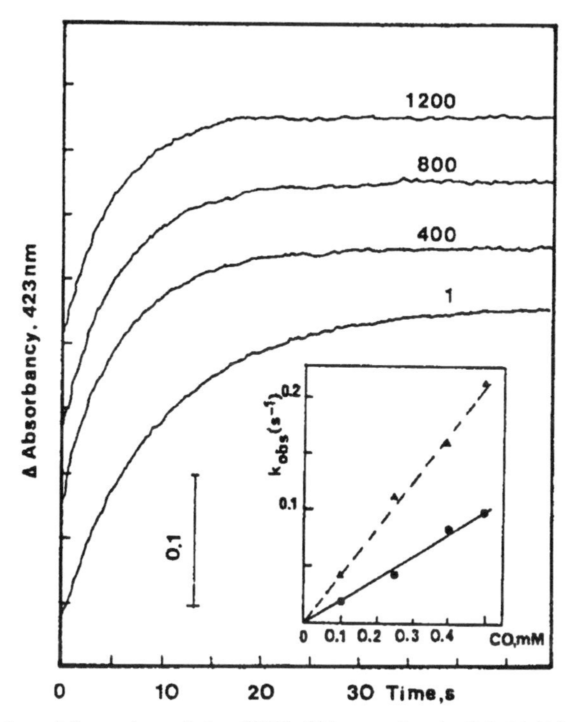

Fig. 8.4. Kinetics of formation of the (HRP-CO) complex in 0.2 M Tris, pH 7.5 in 40% EGOH at different pressures. Inset: dependence of k_{obs} on CO concentration. Temperature: −10°C. Adapted from [48]

pressure can be a very useful physical parameter in mechanistic studies of such a process [50]. Moreover, although even in inorganic chemistry only few results are presently available, they indicate definite trends and shows promising features. My coworkers and I have used a thermodynamic approach for the study of enzymatic electron-transfer processes. Different terms contribute to an observed activation volume ($\Delta V^{\neq}_{obs}$), i.e. ,chemical ($\Delta V^{\neq}_{chem}$), conformational ($\Delta V^{\neq}_{conf}$) and solvation terms ($\Delta V^{\neq}_{sol}$):

$$\Delta V^{\neq}_{obs} = \Delta V^{\neq}_{chem} + \Delta V^{\neq}_{conf} + \Delta V^{\neq}_{solv} \tag{8.3}$$

For the two reactions studied, the chemical term $\Delta V^{\neq}_{chem}$ pertains to the electron transfer itself where we have used, for rate-constant determinations, the following minimum reaction scheme:

$$A_{red} + B_{ox} \overset{K_1}{\leftrightarrow} [A_{red} - B_{ox}] \overset{k_2}{\rightarrow} [A_{ox} - B_{red}] \tag{8.4}$$

where a collision complex is formed in a first step followed by the electron transfer itself and further steps. The observed rate constant is given by:

$$K_{obs} = k_2 \cdot K_1 \cdot [A_{red}] / (K_1 \cdot [A_{red}] + 1) \tag{8.5}$$

HAO Reaction

The reaction of reduced HAO with oxidized mammalian cytochrome *c* is a model reaction for the *in vivo* periplasmic physiological electron transfer from reduced HAO to tetraheme cytochrome *c*554 in *Nitrosomonas europaea*. The reaction has been studied at various temperatures and ionic strengths where the activation volume appeared to be independent of the temperature but strongly dependent on the ionic strength: −25 and −5 ml mol^{-1} at 20 and 420 mM NaCl, respectively. At low ionic strength, the negative activation volume suggests the involvement of an outer-sphere process, as seen for the model reaction of cytochrome *c* with ascorbic acid [51]. At this ionic strength, the contribution of the conformation ($\Delta V^{\ddagger}_{conf}$)and solvation ($\Delta V^{\ddagger}_{sol}$)terms are small. With increasing ionic strength, the outer-sphere process term $\Delta V^{\ddagger}_{chem}$ is compensated by the solvation term $\Delta V^{\ddagger}_{sol}$ having an opposite positive sign. It should be noted that at high ionic strength, where the complex between the two proteins is weak, the observed activation volume is a macroscopic thermodynamic response value close to the expected contribution of solvent reorganization calculated at room temperature [52].

Methanol Dehydrogenase Reaction

For the reaction of methanol dehydrogenase (MDH) with its natural electron acceptor, the cytochrome c_L, both isolated from *Methylophaga marina*, the activation volume was found to be very sensitive both to the temperature and to the ionic strength: at 4°C and 20 mM NaCl, the activation volume was 122 ml mol^{-1} and, at 25°C, 54 and 70 ml mol^{-1} for 10 and 110 mM NaCl, respectively. The Arrhenius plot at atmospheric pressure showed a break at 13°C [53]. In the model reaction of cytochrome c_L with ascorbic acid, the activation volume did not change with temperature or ionic strength, whereas a break could also be observed in the Arrhenius plot at 13°C. The reaction of MDH with Wurster's blue (TMPD) showed no variation of the activation volume with temperature but a dependence on the ionic strength and a linear Arrhenius plot. These finding were interpreted in terms of conformational changes occurring on the cytochrome c_L level. The break observed in the Arrhenius plot suggests the existence of two different conformations. Furthermore, the measurement of the simple model reactions allowed the calculation of a theoretical activation volume which was smaller than the observed value. This discrepancy was interpreted in terms of specific interactions of the two proteins. This specific interaction was also shown by the much smaller activation volume which was measured when a non-specific cytochrome was used. Other experiments are now in progress to clarify these observations.

In regard to these two electron-transfer reactions, HAO and MDH, using the HP-VT-SF apparatus to measure rate constants of electron transfer, my coworkers and I found a very different behavior. In the first case, the reaction of HAO with cytochrome *c*, the changes in the activation volume can be interpreted in terms of solvation changes, whereas in the second example, the reaction of MDH with cytochrome c_L, the activation volume changes considerably with temperature, involving modifications of conformation. With these results, it can be pointed out that the combined use of pressure, temperature and ionic strength is a powerful tool for the investigation of electron-transfer reactions, but it necessitates the use of a stopped-flow apparatus, due to the time constant of the reactions observed. For these reactions, depending on the experimental conditions, k_{obs} could be as fast as 5 s^{-1}, a value which represents a reaction half life-time of 100 ms, a value only accessible using a stopped-flow device (for the case of the electron-transfer reaction HAO cytochrome *c* at 20°C and 20 mM NaCl). However, a clear interpretation of the activation volume for reactions involving biochemical compounds (for the case of electron transfer), is still a problem, as the overall reaction volumes are difficult to quantify.

8.5 Conclusions

The stopped-flow method which has been used for many years in normal conditions [54, 55] cannot be used with all enzyme systems. With it, the system under study must give some optical signal. Another method, the rapid-flow quench, is available, permitting the mixing of two compounds and the analysis of the reaction products when there is not any specific physical signal. It is a sampling technique which has been adapted to sub-zero temperature conditions [2, 56] and to high-pressure experiments [57].

The high-pressure-variable-temperature method is rather new, and few laboratories have the appropriate apparatuses. Why? The main reason is not conceptual, but technical. The apparatus is difficult to use relative to the conventional stopped-flow apparatus. It may be that because the market is limited no commercial devices are developed commercially. The machines actually working are either prototypes or laboratory-made systems. However, the potential is large for the study of enzyme (and chemical) reactions, and to record kinetics without artifacts. Two compounds mix rapidly in set conditions, i.e., at a given temperature and pressure, only these kinds of mixer are accurate.

What improvements can be made to the technique? Clearly, it would be desirable to improve the time resolution. For this, one could take into consideration the continuous-flow capillary mixing apparatus, originally designed 13 years ago [58], but which has been improved very recently with significant advances in mixer design, detection method and data analysis [59]. Test reactions indicate that the mixing is completed within 15 μs of its initiation and that the dead-time of the measurement is 45 μs, which represents an about 30-fold improvement in time resolution over conventional stopped-flow instruments operating at atmospheric pressure. Can this be adapted to extreme conditions ? If the answer is positive, I think that a new future for the understanding of enzyme reactions will be open for mechanistic, thermodynamic and biotechnological studies [60, 61].

Acknowledgements. I am indebted to Professor P. Douzou, the pioneer of cryobiochemistry, for introducing me to the world of extreme conditions. I thank F. Travers and T. Barman for stimulating discussions and my colleagues N. Bec, C. Clery, J. Frank, I. Heiber-Langer, K. Heremans, A. B. Hooper, J. Kornblatt, M. J. Kornblatt, N. Klyachko, S. Kunugi, R. Lange, P. Lemay, P. Masson, V. V. Mozhaev, R. V. Rariy, who carried out most of the experimental work related in this review. I should also like to warmly thank J.-L. Saldana, the laboratory engineer, who played a large part in the development of the HP-VT-SF.

This work was partially financed by the INTAS-93-38 grant project and COST D6.

References

8.1 P. Douzou: *Cryobiochemistry, an Introduction* (Academic Press, London 1977)
8.2 F. Travers, T. Barman, Biochimie **77**, 937 (1995)
8.3 G. S. Greaney, G. N. Somero, Biochemistry **24**, 5322 (1979)
8.4 C. Balny, F. Travers, T. Barman, P. Douzou, Proc. Natl. Acad. Sci. USA **82**, 7495 (1995)
8.5 E. Morild, Adv. Protein Chem. **34**, 93 (1981)
8.6 C. Balny, P. Masson, F. Travers, High Press. Res. **2**, 1 (1989)
8.7 M. Groß, R. Jaenicke, Eur. J. Biochem. **221**, 617 (1994)
8.8 V. V. Mozhaev, K. Heremans, J. Frank, P. Masson, C. Balny, Proteins, Struc. Func. Gen. **24**, 81 (1996)
8.9 R. H. Austin, K. W. Beeson, L. Eisenstein, H. Frauenfelder, I. C. Gunsalus, Biochemistry **14**, 5355 (1975)
8.10 H. A. Kramers, Physica **7**, 284 (1940)
8.11 M. Marden, G. Hui Bon Hoa, Eur. J. Biochem. **129**, 111 (1982)
8.12 P. Butz, K. O. Greulich, H. Ludwig, Biochemistry **27**, 1556 (1988)
8.13 S. D. Hamann, Aust. J. Chem. **37**, 867 (1985)
8.14 D. E. Koshland, Proc. Natl. Acad. Sci. USA **44**, 98 (1958)
8.15 J. L. Markley, F. Travers, C. Balny, Eur. J. Biochem. **120**, 477 (1981)
8.16 R. A. Greiger , C. A. Eckert, AIChE J. 766 (1970)
8.17 M. Sasaki, F. Amita, J. Osugi, Rev. Sci. Instr. **50**, 1104 (1979)
8.18 P. Smith, E. Beile, R. Berger, J. Biochem. Biophys. Methods **6**, 173 (1982)
8.19 Y. Tanigushi, A. Iguchi, J. Am. Chem. Soc. **105**, 6782 (1983)
8.20 K. Heremans, J. Snauwaert, J. Rijkenberg, Rev. Sci. Instr. **51**, 806 (1980)
8.21 C. Balny, J.-L. Saldana, N. Dahan, Anal. Biochem. **139**, 178 (1984)
8.22 C. Balny, J.-L. Saldana, N. Dahan, Anal. Biochem. **163**, 309 (1987)
8.23 R. van Eldik, D. A. Palmer, R. Schmidt, H. Kelm, Inorg. Chim. Acta **50**, 131 (1981)
8.24 H. Kihara, J. Synchrotron Rad. **1**, 74 (1994)
8.25 H. Tsurata, T. Nagamura, K. Kimura, Y. Igarashi, A. Kajita, Z. X. Wang, K. Wakabayashi, Y. Amemiya, H. Kihara, Rev. Sci. Instrum. **60**, 2356 (1989)
8.26 P. Bugnon, G. Laurenczy, Y. Ducommun, P.-Y. Sauvageat, A. E. Merbach, R. Ith, R. Tschanz, M. Doludda, R. Bergbouer, E. Grell, Anal. Chem. **68**, 3045 (1996)
8.27 P. Bugnon, M. Doludda, E. Grell, A. E. Merbach in *High Pressure Research in the Biosciences and Biotechnology*, ed. by K. Heremans (Leuven University Press, Leuven 1997) pp.143
8.28 S. Kunugi, M. Kitayaki, Y. Yanagi, N. Tanaka, R. Lange, C. Balny, Eur. J. Biochem. **248**, 567 (1997)
8.29 I. Heiber-Langer, N. Banzet, J.-L. Saldana, P. Lemay, C. Balny, Biotech. Tech. **7**, 243 (1993)
8.30 P. Degraeve, P. Lemay, Enz. Microbial Tech. **20**, 32 (1997)
8.31 M. J. Kornblatt, R. Lange, C. Balny, Eur. J. Biochem. **251**, 775 (1998)
8.32 R. Lange, J. Frank, J.-L. Saldana, C. Balny, Eur. Biophys. J. **24**, 277 (1996)
8.33 P. Masson, C. Balny, Biochim. Biophys. Acta **874**, 9 (1986)
8.34 C. Clery, P. Masson, I. Heiber-Langer, C. Balny, Biochim. Biophys. Acta **1159**, 295 (1992)

8.35 C. Clery, I. Heiber-Langer, L. Channac, L. David, C. Balny, P. Masson, Biochim. Biophys. Acta **1250**, 19 (1995)
8.36 P. Masson, C. Balny, Biochim. Biophys. Acta **954**, 208 (1988)
8.37 P. Masson, C. Balny, Biochim. Biophys. Acta **1041**, 223 (1990)
8.38 D. McKillop, M. Geeves, C. Balny, Biochem. Biophys. Res. Comm. **180**, 552 (1991)
8.39 N. L. Klyachko, A. V. Levashov, A. V. Khmelnitsky, L. Yu, K. Martinek, in *Kinetics and Catalysis in Microheterogeneous Systems,* ed. by M. Gratzel, K. Kalyanasundaram, Vol. 38 (1991) pp.135
8.40 M. Tuena de Gomez-Puyou, A. Gomez-Puyou, Crit. Rev. Biochem. Mol. Biol. **33**, 53 (1998)
8.41 V. V. Mozhaev, N. Bec, C. Balny, Biochem. Mol. Biol. Inter. **34**, 191 (1994)
8.42 C. Clery, N. Bec, C. Balny, V. V. Mozhaev, P. Masson, Biochim. Biophys. Acta **1253**, 85 (1995)
8.43 R. V. Rariy, N. Bec, J.-L. Saldana, S. Nametkin, V. V. Mozhaev, N. L. Klyachko, A. V. Levashov, C. Balny, FEBS Lett. **364**, 98 (1995)
8.44 R. V. Rariy, N. Bec, N. L. Klyachko, A. V. Levashov, C. Balny, Biotech. Bioeng. **57**, 552 (1998)
8.45 J. Frank, C. Balny, Biochimie **73**, 611 (1991)
8.46 C. Balny, F. Travers, T. Barman, P. Douzou, Eur. Biophys. J. **14**, 375 (1987)
8.47 C. Balny, A. B. Hooper, Eur. J. Biochem. **176**, 273 (1988)
8.48 C. Balny, F. Travers, Biophys. Chem. **33**, 237 (1989)
8.49 R. Lange, I. Heiber-Langer, C. Bonfils, I. Fabre, M. Negishi, C. Balny, Biophys. J. **66**, 89 (1994)
8.50 R. van Zldik, High. Press. Res. **6**, 251 (1991)
8.51 I. Heiber-Langer, A. B. Hooper, C. Balny, Biophys. Chem. **43**, 265 (1992)
8.52 K. Heremans in *Inorganic High Pressure Chemistry*, ed. by R. van Eldik (Elsevier, Amsterdam 1986) pp. 339.
8.53 I. Heiber-Langer, C. Clery, J. Frank, P. Masson, C. Balny, Eur. Biophys. J. **21**, 241 (1992)
8.54 B. Chance, Anal. J. Franklin Inst. **229**, 455 (1940)
8.55 Q. H. Gibson, L. Milnes, Biochem. J. **91**, 161 (1964)
8.56 T. Barman, F. Travers, Meth. Biochem. Anal. **31**, 1 (1985)
8.57 G. Hui Bon Hoa, G. Hamel, A. Else, G. Weill, G. Hervé. Anal. Biochem. **187**, 258 (1990)
8.58 P. Regenfuss, R. M. Clegg, M. J. Fulwyler, F. J. Barrantes, T. M. Jovin, Rev. Sci. Instrum. **56**, 283 (1958)
8.59 M. C. R. Shastry, S. D. Luck, H. Roder, Biophys. J. **74**, 2714 (1998)
8.60 C. Balny, R. Hayashi, K. Heremans, P. Masson, (eds.) *High Pressure and Biotechnology* Vol. 224 (J. Libbey, London 1992)
8.61 R. Hayashi, C. Balny (eds.) *High Pressure Bioscience and Biotechnology,* Progress in Biotechnology Series Vol. 13 (Elsevier, Amsterdam 1996)

9 Pressure Effects on the Intramolecular Electron Transfer Reactions in Hemoproteins

Yoshiaki Furukawa, Yoichi Sugiyama, Satoshi Takahashi, Koichiro Ishimori, and Isao Morishima

Department of Molecular Engineering, Graduate School of Engineering, Kyoto University, Kyoto 606-8501, Japan
E-mail: morisima@mds.moleng.kyoto-u.ac.jp

Abstract. The activation volumes ($\Delta V^{\neq}$) for intramolecular electron transfer (ET) reactions in Ru-modified cytochrome b_5 (Ru-cytb_5) and Zn-porphyrin substituted myoglobins (Ru-ZnMb) were determined to investigate the pressure effects on the ET pathway; this provided us with new insights into the fluctuation—controlled ET reaction mechanism in proteins. Ru-cytb_5, in which the Ru complex was attached at His26, exhibited a large negative activation volume for the ET reaction, although the ET reactions in previous Ru-cytb_5 mutants were almost insensitive to pressurization [14]. The pathway analysis revealed that our Ru-cytb_5 system has a long 'through-space' process on the electron pathway, while the pathway for the previous system consisted of only a 'through-bond' process with covalent bonds or included a single short 'through-space' process. The pressure effects on the ET reaction, therefore, depend highly on the flexibility of the pathway, a flexible 'through-space' or rigid 'through-bond' process, and the ET reaction mediated by the 'through-space' process would be more susceptible to pressurization, implying that the structural fluctuation would preferentially affect the 'through-space' ET reaction. To confirm the prominent effects of pressure on the 'through-space' ET reaction, we also examined the pressure dependence of Ru-ZnMbs, the ET pathways of which have some flexible and long 'through-space' processes. In spite of the same donor–acceptor (D–A) pair, three Ru-ZnMbs, for which D—A distances for the ET reactions are 12.7 (His48Mb), 15.5 (His83Mb) and 19.3 Å (His81Mb), showed different activation volumes for the ET reactions: −1.6 (His83Mb), +3.7 (His81Mb), +6.5 $cm^3 mol^{-1}$ (His48Mb). Since the Marcus theory indicates that the acceleration and deceleration of the ET reaction rates in Ru-ZnMbs would correspond to the shortening and stretching of the D–A distance, respectively, some of the flexible 'through-space' processes in Ru-ZnMbs would reduce the D–A distance and the others would increase the distance by pressurization. In other words, the structural fluctuation affecting the 'through-space' ET process is not isotropic in protein, and local fluctuations count as one of the factors regulating the protein ET reactions.

9.1 Introduction

Electron transfer (ET) reactions play essential roles in several important biological processes, including photosynthesis and respiration [1]. In recent years, interest in the understanding of ET reactions for metal enzymes has significantly increased, since new theories and techniques which provide us with various pieces of information of the factors regulating the ET reaction in proteins have become available [2, 3]. It is in this respect that the effects of protein dynamics on ET processes in protein have received considerable attention [4–8]. As previously reported for many protein systems [2, 3], the static structure of a protein, deduced from crystallographic analysis, would not be enough to characterize the ET reactions. In addition to the static structure of proteins, the contribution of dynamic motions of proteins would be essential for the ET reactions in proteins, because the structural fluctuation by the thermal motions of proteins affects the relative positions of the redox centers and amino acid residues on the ET pathway [9, 10]. Aquino et al. [7] theoretically examined the effects of dynamic fluctuations on the ET reaction between cytochrome c_2 and the photosynthetic reaction center, and suggested that the specific protein fluctuation could enhance the ET rates. Daizadeh et al. [8] also carried out computer simulations to estimate the effects of protein dynamics on ET reactions, and proposed that vibrational motion of the ET pathway regulates the protein ET reaction. Although these theoretical studies predict that the dynamic fluctuation plays a substantial role in the ET reaction, only a few experimental studies paying attention to the effects of dynamic fluctuation on the ET process in proteins have been reported.

To examine the effects of fluctuation on ET reactions, our and other groups have previously investigated the pressure dependence on the intramolecular ET rates for myoglobin (Mb) [11], cytochrome *c* (cyt *c*) [12, 13] and b_5 (cyt b_5) derivatives [14]. Pressure has been considered to perturb the dynamic fluctuation in proteins [15–17]; it increases material density and suppresses the dynamic fluctuation. In cytochromes, the ET rates were accelerated by pressurization or were almost independent of pressure, which has been explained by distance contractions between donor and acceptor sites under high pressure [12, 14]. In sharp contrast to cytochromes, our preliminary study has revealed that a significant deceleration of the ET rates was observed in Mb derivatives, which could not be explained by the distance contraction [11]. Therefore, we introduced the effects of the structural fluctuation into the Marcus equation and proposed the protein fluctuation- controlled ET mechanism.

In this study, we examined the pressure effects on the intramolecular ET rates in two kinds of hemoprotein systems to discuss the contribution of the structural fluctuation to the ET reactions in proteins. One of the two model protein systems is the intramolecular ET reaction system constructed by ruthenium-modified cyt b_5 (Ru-cytb_5). cyt b_5 is one of the ET proteins, the heme of which is coordinated by two histidine residues [18]. The ET reactions in Ru-cytb_5 under high pressure have already been studied, and small or no pressure dependence was reported [14]. The ET pathways in the previous system were mainly constructed with covalent bonds, referred as to the 'through-bond' process, and supposed to be resistant to pressurization [12, 14]; these do not include ET processes mediated by more flexible interactions such as hydrogen bonds ('through-space' ET). Such rigid and pressure-resistant ET pathways would result in a very small pressure dependence. It is, therefore, one of the current discussion points to introduce the flexible 'through -space' ET process into the Ru-cytb_5 system, which would enhance the pressure dependence of the ET reaction.

In order to get further insights into the significance of the 'through-space' process in the pressure dependence of the ET reaction, another system, ruthenium-modified, zinc-substituted myoglobin (Ru-ZnMb) was utilized. Although the ET reactions in Ru-ZnMb are also induced by laser irradiation, they are quite different from those in Ru-cytb_5, in that the ET pathway for Ru-ZnMb contains some long and flexible 'through-space' processes mediated by hydrogen bonds. Since the bond angle and length for hydrogen bonds would be more susceptible to structural fluctuation rather than those for covalent bonds, effects of perturbation in protein fluctuation induced by pressurization on the ET pathway would be different from those in the Ru-cytb_5 model system, leading to the different pressure dependence of the ET rate constants.

By utilizing high-pressure laser flash photolysis, we measured the ET rates of the two model systems, Ru-ZuMb and Ru-cytb_5, under high pressure, up to 100 MPa, and compared their pressure dependence. Particularly, we focused on the significance of the 'through-space' processes in the ET reaction, and discussed the effects of structural fluctuation on the protein ET process in more detail.

9.2 Materials and Methods

9.2.1 Preparation of the Ruthenium-modified Proteins

Preparation of Bis(2,2'-bipyridine) Carbonatoruthenium(II). Bis(2,2'-bipyridine) carbonatoruthenium(II), $Ru(bpy)_2CO_3$, (bpy: bipyridine) was prepared as reported by Johnson et al. [19]. The solid $Ru(bpy)_2Cl_2 \cdot 2H_2O$ (1.0 g) was suspended in 75 mL of deaerated water and refluxed under Ar purging for 15 min. After addition of sodium carbonate (3.3 g), the solution was again refluxed for 2 h. The reaction mixture was filtered and stood at 4°C for overnight, yielding purple, needle-shaped crystals.

Preparation of Aquopentaammineruthenium(II). Aquopentaammine-ruthenium(II), $Ru(NH_3)_5(H_2O)^{2+}$, was prepared by reduction of $[Ru(NH_3)_5Cl]Cl_2$ as previously reported [20]. $[Ru(NH_3)_5Cl]Cl_2$ was reduced by zinc amalgam and the product was precipitated as the hexafluorophosphate salt and stored in a vacuum.

Preparation of Zinc Mesoporphyrin IX Diacid. The zinc(II) complex of mesoporphyrin IX diacid (ZnP) was prepared by the method of Adler et al. [21]. The purity of the product was checked by thin-layer chromatography and ^{1}H NMR spectra (results not shown). No additional bands and signals were observed. The ZnP was stored in a foil-wrapped vial at –80°C.

Preparation of Ruthenium-Modified Rat Hepatic Cytochrome (cyt) b_5. cyt b_5 was prepared as previously described [22]. The purified cyt b_5 gave a single band on SDS polyacrylamide gels and an A_{412}/A_{280} ratio of 5.8.

The $Ru(bpy)_2$(imd) (imd: imidazole) modification of cyt b_5 was followed as previously reported [23]. Although rat hepatic cyt b_5 possesses four exposed histidyl residues (positions 15, 26, 27, and 80), one of the histidine residues, His26, has been proposed to have preferential reactivity to the ruthenium complex

[24—26]. 50-fold excess $Ru(bpy)_2CO_3$ was added to the cyt b_5 solution (2 mg/mL, ca. 10 mL) in 50 mM Tris-HCl, pH 8.0 in an Ar atmosphere, and the reaction mixture was left stood for 17 h at 4°C [23], which was followed by the addition of 50-fold excess imidazole and overnight incubation in Ar at 4°C [23]. After removal of unreacted ruthenium complex and purification by anion-exchange column chromatography, two components were fractionated. The main fraction showed an electronic absorption spectrum similar to those of $Ru(bpy)_2$(dmbpy)-modified cyt b_5 (dmbpy: dimethyl bipyridine) [27] and $Ru(bpy)_2$(imd)-modified cyt *c* [23], which can be assigned to mono-modified $Ru(bpy)_2$(imd)-(His26)-cytb_5 (Fig. 9.1A).

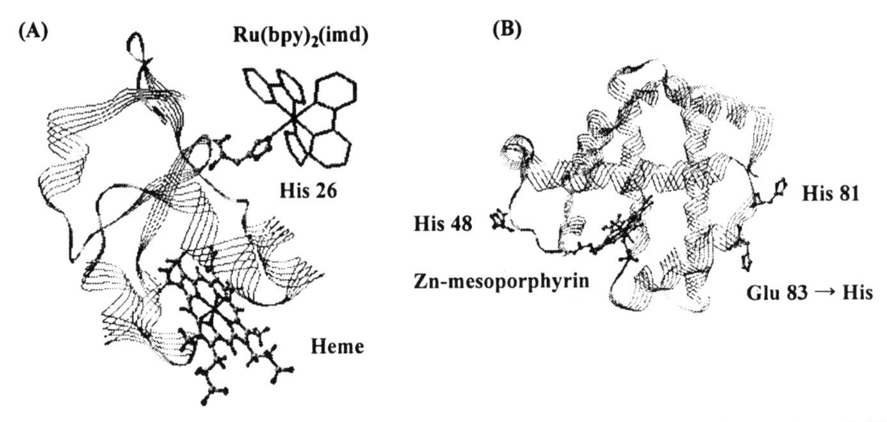

Fig. 9.1. The positions of the Ru-modified histidines in (**A**) rat hepatic cyt b_5 and (**B**) human myoglobin

Preparation of Ru-Modified, Zinc-substituted Myoglobins. Human Mb used in this study has a mutation at Cys 110 (Cys → Ala) to avoid dimer formation during the purification [28]. The procedures for the site-directed mutagenesis and protein purification are described in a previous paper [29]. Human Mb has two accessible histidines on its surface (His48 and His81) against the Ru modification [29]. To construct the singly $Ru(NH_3)_5$-modified Mbs as illustrated in Fig. 1B, the H48N, H81N, and H48N/H81N/E83H human Mb genes were constructed, in which the accessible histidine on the protein surface is limited to His81, His48 and His83, respectively. The donor–acceptor (D–A) distance (distance between Ru complex and ZnP), which is calculated from edge to edge, is 15.5, 19.3, and 12.7 Å for H48N/H81N/E83H Mb (His83Mb), H48N Mb (His81Mb) and H81N Mb (His48Mb), respectively.

After reconstitution of the myoglobin mutant with a heme and its purification, the single histidine residue exposed to the solvent was modified by the Ru complex $Ru(NH_3)_5(H_2O)^{2+}$. The reaction condition and purification for the modification were followed by the procedure reported previously [29].

The heme was removed from Ru-modified Mb to obtain the apo-form of myoglobin [30]. The apoprotein solution was mixed with ZnP to form Ru-ZnMb. The crude sample was purified by cation-exchange column chromatography and sample purity was confirmed by an A_{414}/A_{280} ratio of ca. 16 as previously reported [31]. All manipulations of ZnMb were performed in the dark. The modification of the Ru complex to ZnMb can be confirmed by native-PAGE [32].

9.2.2 Measurements of Flash Photolysis Under High Pressure

High-pressure laser flash photolysis measurements were performed using a steel pressure cell and its inner capsule made of quartz [33]. The experiments under high pressure were performed in 100 mM Tris-HCl, pH 7.4 (Ru-ZnMb), and in 50 mM Tris-HCl, pH 8.0 (Ru-cytb_5). pH of Tris buffer has been shown to be independent of pressure up to 200 MPa [34].

9.3 Results

9.3.1 Electron Transfer in Ruthenium-Modified Cytochrome b_5

Kinetics at Ambient Pressure. The ruthenium complex used in the present Ru-cytb_5 experiments exhibits photoredox behavior very similar to the well-characterized $Ru(bpy)_3^{2+}$ complex [35–37]. A short laser pulse (355 nm; 10 ns pulse width) forms an excited state of the Ru complex which can transfer an electron to the heme Fe(III) to produce Ru(III) and Fe(II), the k_1 process (Scheme 9.1). This reaction is followed by a rapid ground-state ET reaction which returns the system to the initial Ru(II) and Fe(III) redox state, the $k_{ET(b_5)}$ process (Scheme 9.1).

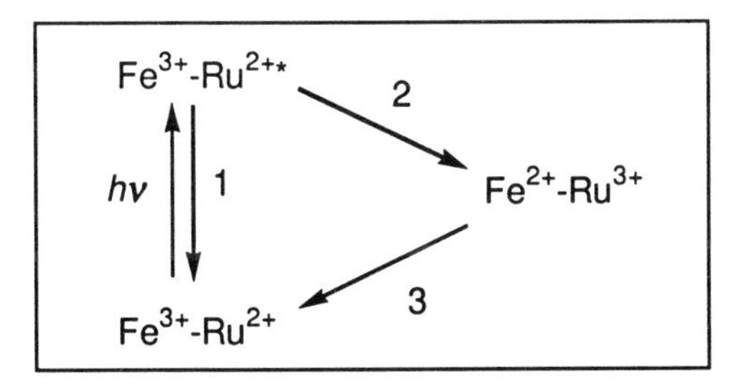

In this ET model system, the intrinsic decay process from the excited state to the ground state of the Ru complex, k_d process (Scheme 9.1), is very fast (k_d = 1.4 × 10^7 s^{-1}) (data not shown) [38], which prevents us from measuring the ET reaction from the excited Rucomplex to the Fe(III) (k_1 process) [39–43] . To obtain the ET rate from Fe(II) to Ru(III), $k_{ET(b_5)}$, the absorbance changes at 424 nm were monitored [14] (Fig. 9.2). Although slight systematic deviations were detected for the residuals, the observed decay kinetics at 424 nm could be fitted to a single exponential function in the range of the sample concentration under 2 μM and the rate constant was estimated as 310 s^{-1}, which was independent of the sample

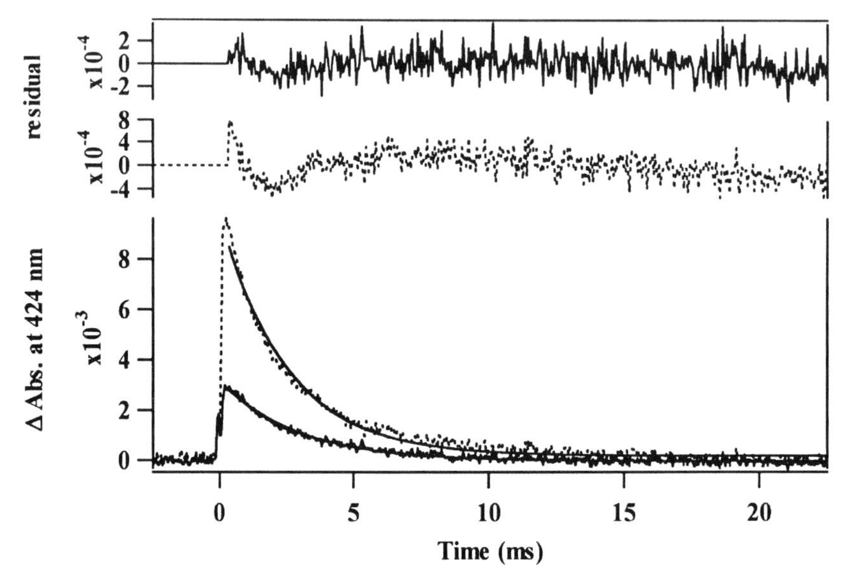

Fig. 9.2. Transient absorbance changes at 424 nm obtained after laser excitation (355 nm; 10ns pulse width) in Ru-cytb_5. The sample concentrations are ca. 2 μM (*solid line*) and ca. 5 μM (*dotted line*). The residuals from the single exponential fitting are shown on the top of each decay signal. The experiments were performed at 298 K, and 0.1 MPa in 50 mM Tris-HCl, pH 8.0

concentration. In the higher concentration of the sample (>5μM), another fast phase probably due to the intermolecular ET clearly appears in the decay (Fig. 9.2, dotted curve).

Kinetics at High Pressure. The ET rate constants in Ru-cytb_5 were measured at 0.1, 20, 50, and 100 MPa. The absorbance decay for the ET reaction at 100 MPa is shown in Fig. 9.3 (solid curve). The sample concentration was ca. 2 μM. Even under high pressure, the residuals from a single exponential fitting were almost random (solid line in the top of Fig. 9.3). The rate constants, $k_{\mathrm{ET}(b_5)}$, under various pressure are summarized in Table 9.1. In order to quantitatively characterize the pressure dependence of the ET reaction in Ru-cytb_5 derivative, we estimated the activation volume ($\Delta V^{\neq}$) for the ET reaction, which is defined by the following equation [17]:

$$\left(\frac{\partial \ln k}{\partial P}\right)_T = -\frac{\Delta V^{\neq}}{RT}, \tag{9.1}$$

where k is the rate constant, P is the pressure, R is the gas constant, and T is the temperature. As shown in Fig. 9.4, the plot of the natural logarithm of the rate constants against the pressure was linear and $\Delta V^{\neq}$ was determined by the linear least-squares fitting. $\Delta V^{\neq}$ for the ET reaction in Ru-cytb_5 was -13 cm^3 mol^{-1}.

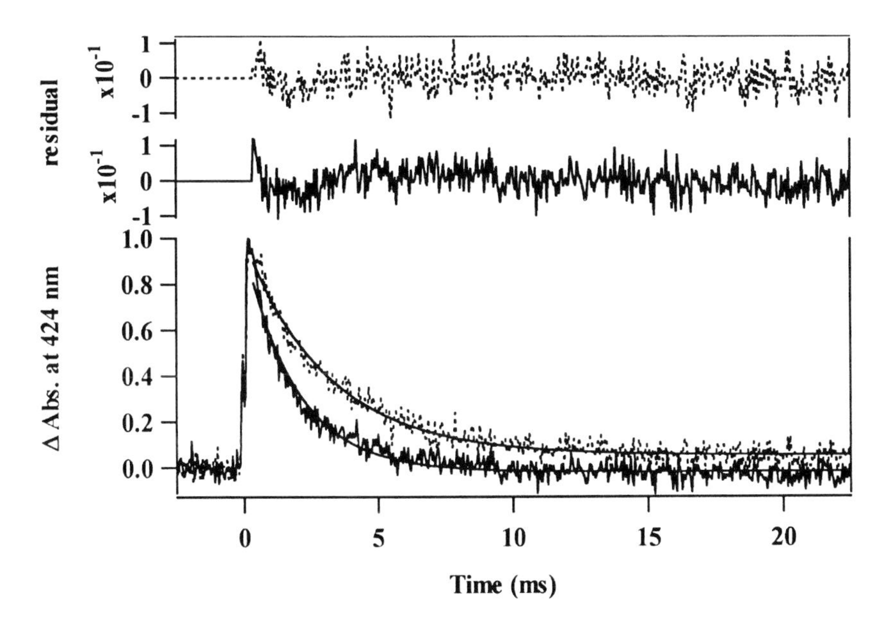

Fig. 9.3. Transient absorbance changes at 424 nm obtained after laser excitation (355 nm; 10ns pulse width) in Ru-cytb_5 under 0.1 MPa (*dotted line*) and 100 MPa (*solid line*). The residuals from the single exponential fitting are shown on the top of each decay signal

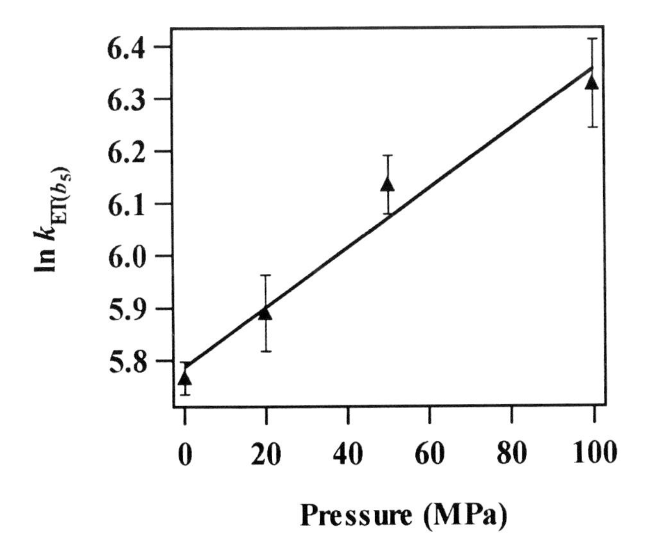

Fig. 9.4. Pressure dependence of intramolecular ET rates in Ru-cytb_5. Linear least-squares fits are shown (*solid lines*)

Table 9.1. Pressure dependence of the ET rate constants of Ru-cytb_5 in 50 mM Tris-HCl, pH 8.0 at 298 K

Pressure (MPa)	$k_{ET(b_5)}$ (s^{-1})
0.1	320
20	360
50	460
100	560

9.3.2 Electron Transfer in Ruthenium-Modified, Zinc-Substituted Myoglobins

Kinetics at Ambient Pressure. The ET reactions in Ru-ZnMbs were also initiated by laser irradiation (532 nm; 10ns pulse width) of the sample solution. The decay of the triplet-excited state in ZnMb ($^3Zn^*Mb$) was monitored at 450 nm for the ET process in Scheme 9.2 [29]. The triplet decay rate, k_{obs}, for Ru-ZnMb after pulsed-laser excitation was the sum of the intrinsic decay rate, k_d, and the ET rate constant, $k_{ET(Mb)}$, from $^3Zn^*P$ to Ru^{3+}.

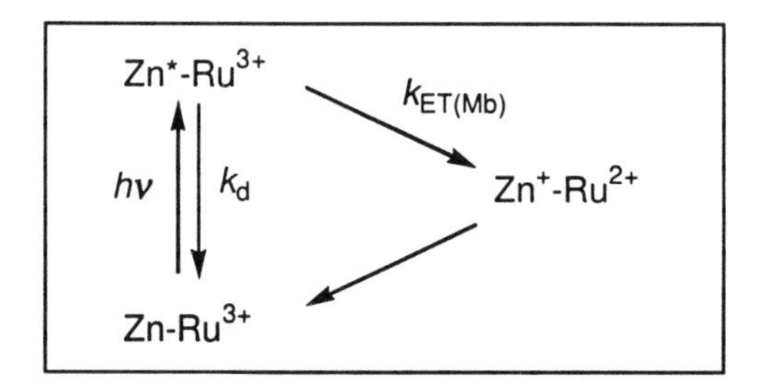

To estimate the intrinsic decay rate for $^3Zn^*Mb$, k_d, the absorbance change at 450 nm was followed in the Ru-unmodified ZnMb sample after the pulsed-laser excitation (Fig. 9.5, dotted curve). The rate constant (k_d), 50 s^{-1}, was obtained at 293 K in 100 mM Tris-HCl, pH 7.4, by fitting a single exponential function; this value is virtually the same as that previously reported [32]. In one of the Ru-ZnMbs, His81Mb, the modification by the Ru complex accelerated the decay rate, as illustrated in Fig. 9.5 (solid curve). Random residuals from a single exponential fitting (solid line in the top of Fig. 9.5) indicate that the decay of the triplet state in this Ru-ZnMb is monophasic. For the other Ru-ZnMbs, His81Mb and His83Mb, the kinetic data can also be fitted by a single exponential function. The rate constant (k_{obs}) obtained was as follows: 500 s^{-1}, 120 s^{-1} and 9.90×104 s^{-1} for His83Mb, His81Mb and His48Mb, respectively.

By subtracting the intrinsic decay rate constant, k_d, from the observed decay rate constants, k_{obs}, the ET rate constant ($k_{ET(Mb)}$) at 293 K was calculated: 450 s^{-1}, 70 s^{-1} and 9.90×10^4 s^{-1} for His83Mb, His81Mb and His48Mb, respectively [29, 30]. The ET rate constants, $k_{ET(Mb)}$, were independent of sample concentration (2 ~5 μM) in all derivatives, showing that the observed ET reactions correspond to the intramolecular ET reactions, not to the bimolecular ET reactions [29, 30].

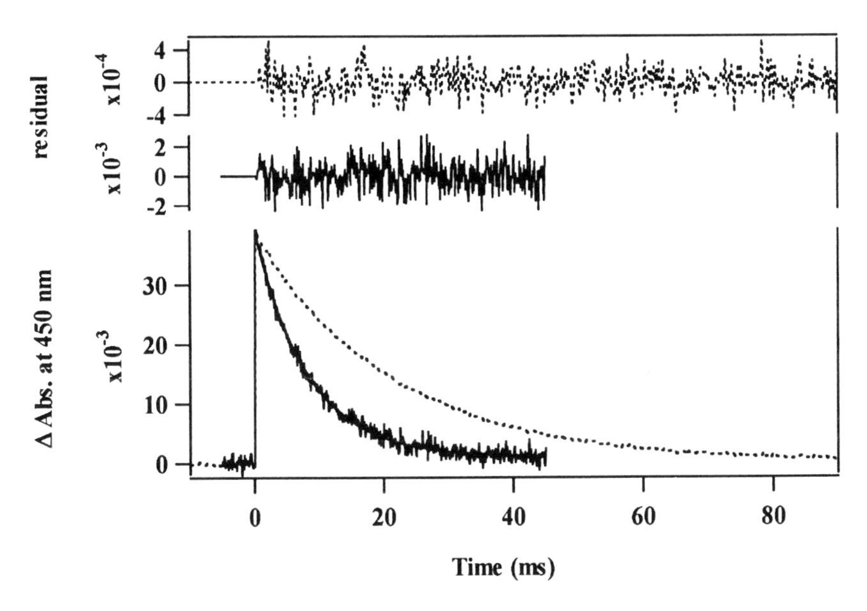

Fig. 9.5. Transient absorbance changes at 450 nm obtained after laser excitation (532 nm; 10ns pulse width) in unmodified ZnMb (*dotted line*) and His81Mb (*solid line*). The sample solutions contained ca. 5 μM of ZnMb derivatives dissolved in 100 mM Tris-HCl, pH 7.4. The residuals shown on the top of each decay signal were obtained by single exponential fitting

Kinetics at High Pressure. The ET rate constants in Ru-ZnMb ($k_{ET(Mb)}$) were measured up to 100 MPa at pressure intervals of 20 MPa. The typical absorbance decay for the ET reaction at 100 MPa for His48Mb are shown in Fig. 9.6 (solid curve). The residuals from a single exponential fitting were also random under high pressure (solid line in the top of Fig. 9.6). The rate constants, $k_{ET(Mb)}$, at 0.1 and 100 MPa for Ru-ZnMb derivatives are summarized in Table 9.2. In the ET reactions of the two Ru-ZnMbs His81Mb and His48Mb the rate constants, k_{ET}, were decelerated by pressurization, while a slight acceleration was found for the reaction rate of the other Ru-ZnMb, His83Mb. Fig. 9.7 shows a plot of the natural logarithm of the rate constants against the pressure. The pressure dependence of the rate constants was linear and $\Delta V^{\neq}$ were determined by the linear least-squares fitting. $\Delta V^{\neq}$ in each ET process are summarized in Table 9.2.

Table 9.2 The ET rate constants of Ru-ZnMbs under 0.1 and 100 MPa, and the activation volumes of each Ru-ZnMb

	$k_{ET(Mb)}$ at 0.1 MPa	$k_{ET(Mb)}$ at 100 MPa	$\Delta V^{\neq}$
	(s^{-1})	(s^{-1})	($cm^3\ mol^{-1}$)
His48Mb	9.9×10^4	8.0×10^4	$+6.5 \pm 1.0$
His81Mb	7.0×10^1	6.0×10^1	$+3.7 \pm 0.1$
His83Mb	4.5×10^2	4.9×10^2	-1.6 ± 0.2

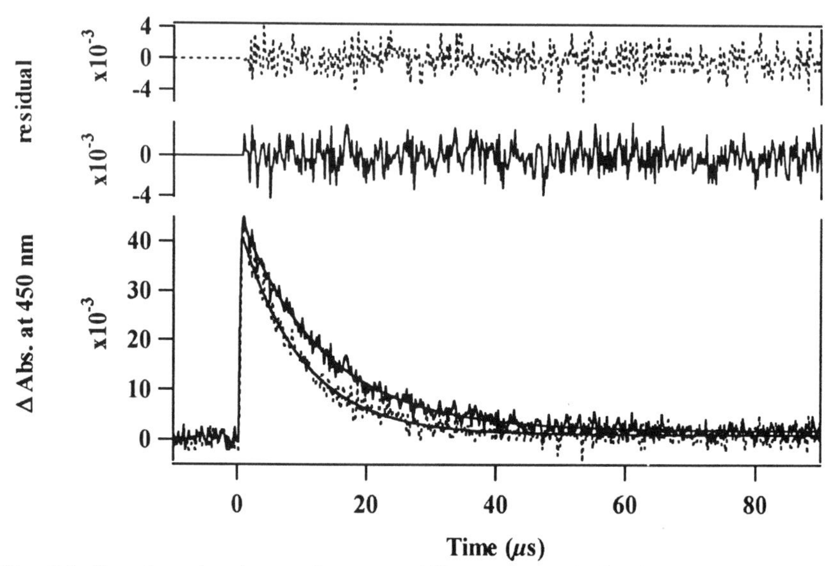

Fig. 9.6. Transient absorbance changes at 450 nm obtained after laser excitation (532 nm; 10ns pulse width) in His48Mb derivative under 0.1 MPa (*dotted line*) and 100 MPa (*solid line*). The residuals shown on the top of each decay signal were obtained by the single exponential fitting

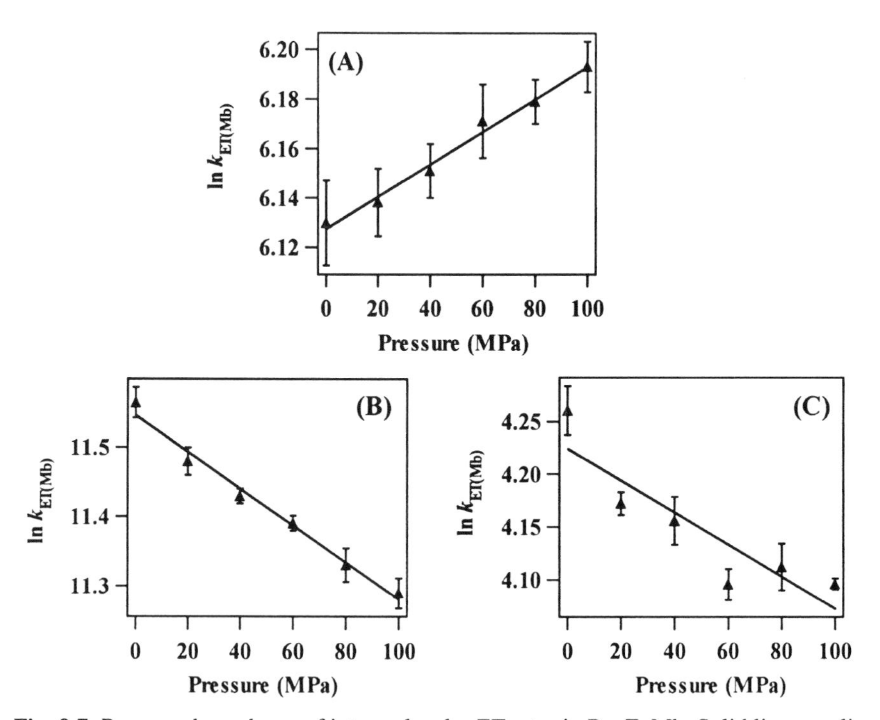

Fig. 9.7. Pressure dependence of intramolecular ET rates in Ru-ZnMb. Solid lines are linear least-squares fitting. (**A**) His83Mb, (**B**) His48Mb, (**C**) His81Mb

9.4 Discussion

9.4.1 Factors Regulating the Electron Transfer Reaction and Their Pressure Dependence

According to the semi-classical ET theory [1], the kinetics for long-range ET reactions have been shown to mainly depend on the following determinants: a D–A distance, d, a redox potential difference, ΔG°, a reorganization energy, λ, electronic coupling at the closest contact (3 Å) between donor and acceptor, $H_{AB}^{\ o}$. Rate constants for long-range protein ET reactions can be described by the following expression, the Marcus equation [1]:

$$k = \frac{4\pi^2 \left(H_{AB}^{\ 0}\right)^2}{h(4\pi\lambda RT)^{1/2}} \times \exp[-\beta(d-3)] \times \exp\left(-\frac{(\lambda + \Delta G^0)^2}{4\lambda RT}\right), \tag{9.2}$$

where β is the distance decay factor, which depends on the mediator between the redox centers. For the factors regulating ET rates, previous studies [12−14, 44] have shown that the reorganization energy, λ, the D–A distance, d, and redox potential difference, ΔG°, may be affected by pressurization.

One of these pressure-dependent parameters, the reorganization energy, λ, corresponds to the energy required to reorganize the solvent or the reactant molecules during the redox reactions, and mainly depends on the local environments around redox centers [2]. Although the estimation of the pressure-induced perturbation on the reorganization energy is not simple, the small contribution to pressure-induced alteration of the ET rates has been reported in the organic molecules. Chung et al. [45] have estimated no pressure effects on the reorganization energy up to 256 MPa by the photoinduced bimolecular ET rates between some anthracene derivatives and simple alkyl-substituted benzene donors. In the present ET reactions, all of the Ru-ZnMb derivatives have the Ru-modified histidine residues on the protein surface and no significant environmental differences of the incorporated zinc porphyrins were found, implying that the reorganization energy would almost be the same in these ET reactions. When the pressure dependence of these ET reactions is assumed to be governed by the pressure-induced perturbation on the reorganization energy, it is likely that the three Ru-ZnMb derivatives exhibit the similar pressure dependence of the ET reactions due to their similar reorganization energies. As clearly shown in our results, however, the pressure dependence of these ET reactions was quite different. In the ET reaction of Ru-cytb_5, a small contribution of the reorganization energy to the pressure dependence has been previously suggested [14]. It would, therefore, be a good approximation to state that the contribution of λ to pressure dependence of ET rates in Ru-modified proteins is small.

Another factor regulating the ET reaction, the redox potential for the D–A pair, has been considered to show significant pressure dependence. Although the activation volume for the ET reactions originated from the pressure-induced redox potential changes has not yet been clarified, cyclic voltumnmetry has revealed that the volume changes for the reduction of $Ru(NH_3)_6^{3+}$ and $Ru(NH_3)_5(isonicotinamide)^{3+}$ are 28.7 $cm^3 mol^{-1}$ [12] and 33 $cm^3 mol^{-1}$ [46], respectively, suggesting a significant contribution of the pressure-induced redox potential change to the activation volume for the ET reaction. However, the effects of the pressure-induced redox potential change would not be dependent on the position of the Ru complex, and the different pressure dependence observed for the Ru-ZnMb derivatives cannot be explained by the pressure-induced redox potential change. Thus, it would be reasonable to assume that the difference in the activation volume for the ET reactions using the same D–A pair can be ascribed to the different perturbation on the ET pathway by pressurization.

9.4.2 The Pathway for Electron Transfer in Ruthenium-Modified Cytochrome b_5

It is quite interesting that the Ru-cytb_5 in this study showed prominent acceleration of the ET reaction rate by pressurization, which is in sharp contrast to the results reported by Scott et al. [14]. Although the previous ET reaction in Ru-cytb_5 has a D–A pair similar to that in our system, the modification site of the Rucomplex on the protein surface is different between the two systems. In the previous Ru-cytb_5 systems, the Ru complexes were covalently attached to Cys65 and Cys73 of the T73C and T65C variants, respectively, while our system has the Ru complex at His26. On the basis of the pathway analysis, the ET reaction in the T65C variant, which is insensitive to pressurization, would be mediated only by covalent bonds and is a typical rigid 'through-bond' ET reaction. Although the another ET in the previous study, the ET from the heme iron to the Ru complex at Cys73, involves a 3.0 Å gap between hydrogen atoms on the α-carbon of Ser71 and a methyl group of the porphyrin in addition to 11 covalent bonds, the coupling through this ET pathway between the Ru complex and the heme would be very similar to the coupling in the T65C-modified Ru-cytb_5 [14]. Significant acceleration was observed for the ET reaction by pressurization in T73C-modified Ru-cyt$b$5, but the activation volume (–1.0 cm^3 mol^{-1}) was quite small, compared to that for His26-modified Ru-cytb_5 (–13 cm^3 mol^{-1}).

On the other hand, the ET pathway of His26-modified Ru-cytb_5 has a large gap (more than 4.4 Å) between hydrogen atoms on the δ-carbon of Leu25 and the α-*meso*-proton of the porphyrin. It is likely that pressure preferentially affects the large and flexible 'through-space' process in His26-modified Ru-cytb_5 and the D–A distance for the process is significantly condensed by pressurization, resulting in an enhanced pressure effect. To confirm the contribution of the 'through-space' process to the ET reaction in His26-modified Ru-cytb_5, we examined the temperature dependence of the ET reaction rates as shown in Fig. 9.8.

By combination of the Marcus equation [1, 47], which can be approximately expressed by (9.3), and temperature dependence of the ET rates, the electronic factor χ_{AB}, defined by $\chi_{AB} = \exp[-\beta(d-3)]$ in (9.2), can be estimated.

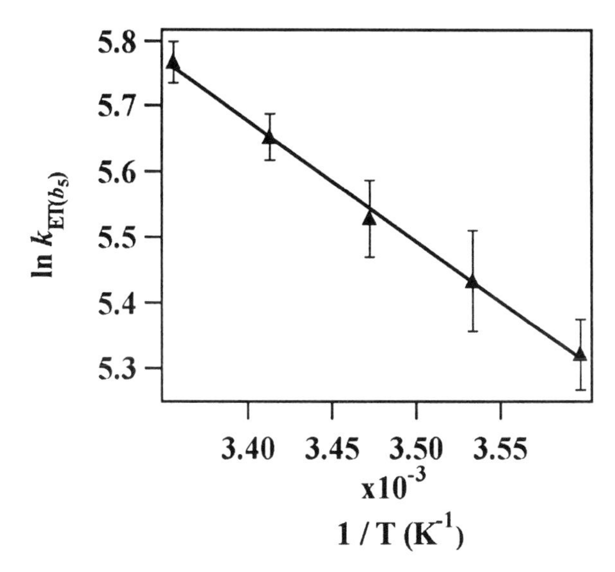

Fig. 9.8. The temperature dependence of the ET rates in Ru-cytb_5. Solid lines are linear least-squares fits

$$k_{ET} = 10^{13} \times \chi_{AB} \times \exp\left(-\frac{(\lambda + \Delta G^0)^2}{4\lambda RT} \right), \tag{9.3}$$

The observed χ_{AB} for the ET reaction in His26-modified Ru-cyt$b$5 was 1.7×10^{-4}, which is closed to the calculated χ_{AB} by the ET pathway including the 'through-space' process (1.8×10^{-4}), but far from that of only the 'through-bond' processes (1.0×10^{-2}). It is, therefore, concluded that the ET reaction in His26-modified Ru-cytb_5 has a flexible 'through-space' process on the ET pathway, unlike the previous T65C- and T73C-modified Ru-cytb_5, which is susceptible to pressurization, leading to the acceleration of the ET rate and a large negative activation volume. Such high sensitivity of the 'through-space' process to pressurization also implies that the structural fluctuation preferentially affects the flexible 'through-space' process in the ET reaction of protein and fluctuation-controlled ET regulation mechanism would be based on the 'through-space' process on the ET pathway.

9.4.3 The Pathway for Electron Transfer in Ruthenium-Modified, Zinc-Substituted Myoglobin

To get further insights into the contribution of the 'through-space' process to the ET in proteins, we also measured the pressure dependence of the ET reaction in Ru-ZnMb systems, in which some 'through-space' processes are involved in the ET pathway. As displayed in Fig. 9.9, the ET pathways for two of the three Ru-

ZnMb derivatives have already been reported [48]. One of the Ru-modified histidine residues, His81, is located near the end of the F-helix and an electron would be transferred along the F-helix to Ser92 at the other end of the helix (Fig. 9.9A). The ET reaction in His81Mb has three 'through-space' processes mediated by hydrogen bonds and 19 'through-bond' processes by the covalent bonds. All of the distances of the 'through-space' process are more than 2.7 Å. In the ET pathway of another Ru-ZnMb, His48Mb (Fig. 9.9B), a single 'through-space' process (2.21 Å) and 17 'through-bond' processes are included. An electron is transferred through the covalent bonds between His48 and Arg45, and then moves to the ZnP via the hydrogen bond between Arg45 and propionate of ZnP.

Unfortunately, quantitative analysis of the pressure-induced changes on the ET pathway has not yet been done, since the contribution of the redox potential change by pressurization to the activation volumes is still unclear. However, it can be safely concluded that the pressure effects on the ET reaction depend on the ET pathway, and a different ET pathway, particularly a difference in the distances and environments of the 'through-space' processes, results in different activation volumes. These observations, therefore, strongly suggest that the pressure effects on the flexible 'through-space' ET reaction are not isotropic and depend on the environmental structure of the 'through-space' processes. The heterogeneous structural changes by pressurization were also reported by high-resolution, two-dimensional NMR spectroscopy for BPTI and lysozyme under high pressure [49, 50].

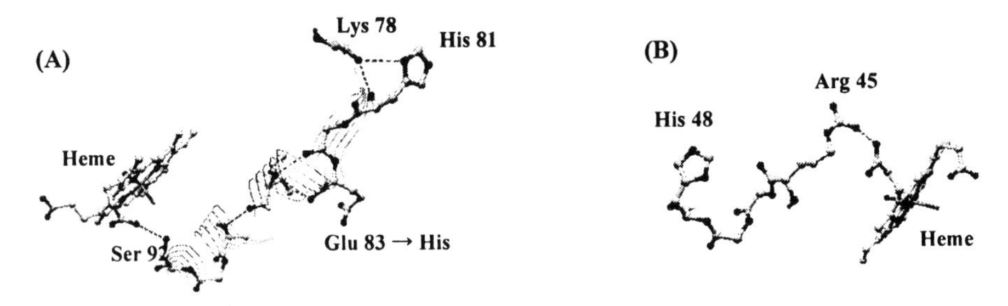

Fig. 9.9. Partial structure of human myoglobin showing heme ring and (**A**) residues 78–92, and (**B**) residues 45–48. Dotted lines show the hydrogen bonds

It should also be noted that the pressure dependence of the His83Mb derivative was opposite to that of the other two derivatives. As shown in Fig. 9A, the ET pathways in His81Mb and His83Mb appear to be quite similar, including three hydrogen bonds. His81 is one of the intrinsic histidine residues on the protein surface forming a hydrogen bond between N_δ of His81 and the carbonyl oxygen atom of the backbone (Lys78), which would fix the orientation of the imidazole ring of His81. Although His83 is located near His81 and would also be exposed to the solvent, position 83 is originally occupied by glutamic acid in native protein and no hydrogen bonds would be formed to fix the imidazole ring of His83. It is more plausible that the hydrogen bond observed for the side chain of His81 restricts the thermal fluctuation around His81. The local structural fluctuation would, therefore, affect the dynamic properties of the ET reaction.

As discussed in this and previous studies [11], the structural fluctuation in proteins may be one of the crucial factors regulating the ET reactions. Our preliminary high-pressure measurement [11] revealed positive activation volumes for the ET reaction in Ru-ZnMbs, and we introduced the Boltzmann factor into the Marcus equation to correlate the positive activation volume with the dynamic fluctuation of the protein. In the present analysis for the activation volume, however, the pressure-induced changes in the flexible 'through-space' process can contribute to the positive activation volume. It is, therefore, still premature to conclude that the introduction of the Boltzmann factor into the Marcus equation allows a description of the dynamic properties of the ET reaction in proteins, particularly the contribution of the dynamic fluctuation to the ET reaction, but the present high-pressure study clearly indicates that the effects of the dynamic fluctuation on the ET reactions are dependent on the ET pathway, a 'through-bond' process or 'through-space' process [51]. Also, it is suggested that the structural fluctuation is not isotropic in the protein structure, and subtle differences in the microenvironments around the ET pathway would lead to significant alteration of the dynamic aspects of the ET reaction. In our laboratory, to gain further insights into the fluctuation-controlled ET mechanism, extensive studies on temperature and viscosity dependence of the ET reaction are now underway.

Acknowledgements. We are grateful to Prof. S. G. Boxer and Dr. R. Varadarajan (Stanford University) for the gift of the expression vector of the human myoglobin gene, and to Prof. S. G. Sligar for the gift of the rat hepatic cytochrome b_5 gene. We are also greatly indebted to Mr. Manabu Teramoto for some protein preparations, and Dr. Shinichi Adachi at RIKEN for some mutant gene constructions. We are also obliged to Mr. Osamu Miyashita at Kyoto University for theoretical discussions.

Note added in proof. Some more detailed descriptions of the pressure dependence study of the ET in Ru-Mb have been published [51].

References

9.1 R.A. Marcus, N. Sutin: Biochim. Biophys. Acta **811**, 265 (1985)
9.2 J.R. Winkler, H.B. Gray: Chem. Rev. **92**, 369 (1992)
9.3 G. McLendon, R. Hake: Chem. Rev. **92**, 481 (1992)
9.4 B.M. Hoffman, M.A. Ratner: J. Am. Chem. Soc. **109**, 6237 (1987)
9.5 B.H. McMahon, J.D. Müller, C.A. Wraight, G.U. Nienhaus: Biophys. J. **74**, 2567 (1998)
9.6 E.S. Medvedev, A.A. Stuchebrukhov: J. Chem. Phys. **107**, 3821 (1997)
9.7 A.J.A. Aquino, P. Beroza, J. Reagan, J.N. Onuchic: Chem. Phys. Lett. **275**, 181 (1997)
9.8 I. Daizadeh, E.S. Medvedev, A.A. Stuchebrukhov: Proc. Natl. Acad. Sci. USA **94**, 3703 (1997)
9.9 H. Frauenfelder, F. Parak, R.D. Young: Annu. Rev. Biophys. Biochem. **17**, 451 (1998)

9.10 A. Ansari, J. Berendzen, S.F. Bowne, H. Frauenfelder, I.E.T. Iben, T.B. Sauke, E. Shyamsunder, R.D. Young: Proc. Natl. Acad. Sci. USA **82**, 5000 (1985)
9.11 Y. Sugiyama, S. Takahashi, K. Ishimori, I. Morishima: J. Am. Chem. Soc. **119**, 9582 (1997)
9.12 M. Meier, R. van Eldik, I.-J. Chang, G.A. Mines, D.S. Wuttke, H.B. Gray: J. Am. Chem. Soc. **116**, 1577 (1994)
9.13 J.F. Wishart, R. van Eldik, J. Sun, C. Su, S.S. Isied: Inorg. Chem. **31**, 3986 (1992)
9.14 J.R. Scott, J.L. Fairris, M. McLean, K. Wang, S.G. Sligar, B. Durham, F. Millett: Inorg. Chim. Acta **243**, 193 (1996)
9.15 K. Heremans, L. Smeller: Biochim. Biophys. Acta **1386**, 353 (1998)
9.16 H. Frauenfelder, N.A. Alberding, A. Ansari, D. Braunstein, B.R. Cowen, M.K. Hong, I.E.T. Iben, J.B. Johnson, S. Luck, M.C. Marden, J.R. Mourant, P. Ormos, L. Reinisch, R. Scholl, A. Schulte, E. Shyamsunder, L.B. Sorensen, P.J. Steinbach, A. Xie, R.D. Young, K.T. Yue: J. Phys. Chem. **94**, 1024 (1990)
9.17 R. van Eldik, T. Asano, W. J. Le Noble: Chem. Rev. **89**, 549 (1989)
9.18 F.S. Mathews, E.W. Czerwinski, P. Argos, in The Porphyrins,Vol.7 ed. by D. Dolphin (Academic Press, New York 1979) p.107
9.19 E.C. Johnson, B.P. Sullivan, D.J. Salmon, S.A. Adeyemi, T.J. Meyer: Inorg. Chem. **17**, 2211 (1978)
9.20 P. Ford, D.F.P. Rudd, R. Gaunder, H. Taube: J. Am. Chem. Soc. **90**, 1187 (1968)
9.21 A.D. Adler, F.R. Longo, F. Kampas, J. Kim: J. Inorg. Nucl. Chem. **32**, 2443 (1970)
9.22 S.B. von Bodman, M.A. Schuler, D.R. Jollie, S.G. Sligar: Proc. Natl. Acad. Sci. USA **83**, 9443 (1986)
9.23 B. Durham, L.P. Pan, S. Hahm, J. Long, F. Millett, in *ACS Advances in Chemistry Series,* Vol. 226 ed. by M.K. Johnson, R.B. King, D.M. Kurtz, C. Kutal, M.L. Norton, and R.A. Scott (American Chemical Society, Washington DC 1990) p.181
9.24 M.P. Jackman, M. Lim, P. Osvath, D.G.A. Harshani de Silva, A.G. Sykes: Inorg. Chim. Acta **153**, 205 (1988)
9.25 J. Altman, J.J. Lipka, I. Kuntz, L. Waskell: Biochemistry **28**, 7516 (1989)
9.26 C.D. Moore, O.N. Al-Misky, J.T.J. Lecomte: Biochemistry **30**, 8357 (1991)
9.27 A. Willie, P.S. Stayton, S.G. Sligar, B. Durham, F. Millett: Biochemistry **31**, 7237 (1992)
9.28 R. Varadarajan, A. Szabo, S.G. Boxer: Proc. Natl. Acad. Sci. USA **82**, 5681, (1985)
9.29 D.R. Casimiro, L. Wong, J.L. Colon, T.E. Zewert, J.H. Richards, I.-J. Chang, J. R. Winkler, H.B. Gray: J. Am. Chem. Soc. **115**, 1485 (1993)
9.30 F.W.J. Teale: Biochim. Biophys. Acta **35**, 543 (1959)
9.31 A.W. Axup, M. Albin, S.L. Mayo, R.J. Crutchley, H.B. Gray: J. Am. Chem. Soc. **110**, 435 (1989)
9.32 J.A. Cowan, H.B. Gray: Inorg. Chem. **28**, 2074 (1989)
9.33 K. Hara, I. Morishima: Rev. Sci. Instrum. **59**, 2397 (1988)
9.34 R.C. Newmann Jr., W. Kauzmann, A. Zipp: J. Phys. Chem. **77**, 2687 (1973)
9.35 B. Durham, J.V. Casper, J.K. Nagle, T.J. Meyer: J. Am. Chem. Soc. **104**, 4803 (1982)
9.36 P.S. Braterman, A. Harriman, G.A. Heath, L.J. Yellowlees: J. Chem. Soc. Dalton Trans. 1801 (1983)

9.37 G.A. Heath, L.J. Yellowlees: J. Chem. Soc. Chem. Comm. 287 (1981)
9.38 K. Sigfridsson, M. Sundahl, M.J. Bjerrum, O.J. Hansson: J. Bioinorg. Chem. 405 (1996)
9.39 K. Sigfridsson, M. Ejdedaeck, M. Sundahl, O. Hansson: Arch. Biochem. Biophys. **351**, 197 (1998)
9.40 L.K. Skov, T. Pascher, J.R. Winkler, H.B. Gray: J. Am. Chem. Soc. **120**, 1102 (1998)
9.41 A.J. Di Bilio, M.G. Hill, N. Bonander, B.G. Karlsson, R.M. Villahermosa, B.G. Malmström, J.R. Winkler, H.B. Gray: J. Am. Chem. Soc. **119**, 9921 (1997)
9.42 A.J. Di Bilio, C. Denninson, H.B. Gray, B.E. Ramirez, A.G. Sykes, J.R. Winkler: J. Am. Chem. Soc. **120**, 7551 (1998)
9.43 D.H. Heacock II, M.R. Harris, B. Durham, F. Millett: Inorg. Chim. Acta **226**, 129 (1994)
9.44 A. Freiberg, A. Ellervee, M. Tars, K. Timpmann, A. Laisaar: Biophys. Chem. **68**, 189 (1997)
9.45 W. Chung, N.J. Turro, I.R. Gould, S. Farid: J. Phys. Chem. **95**, 7752 (1991)
9.46 B. Bänsch, M. Meier, P. Martinez, R. van Eldik, C. Su, J. Sun, S.S. Isied, J.F. Wishart: Inorg. Chem **33**, 4744 (1994)
9.47 B. Durham, L.P. Pan, J.E. Long, F. Millett: Biochemistry **28**, 8659 (1989)
9.48 J.A. Cowan, R.K. Upmacis, D.N. Beratan, J.N. Onuchic, H.B. Gray: Ann. N. Y. Acad. Sci. **550**, 68 (1989)
9.49 H. Li, H. Yamada, K. Akasaka: Biochemistry **37**, 1167 (1998)
9.50 K. Akasaka, T. Tezuka, H. Yamada: J. Mol. Biol. 271, 671 (1997)
9.51 Y. Furukawa, K. Ishimori, I. Morishima: J. Phys. Chem. B 104, 1817 (2000)

10 Marine Microbiology: Deep Sea Adaptations

Chiaki Kato[1], Lina Li[1], Yuichi Nogi[1], Kaoru Nakasone[1], and Douglas H. Bartlett[2]

[1]The DEEPSTAR Group, Japan Marine Science and Technology Center, 2-15 Natsushima-cho, Yokosuka 237-0061, Japan
E-mail: katoc@jamstec.go.jp, lil@jamstec.go.jp, nogiy@jamstec.go.jp, nakasone@jamstec.go.jp
[2]Marine Biology Research Division, 0202, Center for Marine Biotechnology and Biomedicine, Scripps Institution of Oceanography, University of California, San Diego, La Jolla, California 92093-0202, USA
E-mail: dbartlett@ucsd.edu

Abstract. Technological advancements in the recovery and cultivation of deep-sea microorganisms have resulted in the isolation of novel groups of bacteria which are adapted to high pressure. From these collections a number of new species have been identified, including members of the genera *Shewanella*, *Moritella*, *Photobacterium* and *Colwellia*. Some of these isolates have been the subjects of investigations into pressure effects on gene, protein and fatty acid regulation and mechanisms of baro (piezo) adaptation or sensing. Cytochrome *bd* assembly and function appears to be critical to pressure adaptation in at least one *Shewanella* species, and the ToxR/S proteins are required for pressure-responsive regulation in a *Photobacterium* strain.

10.1 Introduction

'I had a talk with a young lady,
on the origin of the gene for organisms
living in the deep sea.'

Tanka poem by K. Suzuki
August 30, 1997

The term 'deep sea' usually refers to waters deeper than 1,000 m, which represents 75% of the total volume of the oceans. The average depth of the oceans, derived by dividing total volume by total area, is around 3,800 m. The maximum depth in the trenches can reach 11,000 m [1]. The deep sea is regarded as an extreme environment with high hydrostatic pressures (up to 110 MPa), predominantly low temperatures (1−2°C), but with occasional regions of extremely high temperature (up to 375°C) at hydrothermal vents, darkness and low nutrient availability. It is accepted that deep-sea microbiology as a definable field did not exist before the middle of this century, and was paid little attention except for the efforts of Certes and Portier [1]. Certes, during the Travaillier and Talisman Expeditions (1882-1883), examined sediment and water collected from depths up to 5,000 m and found bacteria in almost every sample. He noted that bacteria survived at great pressure and might live in a state of suspended animation [2]. In 1904, Portier used a sealed and autoclaved glass tube as a bacteriological sampling device and reported counts of colonies from various depths and locations [3]. In 1949, ZoBell and Johnson started work on the effect of hydrostatic pressure on microbial activities [4]. The term 'barophilic' was first used, defined today as optimal growth at pressure higher than 0.1 MPa or by a requirement of increased pressure for growth. Many microorganisms in the deep sea are extremophiles, such as halophiles, thermophiles, psychrophiles, and barophiles or piezophiles (the term 'piezophile' was proposed as a replacement to barophile as the Greek translations of the prefixes baro and piezo mean weight and pressure, respectively [5]), and some of these microorganisms cannot survive in so-called 'moderate' environments.

In this chapter, we will focus on the taxonomy of microorganisms adapted to the deep sea, and the molecular bases of their high-pressure adaptations.

10.2 Isolation and Taxonomy of Deep-Sea Barophilic (Piezophilic) Microorganisms

10.2.1 Isolation and Growth Properties

The development of manned and unmanned submersibles has supported the approaches used for investigating deep-sea environments and understanding the mechanisms of deep-sea bacterial adaptation. One of these is the manned submersible *SHINKAI 6500*, which is operated by the Japan Marine Science and Technology Center (JAMSTEC). This vehicle has the ability to submerge to a depth of 6,500 m [6]. Recently, an unmanned submersible *KAIKO* was also constructed, which recorded a 10,911 m depth diving point in the Mariana Trench during a dive in March 1995 [7].

Bacteria living in the deep-sea display several unusual features that allow them to thrive in their extreme environment. The first pure culture isolate of a barophilic bacterium was reported in 1979 [8]. The spirillum-like bacterium strain CNPT-3 had a rapid doubling rate at 50 MPa, but did not grow into colonies at atmospheric pressure for several weeks. This bacterium did not lose any of its barophilic characteristics after at least ten transfers, but was maintained at 58 MPa between transfers. Numerous barophilic and barotolerant bacteria have since been isolated and characterized by the DEEPSTAR group at JAMSTEC, from deep-sea sediments at depths ranging from 2,500−11,000 m obtained by the sterilized mud samplers on the submersibles *SHINKAI 6500* and *KAIKO* [9−12]. Some of these strains are listed in Table 10.1. Most of the isolated strains are not only barophilic or barotolerant, but also psychrophilic and cannot be cultured at temperatures above 20°C. Typical growth profiles of such bacteria as a function of several pressures and temperatures are shown in Fig. 10.1. The barophilic bacterium strain DB6705 was not able to grow at atmospheric pressure (0.1 MPa) at temperatures above 10°C;
However, this strain was able to grow better at higher temperatures than lower temperatures when incubated at higher pressures (more than 50 MPa). The growth of the barotolerant bacterial strain DSK1 was also better at a higher temperature (15°C) as

compared -with a lower temperature (10°C) at pressures above 50 MPa. These results illustrate the generalization that deep-sea bacteria show their strongest barophilic responses near their upper temperature limits for growth (typically ~15°C).

Table 10.1. High-pressure-adapted bacterial strains that have been isolated by the DEEPSTAR group at JAMSTEC

Bacterial strain	Optimal growth	Source	Species	Reference
Extremely barophilic bacteria[1]				
DB21MT-2	70 MPa, 10°C No growth at less than 50 MPa	Mariana Trench, at 10,898 m depth	*Shewanella benthica*	12, 28
DB21MT-5	80 MPa, 10°C No growth at less than 50 Mpa	Mariana Trench, at 10,898 m depth	*Moritella yayanosi**	12, 28
Barophilic bacteria[2]				
DB5501	50 MPa, 10°C	Suruga Bay, at 2,485 m depth	*Shewanella benthica*	9, 19, 26
DB6101	50 MPa, 10°C	Ryukyu Trench, at 5,110 m depth	*Shewanella benthica*	9, 19, 26
DB6705	50 MPa, 10°C	Japan Trench, at 6,356 m depth	*Shewanella benthica*	9, 19, 26
DB6906	50 MPa, 10°C	Japan Trench, at 6,269 m depth	*Shewanella benthica*	9, 19, 26
DB172F	70 MPa, 10°C	Izu-Bonin Trench, at 6,499 m depth	*Shewanella benthica*	10, 19, 26
DB172R	60 MPa, 10°C	Izu-Bonin Trench, at 6,499 m depth	*Shewanella benthica*	10, 19, 26
Moderately barophilic bacteria[3]				
DSS12	30 MPa, 8°C	Ryukyu Trench, at 5,110 m depth	*Shewanella violacea**	9, 26
DSJ4	10 MPa, 10°C	Ryukyu Trench, at 5,110 m depth	*Photobacterium Profundum**	22
DSK1	50 MPa, 15°C	Japan Trench, at 6,356 m depth	*Moritella japonica**	9, 34
Barotolerant bacteria				
DSK25	0.1 MPa, 35°C	Japan Trench, at 6,500 m depth	*Sporosarcina* sp.	38

[1]Extremely barophilic bacteria are defined as bacteria that are unable to growth at pressures below 50 MPa but are able to grow well at 100 MPa. [2]Barophilic bacteria are defined as those displaying optimal growth at a pressure of more than 40 MPa. [3]Moderately barophilic bacteria are defined as those displaying optimal growth at a pressure of less than 40 MPa and which are able to grow well at atmospheric pressure. *Novel deep-sea species reported by DEEPSTAR group

The effects of pressure and temperature on cell growth for those deep-sea barophilic strains isolated by Yayanos and the DEEPSTAR group are similar, in that all strains become more barophilic at higher temperatures [9, 13]. These studies indicate that all barophilic isolates are obligately barophilic above the temperature permitting growth at atmospheric pressure. This means that the upper temperature limit for growth can be extended by high pressure. Likewise, barophilic bacteria reproduce more rapidly at a lower temperature (such as 2°C) when the pressure is less than that at its capture depth. It also appears to be true as a general rule that the pressure of maximal rate of reproduction at 2°C may reflect the true habitat depth of an isolate [14]. The reproduction rate of barophiles at pressures near the pressure of the depth of capture increased with increasing temperature over an interval of 6−10°C. Why do barophilic cells grow best at temperatures that on first thought might never be encountered in the cold deep sea? There are three possible explanations. One is that this behavior is simply a consequence of adaptation to high pressure and low temperature. Second, the response may have been inherited by the bacteria from ancestors who lived in a warmer environment [15]. A third explanation is that bacteria of the cold deep sea do periodically encounter warmer temperatures.

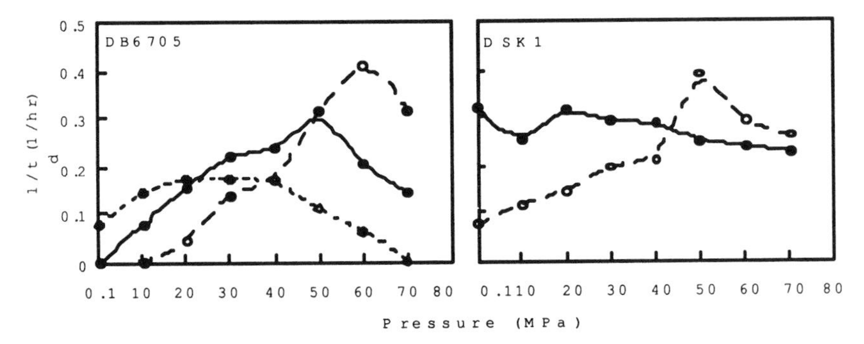

Fig. 10.1. Growth profiles of barophilic bacterium strain DB6705 and barotolerant bacterium strain DSK1 at various pressures and temperatures; Growth at 15°C (open circles), 10°C (solid circles), and 4°C (diamonds) is shown. The growth rate is shown as $1/t_d$, where t_d represents the doubling time (in hours)

10.2.2 Taxonomy

Many deep-sea barophilic bacteria have been shown to belong to the gamma-Proteobacteria through comparison of 5S and 16S rDNA sequences. The G+C content of chromosomal DNA from *Vibrio* spp. that belong to the gamma-Proteobacteria is between 40 and 50% [16] and is considered to be typical of this subgroup. The G+C content of chromosomal DNA from the isolated barophiles in our laboratory was found to be similar, at 40−46%. As a result of a taxonomic study based on its 5S rDNA sequences, Deming reported that, the obligate barophilic bacterium *Cowellia hadaliensis* belongs to the Proteobacteria gamma subgroup [17]. DeLong et al. also documented the existence of barophilic and psychrophilic deep-sea bacteria that belong to this subgroup, as indicated by the 16S rDNA sequence [18]. It is interesting to note that the 16S rDNA sequences of the barophilic strains DB6906, DB172F, and DB172R and the psychrophilic and moderately barophilic strain DSS12 [9] show the highest homology of all, indicating that these strains are very closely related (Fig. 10.2) [19]. The barophilic bacteria reported by Liesack et al. [20] are also included in the same branch of the gamma-Proteobacteria as are the strains isolated by the DEEPSTAR group [9]. These data suggest that most deep-sea high-pressure-adapted (barophilic or piezophilic) bacteria which can be readily cultured belong to the Proteobacteria gamma subgroup and may not be widely distributed within the domain Bacteria.

DeLong et al. reported that 11 cultivated psychrophilic and barophilic deep-sea bacteria are affiliated with one of five genera within the gamma subgroup:

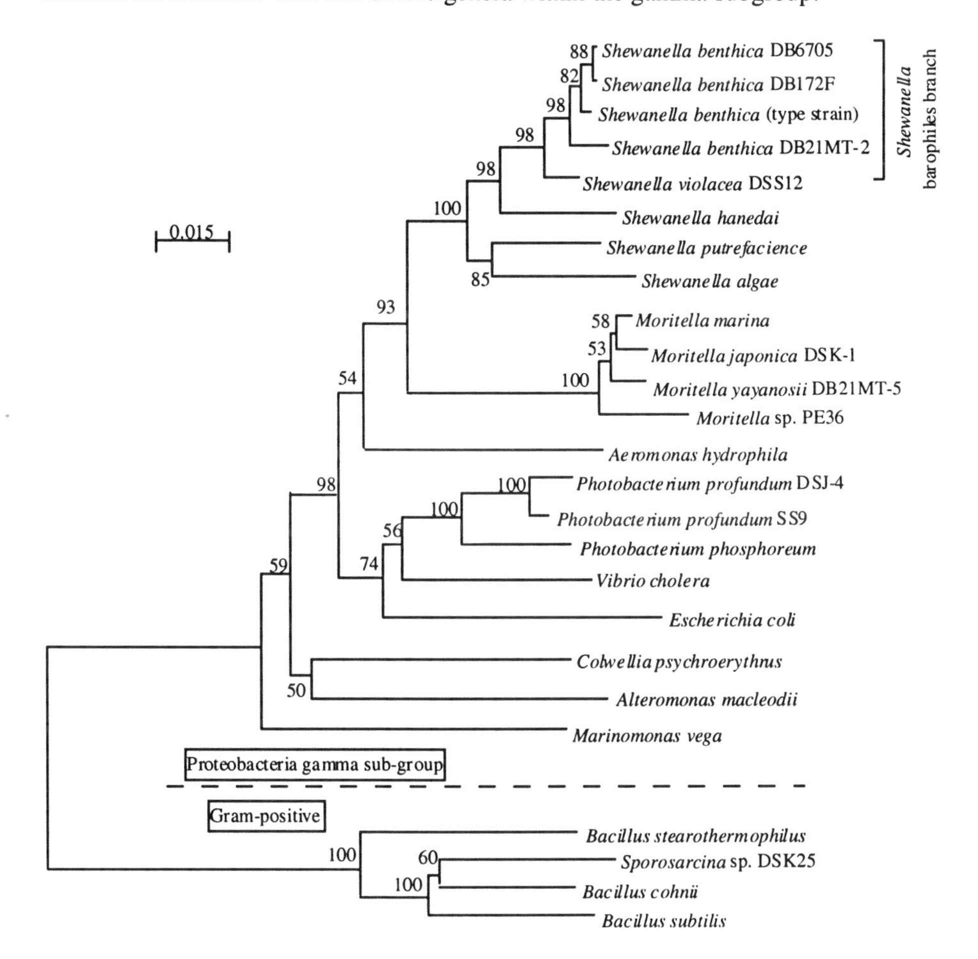

Fig. 10.2. Phylogenetic tree showing the relationships between isolated deep-sea barophilic bacteria within the gamma subgroup of Proteobacteria and genus *Bacillus* as determined by comparing 16S rDNA sequences using the neighbor-joining method. The scale represents the average number of nucleotide substitutions per site. Bootstrap values (%) are shown for frequencies above the threshold of 50%

Shewanella, *Photobacterium*, *Colwellia*, *Moritella*, and an unidentified genus [18]. Only deep-sea barophilic species from two of these genera were reported prior to the reports from the DEEPSTAR group; there were *S. benthica* in the genus *Shewanella* [15, 21] and *C. hadaliensis* in the genus *Colwellia* [17]. The DEEPSTAR group identified four new barophilic species within the above genera based on chromosomal DNA−DNA hybridization studies and several other taxonomic properties. The % values of DNA−DNA relatedness between the deep-sea barophilic strains are shown in Table 10.2.

Table 10.2. Levels of homology for the chromosomal DNAs between barophilic isolates and reference species. **(A) Genus *Photobacterium***

		% Homology with DNA from:						
Strains	GC%	SS9	DSJ4	*P.ang*	*P.dam*	*P.his*	*P.lei*	*P.pho*
SS9	41.5	100	78	27	23	24	17	19
DSJ4	42.0	78	100	28	19	28	18	20
P.ang	39.6	24	21	100	25	30	34	24
P.dam	41.8	20	20	28	100	66	19	19
P.his	41.8	29	16	22	65	100	17	16
P.lei	39.7	20	20	46	23	28	100	25
P.pho	39.1	23	21	43	23	37	24	100

SS9, DSJ4: *P. profundum* strains SS9 and DSJ4; *P.ang: P. angustum; P.dam: P. damselae; P.his: P. histaminum; P.lei: P. leiognathi; P.Pho: P. phosphoreum*

(B) Genus *Shewanella*

		% Homology with DNA from:							
Strains	GC%	DSS12	DB172F	DB6705	DB21MT-2	*S.bent*	*S.peal*	*S.hane*	*S.put*
DSS12	47.0	100	33	35	40	31	21	12	13
DB172F	48.3	39	100	87	77	86	ND*	14	9
DB172R	48.5	28	ND	87	ND	82	ND	14	8
DB5501	49.9	35	98	81	ND	99	ND	9	9
DB6101	49.3	34	100	97	ND	81	ND	12	10
DB6705	48.0	36	79	100	84	87	23	29	12
DB6906	48.0	37	88	75	ND	83	ND	11	10
DB21MT-2	48.7	48	79	78	100	81	ND	20	21
PT-99	47.0	46	80	100	ND	82	ND	16	12
S.bent	47.9	41	79	88	80	100	20	15	7
S.peal	47.0	21	ND	21	ND	20	100	26	ND
S.hane	45.4	15	5	10	10	8	22	100	7
S.alg	51.2	4	7	7	ND	8	ND	6	7
S.put	46.5	13	6	9	9	8	ND	1	100

DSS12:*S. violacea*; DB-series and PT-99: *S. benthica strains: S.bent; S. benthica; S.peal: S. pealeana; S.hane: Shewanella hanedai; S.alg: S. algae; S.put: S. putrefaciens* *ND: not determined

(C) Genus *Moritella*

Strains	GC%	% Homology with DNA from:						
		DSK1	DB21MT-5	*M.mar*	*S.put*	*S.hane*	*S.bent*	*S.viol*
DSK1	45.0	100	22	17	16	23	4	18
DB21MT-5	44.6	41	100	32	10	ND	8	ND
M.mar	42.5	33	28	100	9	12	8	11
S.put	46.5	10	7	9	100	1	8	13
S.hane	45.4	14	ND	9	7	100	8	15
S.bent	47.9	14	12	11	7	15	100	41
S.viol	47.0	18	ND	10	13	12	31	100

DSK1: *M. japonica*; DB21MT-5: *M. Yayanosii*; *M.mar*: *Moritella marina*; *S.viol*: *S. violacea*

Both previously described and new species of bacteria have been identified among barophilic bacterial isolates. *Photobacterium profundum*, a new species, was identified in the moderately barophilic strains DSJ4 and SS9 [22]. *P. profundum* strain SS9 has been extensively studied with regard to the molecular mechanisms for pressure regulation [23–25] as we will describe in section 10.3.3 below. *P. profundum* is the only species to express barophily within the genus *Photobacterium* or to produce the long-chain polyunsaturated fatty acid (PUFA), eicosapentaenoic acid (EPA). No other species of *Photobacterium* have been reported to produce EPA [22].

The moderately barophilic strain DSS12 isolated from the Ryukyu Trench at a depth of 5,110 m [26] was identified to be *Shewanella violacea*, a novel barophilic species belonging within the *Shewanella* barophile branch [19]. Other *Shewanella* barophilic strains, PT-99 [27], DB5501, DB6101, DB6705, DB6906 [9], DB172F, DB172R [10] and DB21MT-2 [12] were all identified to belong to the same species, *S. benthica* [19, 26, 28]. The barophilic and psychrophilic *Shewanella* strains, including *S. violacea* and *S. benthica*, also produce EPA [12, 18, 26]; thus, the occurrence of this PUFA is a property shared by many deep-sea bacteria. *S. violacea* strain DSS12 has been well studied at JAMSTEC, particularly with respect to its molecular mechanisms of adaptation to high pressure. This strain is moderately barophilic, with a fairly constant doubling between 0.1 MPa and 70 MPa, whereas the doubling times of most of the barophilic *S. benthica* strains change substantially with increasing pressure [9–12]. Because there are few differences in the growth characteristics of strain DSS12 under different pressure conditions, this strain is a very convenient deep-sea bacterium for use in studies on the mechanisms of adaptation to high-pressure environments. Studies using this strain include the analyses of the pressure-regulation of gene expression [29, 30] and of the role of *d*-type cytochromes in the growth of cells under high pressure [31, 32]. The molecular mechanisms of gene expression have been analyzed, focusing on a cloned pressure-regulated promoter, and more detailed studies are in progress, as we described in section 10.3.2 [33].

Strain DSK1, which is a moderately barophilic bacterium isolated from the Japan Trench, was identified as *Moritella japonica*. This is the first barophilic species identified in the genus *Moritella* [34]. The type strain of the genus *Moritella* is *M.*

marina (previously known as *Vibrio marinus* [35]), which is one of the most common psychrophilic organisms isolated from marine environments [36]. *M. marina* is closely related to the genus *Shewanella* on the basis of 16S rDNA data and is not a barophilic bacterium [34]. The extremely barophilic bacterium strain DB21MT-5, isolated from the world's deepest trench, Challenger Deep, Mariana Trench at a depth of 10,898 m [12], was identified as a *Moritella* species and designated *M. yayanosii* [28]. The optimal pressure for growth of *M. yayanosii* strain DB21MT-5 is 80 MPa, and this strain is not able to grow at a pressure less than 50 MPa, but is able to grow well at higher pressures, even as high as 100 MPa [12]. Production of the long-chain PUFA docosahexaenoic acid (DHA) is one of the characteristic properties of the genus *Moritella* [12, 18, 34]. DeLong and Yayanos [27, 37] reported that the fatty-acid composition of the barophilic strain changed as a function of pressure, and, in general, greater amounts of PUFAs were synthesized at the higher growth pressures. Approximately 70% of the membrane lipids in *M. yayanosii* is PUFA, which is consistent with its adaption to very high pressures.

From the above studies, four more identified barophilic species could be added to the list of deep-sea barophiles, as shown in Table 10.3.

Table 10.3. List of the identified deep-sea barophilic bacterial species

Genus	Species	Strain	Properties	Reference
Colwellia	*C. hadaliensis*	BNL-1	Extremely barophilic	17
Photobacterium	*P. profundum*	SS9, DSJ4	Moderately barophilic	22
Shewanella	*S. benthica*	PT-99, DB-series	Moderately, obligately and extremely barophillic	12, 15, 19, 21, 26
	S. violacea	DSS12	Moderately barophilic	26
Moritella	*M. japonica*	DSK1	Moderately barophilic	34
	M. yayanosii	DB21MT-5	Extremely barophilic	12, 28

10.3 High-Pressure Sensing and Adaptation in Deep-Sea Microorganisms

10.3.1 Introduction

Before discussing pressure signal transduction and adaptation in deep-sea bacteria, it is useful to consider the effects of hydrostatic pressure on the responses of biological processes in the broader context of mechanostimulation, which is present in all organisms [39, 40]. Pressure differentials across cells exert a variety of effects on different cells and tissues. In mammals, differences in blood pressure of less than 1% of atmospheric pressure can lead to changes in cardiac gene expression, presumably as part of a homeostatic mechanism to maintain blood pressure within a narrow range [41]. Turgor pressure must be sensed and regulated, and in microorganisms with large cell walls, such as Gram-positive bacteria and certain species of algae, it can be as high as 2 MPa [42, 43]. Bioluminescence in dinoflagellates can be greatly increased by relatively small pressure changes associated with laminar fluid shear on the order of 1 dyne cm^{-2} [44, 45].

The sensing of pressure not as a function of space as in the examples above, but as a function of time also exists in marine organisms other than deep-sea bacteria. Many fishes have evolved physiological responses to changes in hydrostatic pressure which influence the concentration of gases within a swim bladder and hence buoyancy [46]. Planktonic crustaceans can display responses to pressure changes of less than 1 kPa (10^{-2} atm) [47]. So, the ability to respond to pressure changes is not unique to deep-sea bacteria; however, the capacity to respond to large changes in pressure in an apparently adaptive manner probably is restricted to those bacteria and higher organisms inhabiting high-pressure environments.

10.3.2 Pressure Regulation in Microorganisms Outside of the Genus *Shewanella*

Pressure-inducible proteins have been observed in *Methanococcus thermolithotrophicus, Rhodotorula rubra* and *Escherichia coli*, representing all three domains of life [48–50]. Even microorganisms which do not come from high- pressure environments may respond to high pressure by inducing the synthesis of certain stress proteins. Overlapping response of organisms to high pressure and high temperature has been observed. The steady-state levels of the heat shock protein GroES appear to be elevated during growth of *E. coli* at 40 MPa [49].

High-pressure effects on the *lac* and *tac* promoters of *E. coli* have also been studied, using promoters fused to the chloramphenicol acetyl transferase (CAT) gene [51–52]. Plasmid-encoded *lac*- or *tac*-CAT activity in *E. coli* cells grown at high-pressure was increased 80–90-fold compared with the level expressed at atmospheric pressure. Quantitative primer extension analysis indicated that growth at 30 MPa and 50 MPa induced high CAT transcript levels in *E. coli* carrying the *lac*-CAT and *tac*-CAT transcriptional fusion plasmids [51].

Another major effect of pressure is on the membrane constituents of bacteria. Elevated pressure represses the synthesis of several outer membrane proteins in *E. coli*, including the OmpC, OmpF, and LamB proteins [53, 54]. One outer membrane protein, OmpH, in the moderately barophilic bacterium *Photobacterium profundum* strain SS9 has been found to be induced 10 –100-fold in mid-log cells grown at the pressure optimum 28 MPa, as compared with cells grown at atmospheric pressure, and the *ompH* regulatory promoter was confirmed [23]. A second OMP of strain SS9, designated OmpL, is repressed at elevated pressures, while a third OMP, designated OmpI, is induced at pressures beyond the pressure optimum of strain SS9, and is the most abundant OMP at 40 MPa [55–57]. It has also been discovered that certain species of marine bacteria, particularly those from the cold deep sea, can produce long-chain and branched PUFAs in their membranes [27, 37, 58, 59]. Increasing the extent of fatty-acid unsaturation can maintain the membrane in a functional liquid crystalline state at high pressures and/or low temperatures [60].

10.3.3 Pressure-Regulated Operons in *Shewanella* Species

A promoter, activated by growth at high pressure, was cloned from the barophilic *S. benthica* strain DB6705 into *E. coli* [61]. Gene expression initiating from this promoter was induced by high pressure in both the barophilic strain DB6705 and in *E. coli* transformants harboring this promoter. Downstream from this promoter, two open reading frames (ORF1 and 2) were identified as one operon, designated as a pressure-regulated operon [62]. The structure of the promoter sequence was similar to that of the promoter of *ompH* from *P. profundum* strain SS9 [63]. The highly conserved pressure-regulated operon from the moderately barophilic bacterium strain DSS12 was also cloned and sequenced [29]. Its sequence is almost identical to the operon from barophilic strain DB6705. Downstream from this operon, another pressure-regulated operon was discovered whose first ORF was designated ORF3 and whose gene expression was also enhanced by high pressure [29]. These two operons represented a pressure-regulated gene cluster (Fig. 10.3). According to the transcriptional analyses, the pressure-regulated operon is expressed at elevated high pressure, and at 70 MPa the largest amount of transcript is present. Based upon its deduced amino acid sequence and heterologous complementation studies in *E. coli*, ORF3 appears to encode the CydD protein [31]. In *E. coli*, CydD is required for the assembly of the cytochrome *bd* complex, one of the components of the aerobic respiratory chain [64]. *E. coli cydD* mutants display increased sensitivity to high pressure, but can be converted to wild-type levels of high-pressure sensitivity if the DSS12 ORF3 gene is present on a plasmid. The cytochrome *bd* protein complex of strain DSS12 was observed spectrophotometrically only in high-pressure cultures. It seems likely that the pressure regulation of this respiratory system in strain DSS12 plays a significant role in cell growth under high pressure [30].

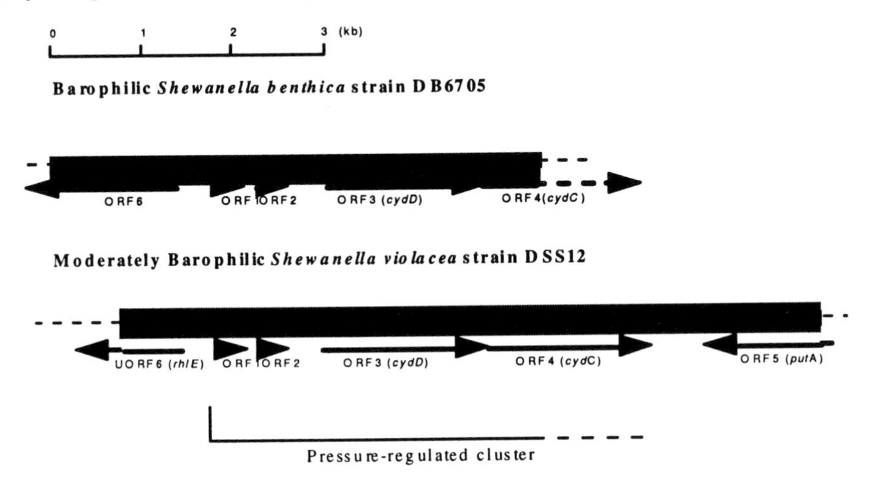

Fig. 10.3. Genetic map of the DNA locus containing the two pressure-regulated operons and flanking ORFs from *Shewanella* strains DB6705 and DSS12. Arrows indicate open reading frames

The molecular details of pressure-regulation in the region upstream of the pressure-regulated operon (ORF 1 and 2) from *S. violacea* strain DSS12 have been analyzed. Five transcription initiation sites have been identified by primer extension analysis in this region, and transcription starting from each of these sites is increased by elevated pressure [29]. Nucleotide sequence analysis of this region indicated the existence of two potential regulatory regions (regions A and B) as follows:
(a) Similarity between a consensus sequence for σ^{54} binding and part of the region designated region A was detected (homology of over 90%) when the region flanking the 5' portion of the second transcription initiation site was compared with several consensus sequences of *E. coli* σ factors. However, typical promoter consensus sequences were not found in the flanking regions of other transcription initiation sites. These findings suggest that this strain not only may have new types of promoter sequences, but perhaps also binding sites for previously unknown types of sigma factors or other regulatory proteins.
(b) A unique octamer motif, AAGGTAAG, was found to be tandemly repeated 13 times in the region designated region B containing the 3rd to 5th transcription initiation sites just upstream of region A. A palindromic sequence, AGTTAAAG ATTAAACT, was found in a region downstream of the tandemly repeated sequence. These regions may have a role in regulating expression of the operon, and it seems possible that some pressure-regulated, DNA-binding proteins might bind there under high-pressure conditions.

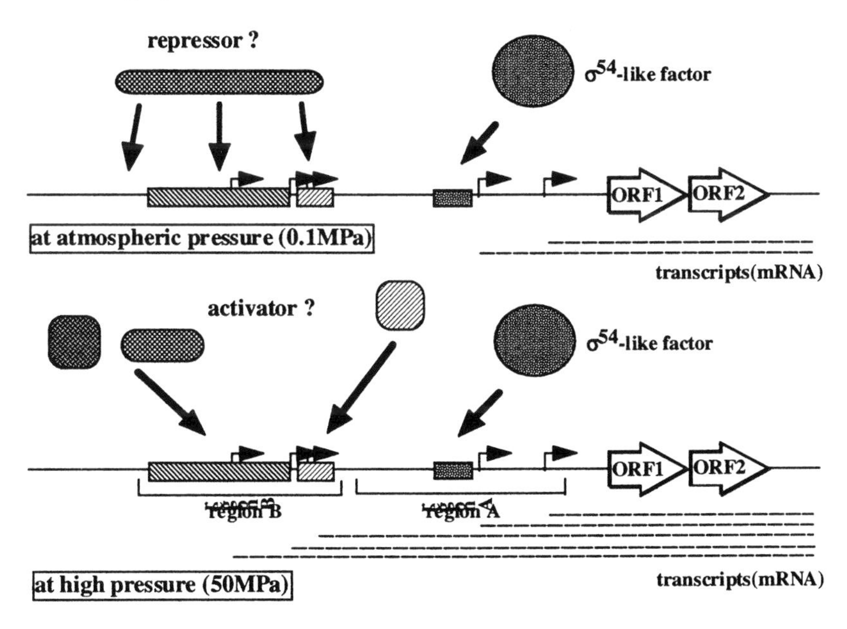

Fig. 10.4. Summary of binding of transacting factors to the region upstream of the pressure-regulated operon of the deep-sea barophilic *Shewanella violocea* strain DSS12 under atmospheric (0.1 MPa) and high-pressure (50 MPa) conditions. The following features are shown: Sequence tandemly repeated 13 times (vertical shading); palindrome structure (diagonal hatching); σ^{54} consensus sequence (*shading*)

In order to analyze whether such DNA-binding proteins exist in strain DSS12, an electrophoretic mobility shift assay (EMSA) was performed [33]. Interactions of factors binding to the upstream region under both atmospheric and high-pressure (50 MPa) conditions considered from the results of EMSA are summarized in Fig. 10.4. They indicated that a σ^{54}-like factor recognizes region A and other unknown factors recognize region B. Different shift patterns of protein-DNA complexes were observed using cell lysates prepared from cells of strain DSS12 cultured at 0.1 MPa or 50 MPa. A factor present in cells grown at 0.1 MPa binds to region B, and a factor in cells grown at 50 MPa binds particularly to the tandem octamer motif (AAGGTAAG; region B2). Since transcripts from region B are not detectable in cells grown at atmospheric pressure (0.1 MPa) [29], the factors binding to region B might function as a transcriptional repressor. In contrast, in cells grown at high pressure (50 MPa), all transcripts are detectable [29], suggesting that the factors expressed under these conditions might act as transcriptional activator. These results indicate that the deep-sea barophilic strain DSS12 expresses different DNA-binding factors under different pressure conditions.

10.3.4 Pressure-Sensing Mechanisms

Additional information on factors controlling gene expression at high pressure has come from studies of *omp* regulation in *P. profundum* strain SS9. Progress in understanding the control of *omp* gene expression in SS9 was made following the isolation of a piezo (pressure)-sensing mutant following transposon mutagenesis. This mutant displayed a constitutive high-pressure-signaling phenotype, producing no OmpL and high levels of OmpH regardless of growth pressure. The operon whose disruption was responsible for this phenotype was identified following transposon cloning and sequencing. The deduced amino acid sequence of its gene products shares extensive similarity to ToxR and ToxS, first discovered as a virulence determinant in the pathogenic bacterium *Vibrio cholerae* [65]. Both are transmembrane proteins, and ToxR is also a DNA-binding transcription factor. Increases in hydrostatic pressure increased ToxR abundance, indicating that it is either autoregulated or under the control of another pressure regulatory mechanism. Overproduction of ToxR by introducing a *toxR*-containing plasmid into SS9 resulted in increased ToxR levels but not increased repression of *ompH* or activation of *ompL*. These results indicated that pressure controls both the amount of ToxR present in the cell and its specific activity. How might pressure control the ToxRS signaling pathway in SS9? Two possibilities, which are not mutually exclusive, are that pressure is affecting the membrane environment in which ToxRS functions or it is directly modulating ToxRS quaternary structure. Elevated pressure is well documented to decrease membrane fluidity and in many instances to promote the dissociation of multimeric proteins [66, 67].

In order to assess the likelihood of a role for membrane structure in ToxRS signaling, the effects of membrane-fluidizing anesthetics were tested. Local anesthetics at concentrations which produce measurable reductions in the lipid chain order of *E. coli* lipid extracts [68] were discovered to reverse the high-pressure inactivation of ToxR in SS9. Antagonism between pressure and anesthetics has been observed in luminous bacteria, tadpoles, newts and mice [69—71]. In all of these instances the counterbalancing influences of pressure and anesthesia have most frequently been posited to arise from their differential effects on motion within the membrane [71]. Thus, membrane fluidity appears to be critical to ToxR responsiveness to pressure changes; however, much more work on the role of the membrane and of the ToxR protein in pressure sensing is needed.

If pressure sensing is indeed a reflection of the fluidity of the membrane, another question arises: Why doesn't the ToxRS system respond to changes in temperature as well as pressure, since temperature changes can also alter membrane fluidity? In other words, why isn't it both a thermometer and a barometer? In *Vibrio cholerae* expression of the *toxRS* operon is decreased at higher temperatures [72]. The basis of this effect is believed to stem from RNA polymerase competition for a heat shock promoter, directing the expression of the *htpG* gene, located immediately upstream of the *toxRS* operator/promoter region. In the case of SS9 a similar situation appears likely to exist, as a homolog of *htpG* is also located upstream of SS9 *toxRS*, and ToxR levels do indeed drop with increases in temperature (Welch and Bartlett, unpublished results). So, perhaps while ToxR (and presumably ToxS) abundance goes down with increases in temperature, membrane fluidity increases with higher temperature result in increased ToxR/S specific activity. In this way the opposing effects of temperature on ToxR/S abundance and specific activity could result in no net change in overall ToxR/S activity, whereas the effects of pressure on ToxR/S abundance and specific activity result in additive changes to the overall activity of this signaling system.

OmpH levels increase with growth into stationary phase as well as in cells incubated at high pressure [54]. However 1 atm stationary-phase production of OmpH can be abolished by the addition one of a variety of sugars to the growth medium. This sugar effect can be largely reversed in many cases by the addition of cyclic adenosine monophosphate (cAMP), and completely reversed if high pressure is applied. Taken together these results imply that *ompH* gene expression, but not *ompL* gene expression, is under the control of the cAMP receptor protein (CRP, also refered to as the catabolite activator protein). Consistent with this suggestion, a possible CRP binding site is present upstream of the *ompH* transcription site [54]. It has been proposed that, under carbon- and energy-limiting conditions, CRP in complex with cAMP is capable of derepressing ToxR repression of *ompH*. Since ToxR repression of *ompH* is low at high pressure, the CRP system has little effect on *ompH* expression at high pressure. Thus, the pressure regulation of *ompH* expression is interwoven with that of another environmental cue: carbon and energy availability.

Figure 10.5 presents a model of pressure regulation of *ompH/L* expression in SS9.

10.4 Concluding Remarks

Although much remains to be learned about the fundamental aspects of high-pressure adaptation in deep-sea microorganisms, the availability of well-defined barophilic strains now provides a solid foundation for future studies. Already evidence has been obtained indicating roles for various membrane components, including unsaturated fatty acids, cytochromes and membrane-localized signaling systems in baroadaptation. There is a need for additonal biophysical inquiries into the functioning of various cell processes at high pressure, the acquisition of genomic

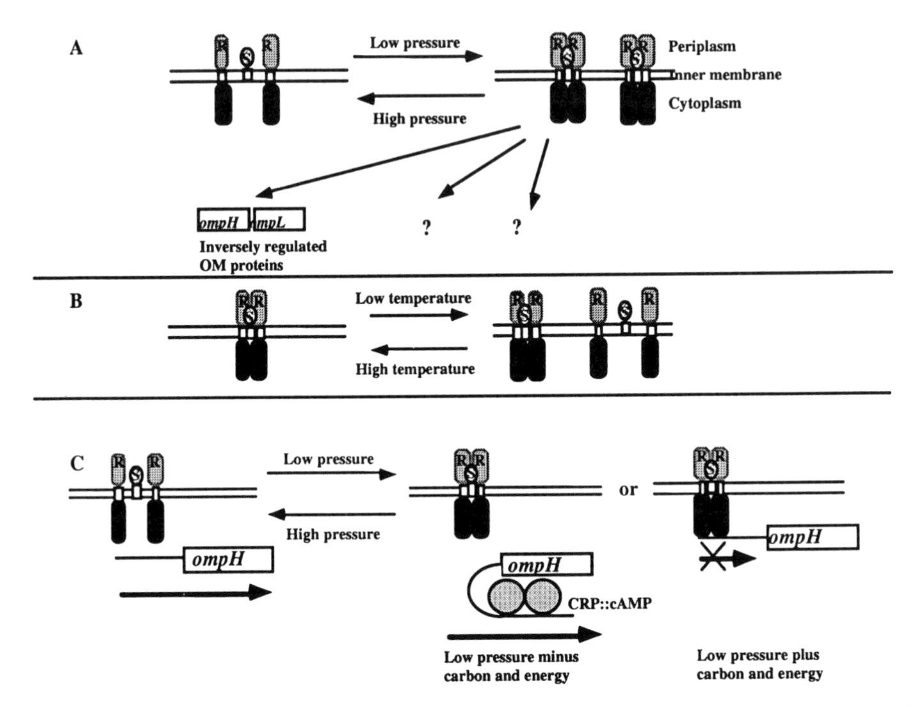

Fig. 10.5. Models depicting the role of ToxR and ToxS proteins in pressure sensing and control of gene expression in the deep-sea bacterium *Photobacterium profundum* strain SS9. A. Pressure influences on the conformational states of the ToxR and ToxS proteins could influence the ability of ToxR to control the expression of numerous pressure-regulated genes, including *ompH* and *ompL*. Low pressure increases both ToxR/S abundance and specific activity (i.e., the ability to repress *ompH* expression and activate *ompL* expression). B. Temperature does not change ToxR/S overall activity. Low temperature increases ToxR/S abundance but decreases its specific activity, resulting in no net change in overall activity. C. The cAMP receptor protein (CRP) interacts with ToxR/S to control *ompH* gene expression. Even at low pressure *ompH* may be transcribed if cells are carbon and energy limited. Under these conditions CRP complexed with its effector molecule cAMP derepresses ToxR/S repression of *ompH*

sequence information from selected barophiles, and the identification of additional genes whose products are needed for life in the deep sea.

Acknowledgements. The members of the DEEPSTAR group are very grateful to Prof. Koki Horikoshi for providing the opportunity to conduct much of the research described in this chapter, as well as for providing encouragement. These authors also thank other colleagues at JAMSTEC for providing deep-sea samples and for additional scientific cooperation. D. H. Bartlett gratefully acknowledges research support from grant MCB96-30546 of the National Science Foundation.

References

10.1 H.D. Jannasch, C.D. Taylor, Annu. Rev. Microbiol. **38**, 487 (1984)
10.2 A. Certes, C. R. Acad. Sci. Paris **98**, 690 (1884)
10.3 J. Richard: L'Oceanographie (Vuibert & Nony, Paris 1907)
10.4 C.E. ZoBell, F.H. Johnson, J. Bacteriol. **57**, 179 (1949)
10.5 A.A. Yayanos, Annu. Rev. Microbiol. **49**, 777 (1995)
10.6 S. Takagawa, K. Takahashi, T. Sano, Y. Mori, T. Nakanishi, M. Kyo, Oceans'89, **3**, 741 (1989)
10.7 M. Kyo, E. Miyazaki, S. Tsukioka, H. Ochi, Y. Amitani, T. Tsuchiya, T. Aoki, S. Takagawa, ROV. OCEANS'95, **3**, 1991 (1995)
10.8 A.A. Yayano, A.S. Dietz, R.V. Boxtel, Science **205**, 808 (1979)
10.9 C. Kato, T. Sato, K. Horikoshi, Biodiv. Conserv. **4**, 1 (1995)
10.10 C. Kato, N. Masui, K. Horikoshi, J. Mar. Biotechnol. **4**, 96 (1996)
10.11 C. Kato, A. Inoue, K. Horikoshi, Trends in Biotechnol. **14**, 6 (1996)
10.12 C. Kato, L. Li, Y. Nakamura, Y. Nogi, J. Tamaoka, K. Horikoshi, Appl. Environ. Microbiol. **64**, 1510 (1998)
10.13 A.A. Yayanos, Proc. Natl. Acad. Sci. USA **83**, 9542 (1986)
10.14 A.A. Yayanos, A.S. Dietz, R.V. Appl. Environ. Microbiol. **44**, 1356, 1982.
10.15 J. W. Deming, H. Hada, R. R. Colwell, K. R. Luehrsen, and G. E. Fox, J. Gen. Microbiol. **130**, 1911 (1984)
10.16 N.R. Kreig, J.G. Holt, Vol. 1 (Williams & Wilkins, Baltimore 1984)
10.17 J.W. Deming, L.K. Somers, W.L. Straube, D.G. Swartz, M.T. Macdonell Appl. Microbiol. **10**, 152 (1988)
10.18 E.F. DeLong, D.G. Franks, A.A. Yayanos, Appl. Environ. Microbiol. **63**, 2105 (1997)
10.19 L. Li, C. Kato, Y. Nogi, K. Horikoshi, FEMS Microbiol. Lett. **159**, 159 (1998)
10.20 W. Liesack, H. Weyland, E. Stackebrandt, Microb. Ecol. **21**, 191 (1991)
10.21 M.T. MacDonell, R.R. Colwell, Appl. Microbiol. **6**, 171 (1985)
10.22 Y. Nogi, N. Masui, C. Kato, Extremophiles **2**, 1 (1998)
10.23 D. Bartlett, M. Wright, A. A. Yayanos, M. Silverman, Nature **342**, 572 (1989)
10.24 D.H. Bartlett, E. Chi, T.J. Welch, High Pressure Bioscience and Biotechnology. eds. by R. Hayashi, C. Balny (Elsevier Science BV, The Netherlands, 1996) p.29
10.25 T.J. Welch, D.H. Bartlett, Mol. Microbiol. **27**, 977 (1998)
10.26 Y. Nogi, C. Kato, K. Horikoshi. Arch. Microbiol. **170**, 331 (1998)
10.27 E.F. DeLong, A.A. Yayanos Appl. Environ. Microbiol. **51**, 730 (1986)
10.28 Y. Nogi, C. Kato, Extremophiles **3**, 71 (1999)
10.29 C. Kato, A. Ikegami, M. Smorawinska, R. Usami, K. Horikoshi, J. Mar. Biotechnol. **5**, 210 (1997)
10.30 C. Kato, L. Li, H. Tamegai, M. Smorawinska, K. Horikoshi, Recent Res. Dev. Agric. Biol. Chem. **1**, 25 (1997)
10.31 C. Kato, H. Tamegai, A. Ikegami, R. Usami, K. Horikoshi, J. Biochem. **120**, 301(1996)
10.32 H. Tamegai, C. Kato, K. Horikoshi, J. Biochem. Mol. Bio. Biophys. **1**, 213 (1998)
10.33 K. Nakasone, A. Ikegami, C. Kato, R. Usami, K. Horikoshi, Extremophiles **2**, 149 (1998)
10.34 Y. Nogi, C. Kato, K. Horikoshi, J. Gen. Microbiol., **44**, 289 (1998)
10.35 H. Urakawa, K. Kita-Tsukamoto, S.E. Steven, K. Ohwada, R. R. Colwell, FEMS Microbiol. Lett. **165**, 373 (1998)

10.36 R.R. Colwell, R.Y. Morita, J. Bacteriol. **88**, 83 (1964)
10.37 E.F. DeLong, A.A. Yayanos Science **228**, 1101 (1985)
10.38 C. Kato, S. Suzuki, S. Hata, T. Ito, K. Horikoshi JAMSTEC R. **32**, 7 (1995)
10.39 P. Blount, S.I. Sukharev, P.C. Moe, M.J. Schreder, H.R. Guy, C. Kung, EMBO J. **15**, 4798 (1996)
10.40 T. Erdos, G.S. Butler-Browne, L. Rappaport, Biochimie **73**, 1219 (1991)
10.41 S. Izumo, B. Nadal-Ginard, V. Mahdavi, Proc. Natl. Acad. U. S. A. **85**, 339 (1998)
10.42 M.A. Bisson, G.O. Kirst, Naturwissenschaften **82**, 461 (1995)
10.43 A. M. Whatmore, R. H. Reed, J. Gen. Microbiol. **136**, 2521 (1996)
10.44 V.D. Gooch, W. Vidaver, Photobiol., **31**, 397 (1980)
10.45 M. I. Latz, J. F. Case, R.L. Gran, Limnol. Oceanogr. **39**, 1424 (1994)
10.46 B. Pelster, P. Scheid, Physiol. Zool. **65**, 1 (1992)
10.47 P. J. Fraser, A. G. Macdonald, Nature **371**, 383 (1994)
10.48 R. Jaenicke, G. Berndardt, H.D. Ludemann, K.O. Stetter, Appl. Environ. Microbiol. **54**, 2375 (1988)
10.49 M. Gross, K. Lehle, R. Jaenicke, K.H. Nierhaus, Eur. J. Biochem. **218**, 463 (1993)
10.50 T.J. Welch, A. Farewell, F.C. Neidhardt, D.H. Bartlett, J. Bacteriol. **175**, 7170 (1993)
10.51 C. Kato, T. Sato, M. Smorawinska, K. Horikoshi, FEMS Microbiol. Lett. **122**, 91 (1994)
10.52 T. Sato, C. Kato, K. Horikoshi, J. Mar. Biotechnol. **3**, 89 (1995)
10.53 K. Nakashima, K. Horikoshi, T. Mizuno, Biosci. Biotechnol. Biochem. **59**, 130 (1995)
10.54 T. Sato, Y. Nakamura, K. Nakashima, C. Kato, K. Horikoshi, FEMS Microbiol. Lett. **135**, 111 (1996)
10.55 D. Bartlett, E. Chi, W.E. Wright, Gene **131**, 125 (1993)
10.56 D. Bartlett, E. Chi, Arch. Microbiol. **162**, 323 (1994)
10.57 E. Chi, D. Bartlett, J. Bacteriol. **175**, 7533 (1993)
10.58 C.O. Wirsen, H.W. Jannasch, S.G. Wakeham, E.A. Canuel, Curr. Microbiol. **14**, 319 (1987)
10.59 K. Kamimura, H. Fuse, O. Takimura, Y. Yamaoka, Appl. Environ. Microbiol. **59**, 924 (1993)
10.60 A.G. MacDonald, A.R. Cossins, Soc. Exp. Biol. Symp. **39**, 301 (1985)
10.61 C. Kato, M. Smorawinska, T. Sato, K. Horikoshi, J. Mar. Biotechnol. **2**, 125 (1995)
10.62 C. Kato, M. Smorawinska, T. Sato, K. Horikoshi, Biosci. Biotech. Biochem. **60**, 166 (1996)
10.63 D.H. Bartlett, T.J. Welch, J. Bacteriol. **177**, 1008 (1995)
10.64 R.K. Poole, F. Gibson, G. Wu, FEMS Microbiol. Lett. **117**, 217 (1994)
10.65 V.L. Miller, R.K. Taylor, J.J. Mekalanos, Cell **48**, 27 (1987)
10.66 A.G. MacDonald, Philos. Trans. Soc. Lond. **B304**, 47 (1984)
10.67 C.A. Royer, Methods Enzymol. **259**, 357 (1995)
10.68 J.A. Killian, C.H. Fabrie, W. Baart, S. Morein, B. de Kruijff, Biochim. Biophys. Acta **603**, 63 (1992)
10.69 F.H. Johnson, D. Brown, D. Marsland, Science **95**, 200 (1942)
10.70 F.H. Johnson, E.A. Flagler, Science **112**, 91 (1950)
10.71 S.M. Johnson, K.W. Miller, Nature **228**, 75 (1970)
10.72 C. Parsot, J.J. Mekalanos, Proc. Natl. Acad. Sci. USA **87**, 9898 (1990)c

11 Submarine Hydrothermal Vents as Possible Sites of the Origin of Life

Kensei Kobayashi[1] and Hiroshi Yanagawa[2]

[1] Department of Chemistry and Biotechnology, Faculty of Engineering, Yokohama National University, Hodogaya-ku, Yokohama 240-8501, Japan
E-mail: kkensei@ynu.ac.jp
[2] Department of Applied Chemistry, Faculty of Science and Technology, Keio University, 3-14-1 Hiyoshi, Japan
E-mail: hyana@applc.keio.ac.jp

Abstract. The latest scenario of life's origin is introduced. Terrestrial organisms are proposed to have been abiogenetically brought into existence on earth about 4 billion years ago, using organic compounds in seawater. Organic compounds could have formed abiotically in both terrestrial and extraterrestrial environments and then become pooled in a primordial sea. Submarine hydrothermal vents are considered to be ideal sites for abiogenic synthesis of the organic compounds preserved to the present day and a possible environment for chemical evolution toward the origin of life in the Archean ocean.

11.1 Introduction

How and when was life begun on earth? This is one of the most fundamental questions remaining for mankind. In the 1920s, Oparin and Haldane independently presented a chemical-evolution hypothesis for the origin of life; they hypothesized that simple compounds gradually and spontaneously 'evolved' into more complicated ones in the primordial ocean. A wide variety of laboratory experiments have been conducted to prove the hypothesis since Miller's historical spark discharge experiment reported in 1953 [1].

In the earlier studies of chemical evolution, the following scheme was widely accepted: (a) Active compounds such as hydrogen cyanide and formaldehyde were abiotically produced from a strongly reduced terrestrial atmosphere by lightening, solar UV radiation, and so on. (b) These active compounds were dissolved into the primordial ocean to give 'primordial soup' and bioorganic monomers like amino acids, purines and pyrimidines were formed. (c) They polymerized to give polypeptides and polynucleotides. (d) Finally 'life' was generated in the soup of biologically important polymers.

Recent findings and theories on the history of earth and other planets have given new insights into the chemical evolution and the origin of life. Particularly, the discovery of submarine hydrothermal vents in the late 1970s [2] has drastically changed the image of the primordial sea where life was believed to have begun. In this chapter, we would like to introduce the latest idea of the origin of life, centering around the proposed roles of 'hot' environments of the primordial sea suggested by submarine hydrothermal vents.

11.2 Abiotic Formation of Bioorganic Compounds in Planetary Atmospheres

Bioorganic compounds such as amino acids are believed to have been supplied to the primitive earth before the origin of life on earth. Were the bioorganic compounds formed on the primitive earth or formed in extraterrestrial environments and delivered to the earth? A large number of experiments have be performed to obtain evidence for amino acid formation on the earth or in space.

From the point of view of chemical evolution, reducing environments are attractive because amino acids can be synthesized abiotically from reduced gas mixtures by spark discharge [1], heat [3], ultraviolet light [4], and shock waves [5]. These experiments demonstrate that if the primitive earth atmosphere was reduced, consisting of a mixture of methane, ammonia and water, amino acids and other organic compounds could have been easily obtained using the available energies on the early earth.

Recent studies on the formation of planets have suggested, however, that high-velocity impacts of planetesimals onto a growing planet can result in the impact-degassing of volatiles and the formation of an impact-induced atmosphere at high temperature [6−8]. According to these hypotheses, the resulting impact-induced atmosphere contained carbon monoxide or carbon dioxide as a major carbon source. In addition, results of photochemical studies of the early earth's atmosphere show that the presence of hydroxyl radicals from the photodissociation of H_2O, together with the greater flux of UV radiation from the young sun, would have limited the half lives of reduced gases such as methane and ammonia to about 50 yr and about 1 week, respectively [9, 10]. Therefore, for a variety of reasons, the primitive earth atmosphere may have been only 'mildly reduced'. Some investigators view this as a problem for conventional chemical-evolution theories because only a trace of amino acids is obtained in simulated primitive atmosphere experiments when carbon monoxide or carbon dioxide are used as a carbon source [11, 12].

Kobayashi et al. [13, 14] examined possible roles of cosmic rays in the abiotic formation of bioorganic compounds. Figure 11.1 shows a typical experimental irradiation set-up. A mixture of carbon monoxide, carbon dioxide, nitrogen and water was irradiated with high energy (3−40 MeV) protons generated from a van de Graaff accelerator or an SF cyclotron. Figure 11.2 shows a chromatogram of the product after acid hydrolysis. It was proved that a number of proteinaceous and nonproteinaceous amino acids such as glycine, alanine, aspartic acid, and β-alanine, were abiotically produced from the gas mixture with a high yield. The G-value (number of formed molecules per 100 eV) of glycine was calculated to be about 0.02. The *G*-value of glycine was also ca 0.02 when methane was used in place of carbon monoxide [15]. It should be noted that the *G*-value of glycine is 0.002 for methane [16] and 10^{-6} for carbon monoxide [17] in the discharge experiments. The latest estimation of cosmic ray energy, 0.011 cal/cm^{2}yr 2π [14], gives a production rate of glycine of 1.0 μmol/m^2 yr in a 1:1 mixture of carbon monoxide and nitrogen with water vapor.

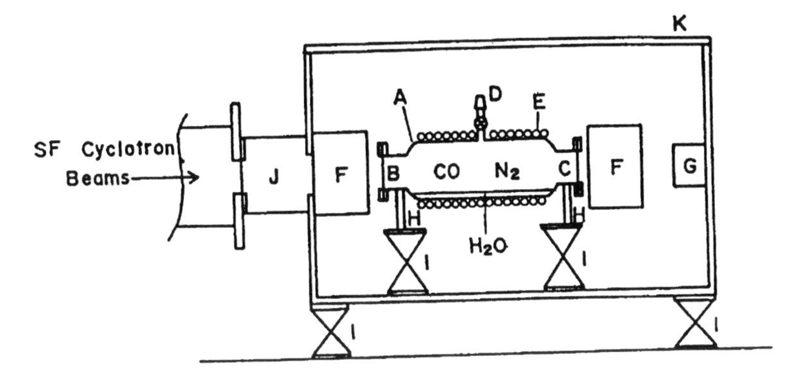

Fig. 11.1. Cross section of an experimental apparatus for the irradiation with high-energy particles when an SF cyclotron is used. A: Pyrex tube (840 mL), B: Havar foil (10 μm), C: Havar foil (40 μm), D: gas inlet, E: water jacket, F: ion chamber, G: Faraday cup, H: supporter, I: jack, J: shield tube, K: shield box

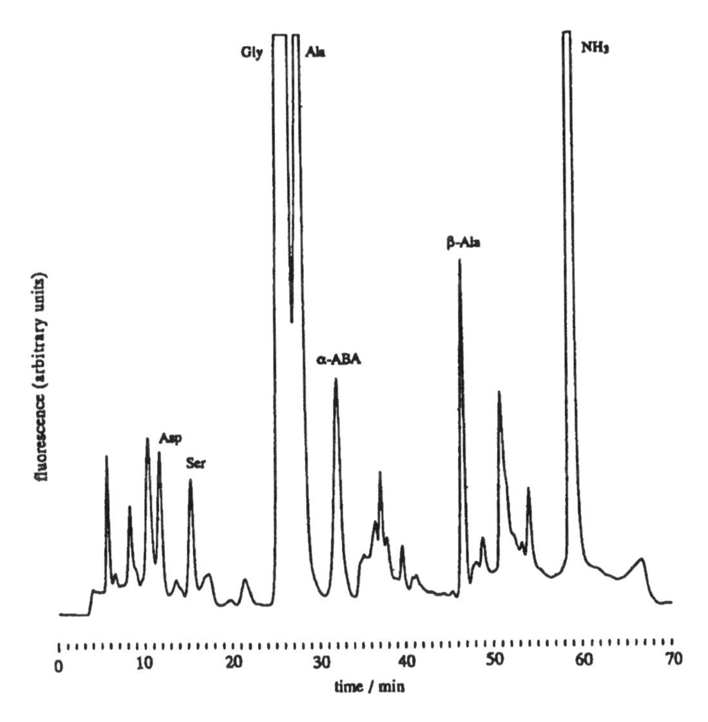

Fig. 11.2. Ion exchange chromatogram of a proton irradiation product after acid hydrolysis. A mixture of carbon monoxide (280 torr), nitrogen (280 torr) and water (20 torr) was irradiated with 3 MeV protons. D: aspartic acid, S: serine, Sa: sarcosine, E: glutamic acid, G: glycine, A: alanine; B: aminobutyric acid

If the composition of the gas mixture is changed, the *G*-value of glycine is also changed. When carbon dioxide is added to the mixture, for example, the *G*-value of glycine would be decreased by a factor equivalent to the $CO/(CO+CO_2)$ ratio in gas mixtures [13]. Thus, the estimated production rate of amino acids in the primitive earth atmosphere is greatly affected by the composition of the atmosphere.

11.3 Abiotic Formation of Bioorganic Compounds in Space

It is known that a wide variety of organic compounds are present in extraterrestrial environments; infrared spectrometers on board Voyagers 1 and 2 detected the presence of organic compounds in such solar system bodies as Jupiter and Titan (a moon of Saturn) [18]. Mass spectrometers on board Vega 1 and Giotto showed that cometary dusts contained a wide variety of complex organic compounds [19]. Microwave radiotelescopes have given evidence that a great number of organic molecules exist in molecular clouds which would evolve to new star systems [20].

Among them, cometary organic compounds have been regarded as an important source of those used for the generation of life on earth: it is believed that comets bombarded earth quite often in the early history of the earth (before ca 4 billion years ago), and brought water and organic compounds. It is supposed that organic compounds in comets were formed in their precursor bodies, interstellar dust particles (ISDs). ISDs in dense clouds were covered with ice mantles containing such volatiles as water, carbon monoxide, ammonia. In order to simulate reactions in ISDs, ice mixtures of carbon monoxide (or methane), ammonia and water made in a cryostat were irradiated with high-energy protons or ultraviolet light; organic compounds such as nitriles were detected in the products by infrared spectroscopy, but bioorganic compounds could not be detected in them [21, 22]. Briggs et al. [23] suggested that glycine was formed after ultraviolet irradiation of an ice mixture of carbon monoxide, methane, ammonia and water, but the experiments were not conducted quantitatively.

Abiotic formation of amino acid precursors in ice mantles of ISDs was quantitatively estimated for the first time through laboratory simulation. Mixtures of carbon monoxide (or methane, or methanol), ammonia and water were irradiated with protons, electrons or γ-rays. Carbon monoxide, ammonia and water were selected as starting materials since they are strongly suggested to be present in comets and ISDs as major volatile constituents [24–26]. Methane and methanol, which are also suggested to be in ISDs [22, 24], were used in some experiments. We used high-energy protons as energy sources, since major energy sources in space are ultraviolet light and cosmic rays, and major components of the latter are high-energy protons. Then, ice mixtures of carbon monoxide (or methane or methanol), ammonia and water ('simulated ISD ices') were irradiated with high-energy protons.

Figure 11.3 shows a schematic of an apparatus for proton irradiation of simulated ISDs. A gas mixture of carbon monoxide (or methane), ammonia and water vapor (1:1:1 in volume) was made with the gas mixer and propelled against the copper block located in the cryostat evacuated to $10^{-6}-10^{-7}$ torr, in order to make the simulated ISD ices. The ice mixtures were irradiated with 2.8 MeV protons from a Van de Graaff accelerator of Tokyo Institute of Technology.

After irradiation, the temperature of the metal block was naturally raised to room temperature, while generated gases were analyzed with the quadrupole mass spectrometer located between the cryostat and the turbo molecular pump. After analysis of volatile products, residual matter on the metal block was recovered with water, which was subjected to amino acid analysis.

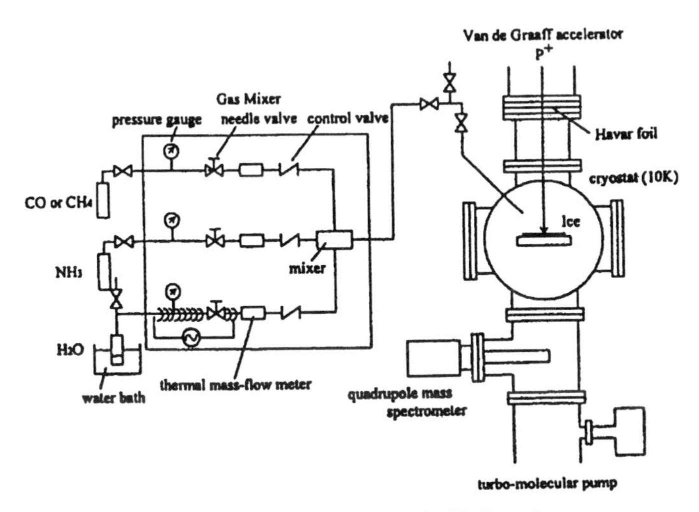

Fig. 11.3. Apparatus for proton irradiation of simulated ISD ice mixtures

Amino acid precursors, which gave amino acids such as glycine after acidhydrolysis, were detected in each product. Using the *G*-values of glycine experimentally obtained, a *G*-value of glycine in ice mantles of ISDs, whose major components were water, carbon monoxide and ammonia (100:10:1) was estimated to be ca 6×10^{-6} [28]. It is suggested that comets, aggregates of ISDs, contain ca 20 nmol/cm^3 of glycine precursors [29]. Bioorganic compounds delivered to the primitive earth by comets or their fragments (interplanetary dusts) are quite important sources for the building blocks of the first life on earth, as well as endogenous bioorganic compounds.

11.4 The Primeval Ocean as a Cradle of Life on Earth

Where did life start on earth? Some scientists insist that life was begun extraterrestrially and that seeds of the life were carried to the earth, which is known as the 'panspermia theory'. Most scientists believe that life was started in the oceans of our earth. One of the reasons is a close correlation between elemental composition of sea water and that of organisms. Nine of the top ten major elements found in living organisms, H, O, C, N, Na, Ca, S, K, Cl are also among those found in sea water (the exception being P). Not only the major elements, but also trace metal elements have the same tendency: transition elements relatively abundant in sea water such as molybdenum, iron, and zinc must have played important roles in the course of chemical evolution in the primeval sea [30]. Egami proposed that the inorganic composition of the primeval sea was essentially similar to that of the present sea [31].

As described in the previous sections, organic compounds could be formed both in the primitive earth atmosphere and in the interstellar media. Those of the former were directly collected in sea water, and those of the latter could have been brought to the ocean by comets or meteorites. Life is believed to have been born in sea water using these organic molecules. Then what was the primeval ocean like? We do not have solid evidence for how it was, but it was supposed to be something like 'cold consomme' or 'warm potage'. Recently, however, it is strongly suggested that bolide impacts of comets and meteorites on the primitive earth [32] kept the earth's surface environment 'hellishly' hot until the end of the late heavy bombardment era.

In the late 1970s, very hot water over 300°C was found venting at the Galapagos spreading center [2]. Such places were named 'submarine hydrothermal vents'. Submarine hydrothermal vents have been found at various sites along plate boundaries all over the world, which stimulated geophysical, geochemical, microbiological, ecological and ore-deposit research on submarine hydrothermal systems. They were also regarded as models of the primeval ocean where chemical evolution toward the generation of life occurred [33].

11.5 Implication of the Present Hydrothermal Systems for the Condition of the Primeval Ocean

Since submarine hydrothermal vent systems represent reducing, energy-rich, and metal-ion-rich conditions in the present marine environment, they are considered to be ideal sites for present-day abiogenic synthesis of organic compounds and have been suggested as a possible environment for chemical evolution toward the origin of life [2, 34]. It had also been suggested before the discovery of submarine hydrothermal vents that an important relationship may exist between the evolution of the earth's crust such as the hydro-geothermal zone associated with axes of plate spreading and the process of chemical evolution [35, 36].

Submarine hydrothermal vent systems have the following interesting characteristics: (a) thermal energy can be applied to synthesize organic compounds from simple starting materials; (b) they represent reducing conditions, which are favorable for organic syntheses; (c) the thermal fluid contains a quite high concentration of transition metal ions which catalyze chemical and biochemical reactions. We would like to examine these characteristics from the point of view of chemical evolution.

11.5.1 Heat Energy and Quenching

Since Miller's historical discharge simulating lightening [1], there have been various types of experiments simulating primitive earth environments. Besides lightenng, a number of energy sources have been simulated, including UV light from the sun [4], volcanic heat [3], radiation from radioisotopes in the crust [37], shockwaves after bombardment of meteorites [5] and cosmic rays [13]. It has been shown that bioorganic compounds such as amino acids are easily formed by using any of the energy sources above, if the starting gas mixture is a 'strongly' reduced atmosphere containing methane and/or ammonia.

In the case of hydrothermal vents, quite a large amount of energy is usual available from the magma underground. As shown in Fig. 11.4, sea water reaches to a deep reaction zone, where the water-rock interaction occurs at high temperature (over 300°C). Thus, heat energy can be used to form organic compounds. The heated fluid rises in an up-flowing zone, where a secondary modification process occurs at various temperatures. If the reaction product resides in the high temperature area, thermal reaction products will be decomposed by further heating. In the case of hydrothermal vents, however, the fluid in the high temperature area quickly rises and is discharged into cold sea water. Such a rapid cooling of the heat products is called 'quenching', which makes it possible for the organic products not to be decomposed by further heating.

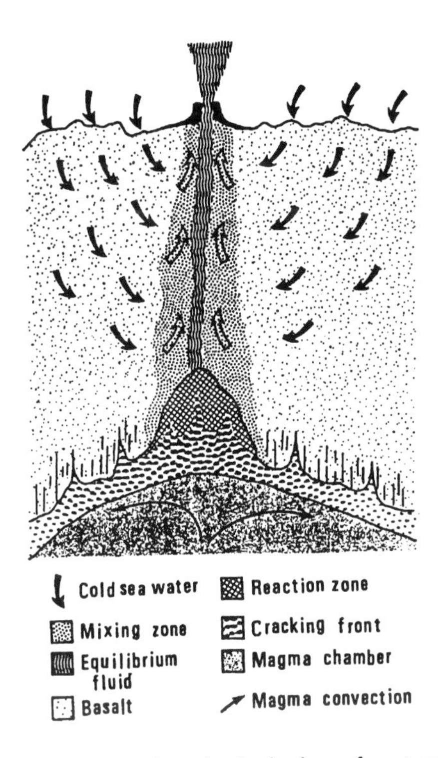

Fig. 11.4. The submarine hydrothermal vent reactor

11.5.2 Reducing Environments

From the point of view of chemical evolution, reducing environments are attractive because amino acids can be synthesized abiotically from reduced gas mixtures by spark discharge or other energy sources. These experiments demonstrate that if the starting materials were reduced, e.g., contained methane, ammonia or hydrogen in the mixtures, amino acids and other organic compounds could have been easily obtained using various types of energy, including heat. On the other hand, formation of organic compounds, particularly that of nitrogen-containing ones, was proved to be quite difficult to achieve from non- or slightly reduced starting materials [11, 12], with the exception of irradiation with high-energy particles [14].

It is strongly suggested that the oxidation states prevailing during large impacts will be strong functions of the composition of the target, which in the case of the earth would likely lead to oxidized conditions [38]: the primitive atmosphere of the earth was composed of carbon dioxide, carbon monoxide, nitrogen and water [8], that is, the slightly reduced gases. In a 'global' mildly reduced environment, organic molecules required for the origin of life could have been formed in 'local' reduced environments. Submarine hydrothermal systems may provide the type of reduced environment needed for the abiotic synthesis of various bioorganic compounds. In the present-day oceans, hydrothermal vents are more reduced than their immediate vicinity, since hydrothermal fluids contain reduced gases such as methane, hydrogen, hydrogen sulfide, and ammonia [39].

11.5.3 High Concentration of Trace Metal Ions

Submarine hydrothermal fluids contain high concentrations of metal ions, such as iron, manganese, copper, and zinc, as shown in Table 11.1 [40, 41], They may be able to catalyze organic synthesis reactions, which are indeed essential to many present-day biochemical processes.

Table 11.1. Concentration of metal ions in normal deep-sea water and 350°C hydrothermal fluids at 21° N, on the East Pacific Rise

	Units	Seawater	350°C fluid	Reference
pH		7.9	3.6	40
SiO_2	mmol/kg	0.16	21.5	40
Mg	mmol/kg	52.7	0	40
Ca	mmol/kg	10.3	21.5	40
Ba	mmol/kg	0.145	35−95	40
K	mmol/kg	10.1	25	40
SO_4	mmol/kg	28.6	0	40
Fe	mmol/kg	≈0	1.8	40
Mn	mmol/kg	0.002	610	40
Zn	mmol/kg	≈0	110	41
Cu	mmol/kg	≈0	15	42

Based upon this idea, Egami, Yanagawa and coworkers have been trying an experimental approach to the chemical evolution in the primeval sea. A 'modified sea medium (MSM)' was designed to accelerate possible reactions in chemical evolution in the laboratory. The MSM was enriched with six transition elements, iron, molybdenum, zinc, copper, cobalt, and manganese, by 1,000 – 100,000 times compared with present-day sea water. As mentioned above, it was found that amino acids and related compounds could be formed from formaldehyde and hydroxylamine in the MSM [42]. Furthermore, it was found that 'marigranules', highly organized particles, were produced from a reaction of glycine, and acidic, basic, and aromatic amino acids at 105°C in the MSM [43, 44]. We believed that metal ions in the modified sea medium played important roles in the formation of amino acids, peptides, and peptide-like polymers. When Egami and co-workers designed the MSM enriched with high concentrations of six metal ions in 1973, they were unaware of a natural, near-surface environment containing such high concentrations of metal ions. Submarine hydrothermal vents were discovered later and their metal-ion concentrations are similar to those of the MSM.

11.6 Experiments in Simulated Hydrothermal Vent Environments

11.6.1 Synthesis of Amino Acids

We performed experiments simulating submarine hydrothermal vent environments in order to examine the possible formation and alteration of amino acids. A medium that approximated hydrothermal vent sea water, designated as modified hydrothermal vent medium (MHVM), contained 2 mM $Fe(NH_4)_2(SO_4)_2$, 0.6 mM $MnCl_2$, 0.1 mM $ZnCl_2$, 0.1 mM $CuCl_2$, 20 mM $CaCl_2$, 0.1 mM $BaCl_2$, and 50 mM NH_4Cl. Its pH was adjusted to 3.6 with 1 N HCl [45]. The metal ion concentrations and pH were set close to the reported value of 350°C for hydrothermal fluid at 21° N on the East Pacific Rise [40, 41]. The MHVM was put into a Pyrex glass tube, which was placed in a stainless steel autoclave. After being purged with nitrogen five times, a gas mixture of methane (40 kg/cm^2) and nitrogen (40 kg/cm^2) was put into the autoclave at room temperature, which was then heated at 325°C for 1.5 – 12 h. After cooling to room temperature, the resulting aqueous solution was filtered through a membrane filter, and an aliquot of the solution was subjected to amino acid analysis after acid hydrolysis.

Figure 11.5 shows an ion exchange chromatrograms of (a) the amino acid fraction of the product compared to (b) a procedural blank. Amino acids such as glycine, alanine, aspartic acid, serine, and glutamic acid, together with non-proteinaceous amino acids (e.g., 2-aminobutyric acid and sarcosine) were detected in the chromoatogram. The major amino acid products, glycine, alanine, and sarcosine, were also identified by gas chromatography and/or gas chromatography combined with mass spectrometry. Evidence that these amino acids were synthesized abiotically from methane, includes: (a) the gas chromatographic analysis with an optically-active column indicated that the resulting amino acids were racemic mixtures; (b) non proteinaceous amino acids such as sarcosine and 2-aminobutyric acid were detected; and (c) the procedural blank gave much lower amounts of amino acids. Only trace amounts of aspartic acid, serine, and glutamic acid were detected in the products. Table 11.2 shows the yield of the major amino acids synthesized in the present experiments. Larger amounts of glycine, alanine, and sarcosine were obtained when a longer heating period was applied. After the reaction, silica was found with Fe(III) and Mn(IV) in the precipitate. The silica was apparently derived from the Pyrex glass wall. Effects of metal ions on the formation of amino acids was examined by removing metal ions from the MHVM, in which case the yield of amino acids was remarkably decreased. Some trace metals are now known to be essential to present-day life, and they may act as catalysts for prebiotic reactions in the course of chemical evolution [30].

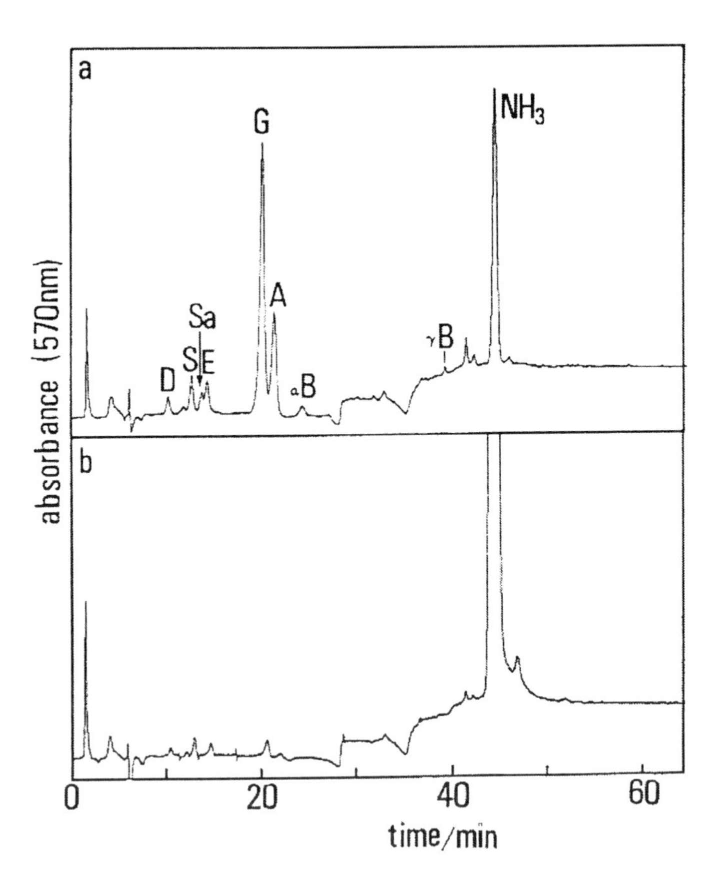

Fig. 11.5. Amino acid analysis of the products formed from methane and nitrogen in a modified hydrothermal vent medium by heating at 325°C and 200 kg/cm^2 for 1.5 h (***a***) and a procedural blank (***b***). Abbreviations: D, aspartic acid; S, serine; Sa, sarcosine; E, glutamic acid; G, glycine; A, alanine; αB, 2-aminobutyric acid; γB, 4-aminobutyric acid

Table 11.2. Formation of amino acids in hydrothermal experiments

	Reaction system					
Reaction conditions	Blank 1	Bank 2	Run 1	Run 2	Run 3	Run 4
Medium	MHVM[a]	MHVM[a]	H_2O	MHVM[a]	H_2O	MHVM[a]
Volume (mL)	40	40	50	50	50	30
CH_4 (kg/cm^2)	40	0	40	40	40	40
N_2 (kg/cm^2)	40	80	40	40	40	20
CO_2 (kg/cm^2)	0	0	0	0	0	12
Temperature (°C)	25	325	325	325	325	260
Pressure (kg/cm^2)[b]	80	200	200	200	200	140
Time (h)	12	3	12	1.5	12	6
Major amino acids produced (pmol/mL)						
Glycine	54	47	57	1190	2080	2400
Alanine	29	21	trace	600	980	630
Sarcosine	0	0	trace	640	1960	c

[a] Modified hydrothermal vent medium.
[b] Total pressure at the reaction temperature.
[c] Not separated from a large peak of glutamic acid

The results summarized here support the possibility that amino acids such as glycine and alanine are produced abiotically in present-day submarine hydrothermal systems [45]. The yield of amino acids was not high, because the experimental set up is a closed system, and amino acids synthesized in the thermal reaction would be thermally decomposed. If reactions were carried out in flow reactors resembling actual submarine hydrothermal systems [46], the yield would be expected to increase since the products would be moved away from the threat of thermal decomposition.

11.6.2 Stability of Amino Acids in Vent Environments

It has been, however, pointed out that organic compounds such as free amino acids are unstable in aqueous solution at high temperature. Miller and Bada [47] demonstrated that most amino acids were decomposed when heated at 250°C. Their experiments, however, were criticized by many investigators [48]: they did not consider the pH, oxidation states and concentration of dissolved species which are strongly influenced by fluid rock reactions.

We performed experiments simulating submarine hydrothermal vent environments in order to examine the stability and possible reactions of amino acids. Twenty cm^3 of 50 mM ammonium chloride-hydrochloric acid aqueous solution of amino acids (pH 3.6) were put into a Pyrex glass tube, which was placed in an autoclave. The autoclave was purged with nitrogen five times, and pressurized with nitrogen or a mixture of nitrogen (99 vol%) and hydrogen (1 vol%) with a pressure of 80 atm at room temperature. The latter was used to set the fugacity of hydrogen close to that in submarine hydrothermal vent systems [49]. Then the autoclave was heated at 200 350°C for 2 h. After cooling to room temperature, the resulting aqueous solution was filtered through a membrane filter (pore size: 200 nm) and lyophilized, and a part of the resulting product was hydrolyzed with 6 M hydrochloric acid at 110°C for 24 h. Amino acids in both the hydrolyzed and unhydrolyzed solutions were determined.

When an aqueous solution of amino acids was heated, the amount of amino acids was decreased. It was, however, increased after acid hydrolysis, particularly in the case of glutamic acid. It is suggested that the decrease in the concentration of amino acids was caused not only by decomposition, but also by formation of heat-resistant compounds which yielded amino acids after hydrolysis. Pyroglutamic acid was identified as a major product by RPC and LI-MS, when glutamic acid solution was heated (11.1).

$$NH_2CH(CH_2CH_2COOH)COOH \rightarrow CH_2\,CH_2CONHCHCOOH + H_2O \qquad (11.1)$$

Simple amino acids such as glycine and alanine were found in the heated solution of glutamic acid.

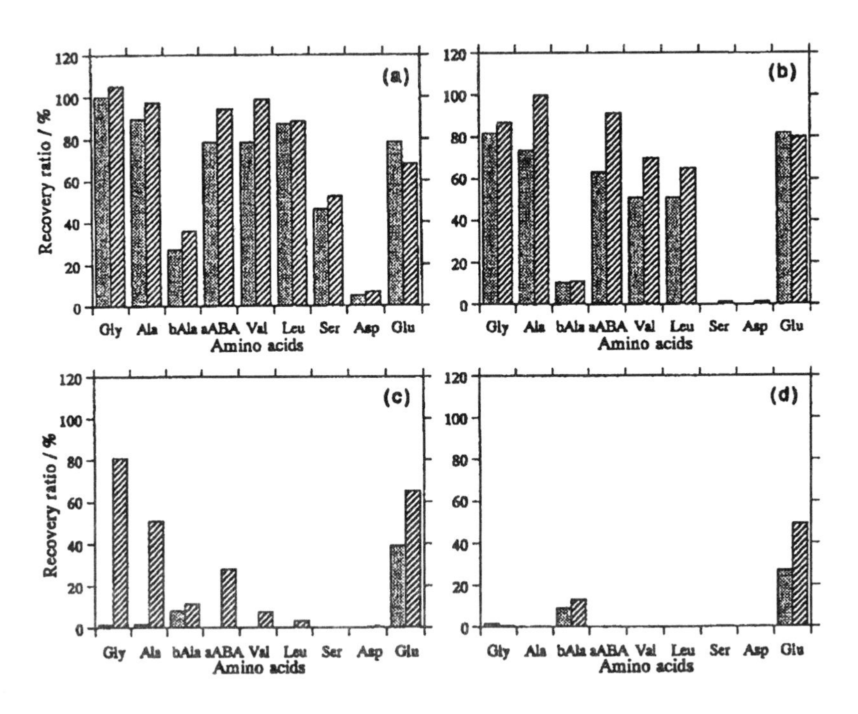

Fig. 11.6. Recovery ratio (%) of amino acids after acid hydrolysis. Aqueous solution of amino acids or amino acid amides were heated at (*a*) 200, (**b**) 250, (**c**) 300 and (**d**) 350 °C for 2 h after being pressurized with nitrogen or a mixture of nitrogen and hydrogen. Abbreviations are as follows: bAla: β-Alanine; aABA: α-aminobutyric acid.

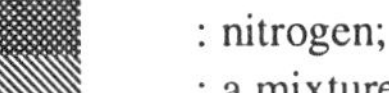
: nitrogen;
: a mixture of nitrogen (99 vol%) and hydrogen (1 vol%)

Figure 11.6 shows the recovery ratio of amino acids after acid hydrolysis when an aqueous solution of amino acids was heated at 200−350°C for 2 h. At each temperature, the recovery ratio of amino acids was improved when hydrogen was added to the pressurizing gas; it was drastically increased particularly at 300°C [50]. It was suggested that amino acids could be more stable than Miller and Bada [47] insisted, if the fugacity of hydrogen in the system was as high as that in the actual submarine hydrothermal systems.

Stability of organic compounds in hydrothermal systems has been discussed by using such equilibria as (11.2) [49].

$$2CO_2 + 1/2N_2 + 9/2H_2 = \text{Glycine } (C_2H_5O_2N) + 2H_2O \tag{11.2}$$

The equilibria above suggested that destruction of organic compounds such as amino acids was suppressed by the high fugacity of hydrogen. In addition, our results showed that polymerization reactions might occur among amino acids in hydrothermal fluids [50]. In such case, activities of dissolved organic species cannot be calculated by using the simple equilibria used in previous works [47,49].

11.6.3 Formation of Microspheres and Oligomers

The experimental results summarized above indicate that abiotic synthesis of amino acids and other organic compounds from inorganic starting materials may be possible in submarine hydrothermal systems. Questions as yet unanswered concern the role of hydrothermal environments in the abiotic synthesis of oligomers from amino acids. Results summarized below show that extremely thermophilic microspheres and amino acid oligomers can be formed in aqueous solution at 250−350°C.

When an aqueous solution (15 mL) containing 0.3 M glycine, 0.1 M L-alanine, 0.3 M L-valine, and 0.1 M L-aspartic acid was put into a glass tube and heated at 250°C and 134 atm for 6 h, numerous microspheres were formed [51]. These are organized particles of 1.5 to 2.5 μm in diameter (Fig. 11.7). They were also obtained in reactions at 300°C, though their spherical structures were deformed. Microspheres were not obtained in the reaction at 200°C. However, IR and NMR studies on the heated mixture revealed the presence of amide bonds and the near absence of imide bonds. After hydrolysis of products, glycine, alanine, valine, and aspartic acid were found in nearly equal amounts. When the reaction mixture was heated at 250°C and 134 atm for 6 h in a stainless steel vessel, no microspheres were observed. Microspheres were not obtained from mixtures of glycine alone, of glycine−alanine or of glycine−alanine−valine heated at 250°C for 6 h, suggesting that the presence of aspartic acid is crucial for microsphere production. Other polar amino acids such as glutamic acid, lysine, arginine, histidine, serine, threonine, and 4-hydroxyproline could be used instead of aspartic acid to form microspheres. Moreover, basic amino acids (lysine, arginine, and histidine) resulted in the formation of larger microspheres (4−8 μm in diameter). These results indicate that a polar amino acid, a glass tube, and a reaction temperature of 250°C or above are necessary for the formation of the microspheres.

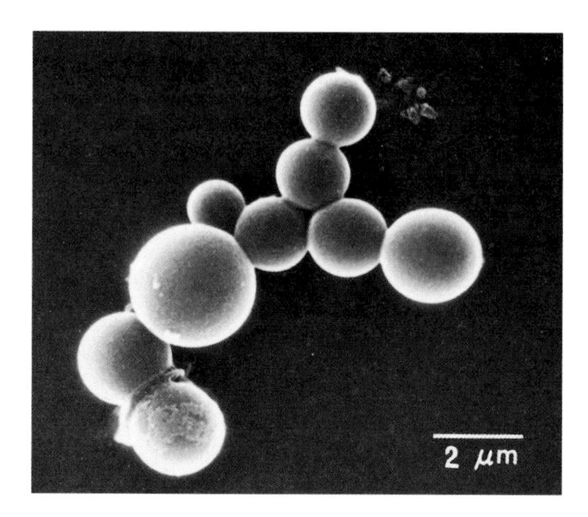

Fig. 11.7. Formation of microspheres in a reaction of mixture containing glycine, alanine, valine, and aspartic acid at 250°C under a hydrostatic pressure of 134 atm. A reaction mixture (15 mL) containing 0.3 M glycine, 0.1 M L-alanine, 0.3 M L-valilne, 0.1 M L-aspartic acid, and 0.1 M $KHCO_3$-NaH_2PO_4 buffer (pH 7.2) was put into a pyrex tube (20 x 105 mm), which was capped and placed in a stainless steel autoclave (70 mL) that was encased in an aluminum heating block. The autoclave was heated at 250°C for 6 h. The resulting solution was centrifuged and the precipitate was placed on a clean glass coverslip. The dried specimens were examined under a JEM 100 CX-ASID

The microspheres were dissolved with 1% SDS-8 M urea or 0.5% hydrofluoric acids. The microspheres completely disappeared on treatment with 1% SDS-8 M urea at 100°C for 10 min or 0.5% hydrofluoric acid at room temperature for 10 min. Polyacrylamide gel electrophoresis of the 1% SDS-8 M urea-solubilized components of the microspheres gave, after silver staining, a broad band with molecular weights of 1,000−2,000 daltons. The microspheres were partially hydrolyzed with 6 N HCl at 110°C for 72 h. After the hydrolysis, 10% of the total amide bonds were cleaved and silica was found in the hydrolyzate. These results suggest that the microspheres are made of oligomers with the following characteristics: (a) resistance to acid hydrolysis, (b) silicon bonds are included, and (c) molecular weights of 1,000−2,000 daltons.

Oligomers were also present in the supernatant of the reaction mixture, and they were found to remain on an Amicon YM-2 membrane after ultrafiltration. Polyacrylamide gel electrophoresis of oligomers in the supernatant gave, after silver staining, a broad band with molecular weights of 1,000−2,000 daltons. The electrophoretic pattern was very similar to that of the 1% SDS-8 M urea-solubilized component of the microspheres.

Figure 11.8 shows gel chromatograms of the products of several reaction mixtures heated at different temperatures. The reaction mixture at 200°C gave a single peak with a molecular weight of 1100 daltons. This peak can barely be detected at 100°C, increases at 150°C and is the sole product at 200°C (Fig. 11.8a). At 250°C several peaks with molecular weights up to 4000 daltons appear in the chromatogram (Fig. 11.8b). A gel chromatogram of the product of the reaction mixture heated at 250°C in a stainless steel vessel was much the same as that of the reaction mixture heated at 250°C in a glass tube. At 300°C these peaks disappeared and only a peak with a large molecular weight and a peak with a small molecular weight remained. (Fig. 11.8c). Judging from the fact that polyacrylamide gel electrophoresis of the large molecular weight peak gave a broad band with a molecular weight of 1000–2000 daltons, this large molecular weight peak seems to be an aggregated form of oligomers. Apparently, at 250°C and 300°C aggregation and degradation of the oligomers occurs simultaneously. At 350°C these peaks decreased considerably (Fig. 11.8d).

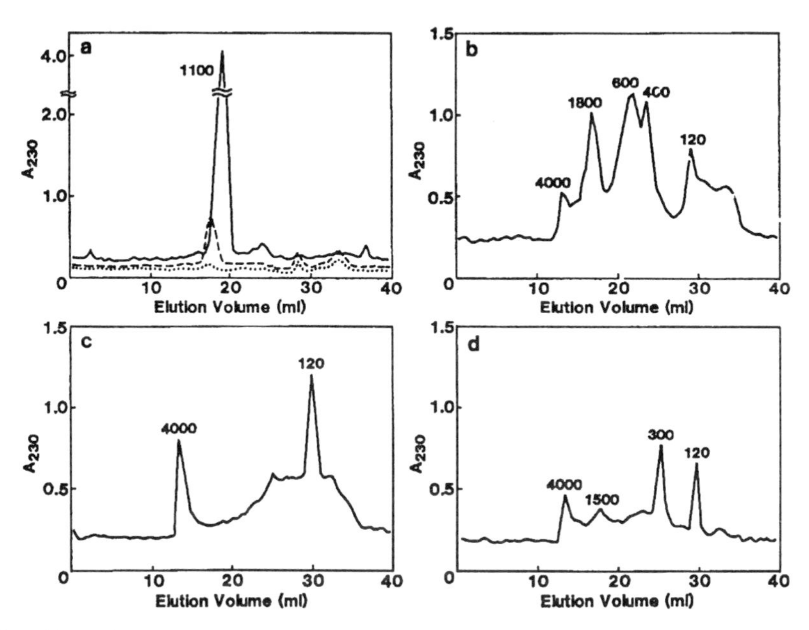

Fig. 11.8. Bio-Gel P-4 chromatography of supernatants of a reaction mixtures heated at different temperatures. (**a**), Supernatants of reaction mixtures heated at 100 (...), 150 (---), and 200°C(-) for 6 h; (**b**), (**c**), and (**d**): supernatants of reaction mixtures heated at 250°C, 300°C and 350°C for 6 h, respectively. Numerical values over peaks represent molecular weight of the peaks

11.7 Conclusion

In our experiments, 'modified hydrothermal vent medium' was pressurized and heated. There are, however, several differences between our simulation system and the actual hydrothermal system. First, we heated aqueous fluid in a closed autoclave. The actual hydrothermal system is, however, considered as 'a flow system' where quenching of the heated fluid by cold seawater occurs. Second, we did not use hydrogen sulfide, one of the characteristic gaseous components in hydrothermal system, due to restriction of our experimental system. Third, the actual pH and redox condition of hydrothermal fluid is controlled with mineral assemblage buffers [49], but we did not use them. Further experiments on flow systems in which the pH, oxidation state and activities of inorganic ions are buffered to values consistent with observations of submarine hydrothermal systems will help clarify the extent to which the present results reflect what can happen in natural systems.

These experiments may also have implications for ancestral and contemporary life at high temperatures. A variety of micro-organisms are now known which grow optimally above 65°C. Since 1982, when Stetter isolated an organism capable of growing at 105°C from the hot seafloor of a submarine solfatara field [52], several extremely thermophilic organisms have been found with optimum growth temperatures at 100°C or above [53,54] and a few are capable of growth at 113°C [55]. All thermophilic bacteria are eubacteria or archaebacteria, and no eucaryotes are known which can grow at these temperatures. On taxonomic grounds there is good support for the ancestral organism to have been thermophilic [56]. Specifically, extreme thermophily is most strongly represented towards the root of the archaebacteria tree. Therefore, the ancestral archaebacterium might have been an extremely thermophilic anaerobe dependent on sulfur reduction.

Our experiments indicated that amino acids and their oligomers are stable at temperatures up to 250°C. However, so far there is no evidence that micro-organisms can grow at 113°C or above [55]. An archaebacterium, *Pyrolobus fumarii*, holds the current record for life at high temperature, growing at temperatures as high as a scalding 113°C. This may suggest that the primary structure of proteins is relatively stable at temperatures up to 250°C, but their secondary and tertiary structures are much unstable at 113−250°C. Our experiments suggest that extremely thermophilic cellular structures capable of growing at 113°C or above may be possible if they are somehow analogous to the microspheres produced from amino acids at 250°C.

References

11.1 S.L. Miller: Science **117**, 528 (1953)

11.2 J.B. Corliss, J. Dymond, L.I . Gordon, J.M. Edmond, R.P. von Herzen, R.D. Ballard, K. Green, D. Williams, A. Bainbridge, K. Crance, T.H. van Andel, Science **203**, 1073 (1979)

11.3 K. Harada, S. W. Fox, Nature **201**, 335 (1964)

11.4 C. Sagan, B. N. Khare, Science **173**, 417 (1971)

11.5 A. Bar-Nu, N. Bar-Nun, S. H. Bauer, C. Sagan, Science **168**, 470 (1970)

11.6 T. Matsui, Y. Abe, Nature **319**, 303 (1986)

11.7 T. Matsui, Y. Abe, Nature **322**, 526 (1986)

11.8 J. Kasting: Origins Life Evol. Biosphere **20**, 199 (1990)

11.9 J. S. Levine, J. Mol. Evol. **18**, 161, (1982)

11.10 J. F. Kasting, K.J. Zahnle, J. C. G. Walker, Precambrian Res. **20**, 121 (1983)

11.11 G. Schlesinger , S. L. Miller, J. Mol. Evol. **19**, 376 (1983)

11.12 A. Bar-Nun, S. Chang, J. Geophys. Res. **88**, 6662 (1983)

11.13 K. Kobayashi, M. Tsuchiya, T. Oshima, H. Yanagawa, Origins Life Evol. Biosphere **20**, 99 (1990)

11.14 K. Kobayashi, T. Kaneko, T. Saito, T. Oshima, Origins Life Evol. Biosphere **28**, 155 (1998)
11.15 K. Kobayashi, T. Kaneko, M. Tsuchiya, T. Saito, T. Yamamoto, J. Koike, T. Oshima, Adv. Space Res. **15**, 127 (1995)
11.16 Y. Honda, R. Navarro-Gonzarez, K. Kobayashi, C. Ponnamperuma, Viva Origino **19**, 18 (1991)
11.17 K. Kobayashi, T. Kaneko, C. Ponnamperuma, T. Oshima, H. Yanagawa, T. Saito, Nippon Kagaku Kaishi, **1997**, 823
11.18 G. Carle, D. Schwartz, J. Huntington (eds.), Exobiology in Solar System Exploration, NASA SP 512 (1992)
11.19 J. Kissel, F. R. Krueger, Nature **326**, 755 (1987)
11.20 L.E. Snyder: Origins Life Evol. Biosphere **27**, 115 (1997)
11.21 M.H. Moore, B. Donn, R. Khanna, M.F. A'Hearn, Icarus **54**, 388 (1983)
11.22 L.J. Allamandola, S.A. Sandford, and G.J. Valero, Icarus **76**, 225, (1988)
11.23 R. Briggs, G. Ertem, J.P. Ferris, J.M. Greenberg, P. J. McCain, C. X. Mendoza-Gomez, W. Schutte, Origins Life Evol. Biosphere **22**, 287 (1992)
11.24 T. Yamamoto, in Comets in the Post-Halley Era, ed. by R. L. Newburn, Jr., et al. Vol. 1 (Kluwer Academic, Dordrecht 1991) pp.361
11.25 G. Strazzulla, R.E. Johnson, in Comets in the Post-Halley Era, ed. by R. L. Newburn, Jr., et al. Vol. 1 (Kluwer Academic, Dordrecht 1991) pp.243
11.26 J.M. Greenberg et al., in The Chemistry of Life's Origins, ed. by G. M. Greenberg et al. (Kluwer Academic Publishers, Dordrecht 1993) pp.1
11.27 K. Kobayashi, T. Kasamatsu, T. Kaneko, J. Koike, T. Oshima, T. Saito, T. Yamamoto, H. Yanagawa, Adv. Space Res. **16**, 21 (1995)
11.28 T. Kasamatsu, T. Kaneko, T. Saito, K. Kobayashi, Bull. Chem. Soc. Jpn. **70**, 1021 (1997)
11.29 T. Kasamatsu, T. Kaneko, T. Saito, K. Kobayashi, Chikyukagaku (Geochemistry) **31**, 191 (1997)
11.30 K. Kobayashi, C. Ponnamperuma, Origins Life Evol. Biosphere, **16**, 41 (1985)
11.31 F. Egami: J. Mol. Evol. **4**, 113 (1974)
11.32 N.H. Sleep, K.J. Zahnle, J.F. Kasting, H.J. Morowitz, Nature **342**, 139 (1989)
11.33 N.G. Holm (ed.), Marine Hydrothermal Systems and the Origin of Life (Kluwer, Dordrecht 1992)
11.34 J. A. Baross, S. E. Hoffman, Origins Life Evol. Biosphere **15**, 327 (1985)
11.35 D.E. Ingmanson, M. J. Dowler, Origins of Life **8**, 221 (1977)
11.36 E.T. Degens, in The Global Carbon Cycle, ed by B. Bolin, E.T. Degens, S. Kemper and P. Ketner, P. (John Wiley & Sons, New York 1979) pp.57
11.37 W.M. Garrison, D.C. Morrison, J.G. Hamilton, A.A. Benson, M. Calvin, Science **128**, 214 (1951)
11.38 B. Fegley, Jr., R.G. Prinn., H. Hartman., G.H. Watkins, Nature **319**, 305 (1986)
11.39 M.D. Lilley, J.A. Baross, L.I. Gordon, in Hydrothermal Processes at Seafloor Spreading Centers, NATO Conference Series, IV: Marine Sciences, Vol. 12, ed. by P. A. Rona, P.A. et al. (Plenum, New York, 1983) pp.411
11.40 J. M. Edmond, K.L. Von Damm, R.E. McDuff, C.I. Measures, Nature **297**, 187 (1982)
11.41 K.L. Von Damm, B. Grant, J.M. Edmond, in Hydrothermal Processes at Sea Floor Spreading Centers, NATO Conference Series, IV: Marine Sciences, ed. by P.A. Rona, K. Bostrom, L. Laubier, K.L. Smith, Jr. Vol. 12 (Plenum, New York 1983) pp.391

11.42 H. Hatanaka, F. Egami, J. Biochem. **82**, 499 (1977)
11.43 H. Yanagawa,F. Egami, Biosystems **12**, 147 (1980)
11.44 H. Yanagawa, K. Kobayashi, F. Egami, J. Biochem. **87**, 855 (1980)
11.45 D.W. Ingmanson, M.J. Dowler, Nature **286**, 51 (1980)
11.46 J.B. Corliss,:Origins Life Evol. Biosphere **16**, 381 (1986)
11.47 S.L. Miller, J.L. Bada, Nature, **334**, 609 (1988)
11.48 E.L. Shock: Origins Life Evol. Biosphere **20**, 331 (1990)
11.49 E.L. Shock,: Origins Life Evol. Biosphere **22**, 67 (1992)
11.50 M. Kohara, T. Gamo, H. Yanagawa, K. Kobayashi, Chem. Lett. **1997**,1053
11.51 H. Yanagawa, K. Kojima, J. Biochem. **97**, 1521 (1985)
11.52 K.O. Stetter, Nature **300**, 258 (1982)
11.53 G. Fiala, K.O. Stetter, Arch. Microbiol. **145**, 56 (1986)
11.54 H.W. Jannasch, C.O. Wirsen, S.J. Molyneaux, T.A. Langworthy, Appl. Environ. Microbiol. **54**, 1203 (1988)
11.55 E. Blochl, R. Rachel, S. Burggraf, D. Hafenbradl, H.W. Jannasch, K.O. Stetter, Extremophiles **1**, 14 (1997)
11.56 C.R. Woese, O. Kandler, M.L. Wheelis, Proc. Natl. Acad. Sci. USA **87**, 4576 (1990)

12 The Effect of Hydrostatic Pressure on the Survival of Microorganisms

Horst Ludwig, Günter van Almsick, and Christian Schreck

Institute for Pharmaceutical Technology and Biopharmacy, Section Physical Chemistry, Im Neuenheimer Feld 346, 69120 Heidelberg, Germany
E-mail: horst.ludwig@urz.uni-heidelberg.de

Abstract. The sensitivity to elevated hydrostatic pressure is investigated for different bacterial species. Dependent on the species, a minimal pressure between 100 and 350 MPa is necessary for inactivation. The most sensitive cells are bacteria of oblong shape, the most resistant are cocci. The cell wall does not stabilize vegetative bacteria against pressure. Details of the kinetics of inactivation point to proteins as being the targets of pressure's action. Other features suggest that the membrane plays a role. The latter aspect is substantiated by staining experiments and electron microscopy. Thus, it seems a likely supposition that inactivation of vegetative bacteria is caused by the damage of membrane proteins.

12.1 Introduction

Inactivation of microorganisms by high hydrostatic pressure has a long tradition of more than 100 years. In 1883 Certes found living bacteria in water samples from the deep sea [1]. They were collected at a depth of 5100 m, thus corresponding to a pressure of 50 MPa. One year later Certes built an apparatus to produce pressures up to 600 MPa to continue his studies on the viability of microorganisms under high hydrostatic pressure [2]. Subsequent investigations have shown that all kinds of microorganisms, including viruses, can in principle be inactivated using pressures between 100 and 1000 MPa [3–7]. Some difficulties have arisen in the case of bacterial endospores [8]. They were found to be much more resistant to high hydrostatic pressure than vegetative cells. The role of pressure in the stabilization and instabilization of bacterial spores was discussed by Gould and Sale [9], and strategies for their inactivation by pressure were developed recently [10–12].

For many years high-pressure work on biological systems was mainly directed towards marine biology, deep-sea diving, and hyperbaric medicine. Lethal and sublethal pressure effects on living cells have been reported [13–16]. It was only in the last decade that a new stimulus was given in Japan by emphasizing the prospective benefits of pressure-treated foods [17]. This resulted in a tremendous increase of high-pressure studies, stimulating also investigations in the fields of medicine and pharmacy [18–22]. Nevertheless, there are no systematic studies available regarding the pressure resistance of different bacterial species. It is the aim of this study to compare the barotolerances and kinetics of different vegetative bacteria to find out general rules. The combined results of kinetic experiments and microscopic examinations of pressure-treated cells are used to shed light on the mechanisms of pressure-induced inactivation.

12.2 Experimental Methods

The bacteria were always freshly prepared just before the high-pressure experiments. In order to obtain genetically homogeneous populations they were grown up from single colonies in suitable media and were then harvested in the early stationary phase when they approached their maximal pressure resistance. The bacteria were usually pressurized in their growing media, the pH at normal pressure being 7 to 7.5. The surviving organisms were counted as colony forming units (cfu) on suitable agar plates. All experiments were repeated at least once. The relative standard deviation of the bacterial counts was about 15%, this can hardly be seen in the logarithmic plots. Only in case of very small numbers of surviving bacteria can the relative deviation rise to 50% and more.

12.2.1 Microorganisms

The bacteria were obtained from the Deutsche Sammlung von Mikroorganismen und Zellkulturen GmbH in Braunschweig, Germany. For the kinetic experiments the following strains were used: *Corynebacterium renale* ATCC 10848, *Escherichia coli* ATCC 11303, *Kurthia zopfii* ATCC 33403, *Micrococcus luteus* ATCC 4698, *Paracoccus denitrificans* ATCC 17741, and *Spirillum* DSM 9662, whose classification is not yet concluded. In all cases the bacteria were grown in Standard I medium from Merck (Darmstadt, Germany).

Comparable results of additional bacterial species are described in earlier publications. Put together, a broad enough spectrum of different bacteria is provided to compare their pressure tolerance.

12.2.2 High-Pressure Experiments

The high-pressure device consisted of ten small pressure vessels which could be thermostated and pressurized simultaneously. The pressure-transmitting medium was water, the maximal pressure 700 MPa [23]. The rate of pressure increase could be varied from very slow to about 10 s for 100 MPa. Thus it was possible to control adiabatic temperature effects. The bacterial suspensions were enclosed in polyethylene tubes (sample size about 1 mL) and placed into the pressure medium water inside the vessels. After simultaneous pressurization the single vessels could be depressurized and opened at different times in order to measure the kinetics of inactivation.

12.2.3 Staining of *E. coli* Cells with Fluorescent Dyes

Cells were stained using ethidium bromide or propidium iodide. Both dye-stuffs intercalate into nucleic acids, with this increasing the fluorescence intensity by a factor of 50-100. This does not work in the case of living cells because the dye cannot permeate the membrane or it is actively removed to the outside.

E. coli from the exponential phase were pressure treated at 25°C using 300 or 500 MPa for 10 min. The staining procedure was the same with both substances and followed the method given in [24]. The fluorescence of cells stained with ethidium bromide was controlled by viewing them in the light microscope (Zeiss 473028, magnification 1000×) using the 485 nm line from a mercury lamp (HBO 50, Osram) with filter (Zeiss H485) for excitation.

Stainability with propidium iodide was investigated using a flow cytometer (FACScan, Becton Dickinson, Heidelberg). About 50,000 cells were measured the flow rate being ≤5000 cells s^{-1}. Excitation was done by the 488 nm line of an argon laser, the sideward scatter (90°) of the emission between 600 and 650 nm was analyzed. The amplifying factor was the same for all fluorescent signals.

12.2.4 Transmission Electron Microscopy of *E. coli* Cells

The bacterial suspensions were fixed with diglutaraldehyde, contrasted with osmiumtetroxide, and cautiously dehydrated starting with 30% acetone and finishing with 100% acetone. The bacteria were then transferred into synthetic resin, polymerized, and sliced into 40-90 nm thin sections. These were contrasted using uranyl acetate and lead citrate and thus studied in a transmission electron microscope (Philips CM10). A detailed description of the procedure is given elsewhere [24, 38]. The magnification was 25,000× to 40,000×.

12.3 Results and Discussion

12.3.1 Barotolerance of Bacteria

Figures 12.1 and 12.2 show the sensitivity to pressure of six bacterial species. In all cases the pressure given on the abscissa lasted for 5 min, the temperature was 25°C. The ordinate gives the decrease of the number of living bacteria on a logarithmic scale. The most sensitive species is the *Spirillum*. Its inactivation starts at 150 MPa, *E. coli* and *K. zopfii* follow above 200 MPa (Fig. 12.1). More stable are *C. renale*, *P. denitrificans*, and *M. luteus* with starting pressures of 250, 300, and 350 MPa, respectively (Fig. 12.2). Beyond the starting pressure the rate of inactivation increases exponentially during the first 100 to 150 MPa, but then it accelerates more slowly and finally it seems to approach a limiting value. This is clearly recognized for *E. coli* (Fig. 12.1) and is indicated in the cases of *P. denitrificans* and *M. luteus* (Fig. 12.2). This supposition is substantiated for *P. denitrificans* in Fig. 12.3. The figure shows the inactivation effects which occur during the time of pressure build-up as well as the raw data. The raw data were obtained in the following way: The samples were pressurized to the nonlethal pressure of 250 MPa and equilibrated for 2 min, which is enough to remove adiabatic heat effects. This was followed by a fast pressure rise of 50 MPa to 300 MPa (with negligible adiabatic heating). Then the effect of 2.5 min at 300 MPa was determined. To get the values at 350 MPa, after equilibration at 250 MPa and the fast pressure rise to 300

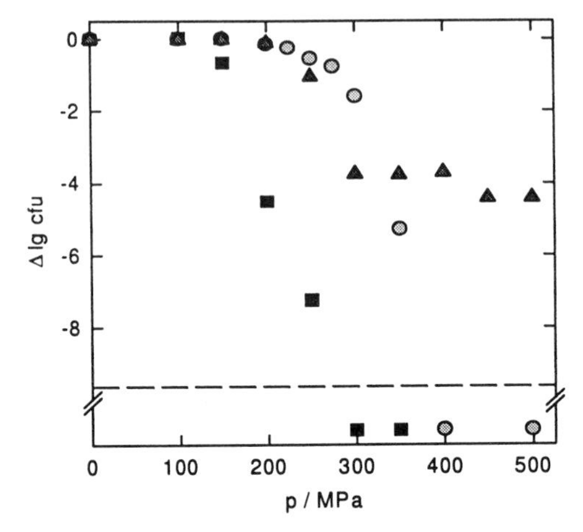

Fig. 12.1. Barotolerance of ■ *Spirillum* DSM 9662, ● *K. zopfii*, and ▲ *E. coli*. Experimental conditions: 5 min at 25°C and different pressures. The detection limit is shown (dashed line)

MPa, the system was equilibrated for 30 s and then again pressurized by another 50 MPa. The 2.5 min effect at 350 MPa could then be measured (raw data). This was corrected by using one-fifth of the effect determined at 300 MPa.

This procedure was continued in 50 MPa steps to higher pressures. The resulting corrected data give the real effect of 2.5 min under the desired pressure. Now the limiting rate can clearly be seen, 2.5 min at 450 or 500 MPa causes the same inactivation of *P. denitrificans*. Pressure limits of the inactivation rate could also be proved in case of *Spirillum* [25] and in the cases of *E. coli* and *Pseudomonas*

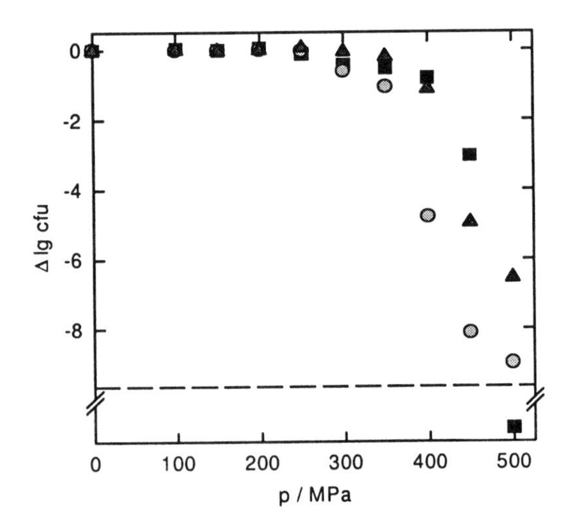

Fig. 12.2. Barotolerance of ■ *C. renale*, ● *P. denitrificans*, and ▲ *M. luteus*.Experimental conditions: 5 min at 25°C and different pressures. The detection limit is shown (dashed line)

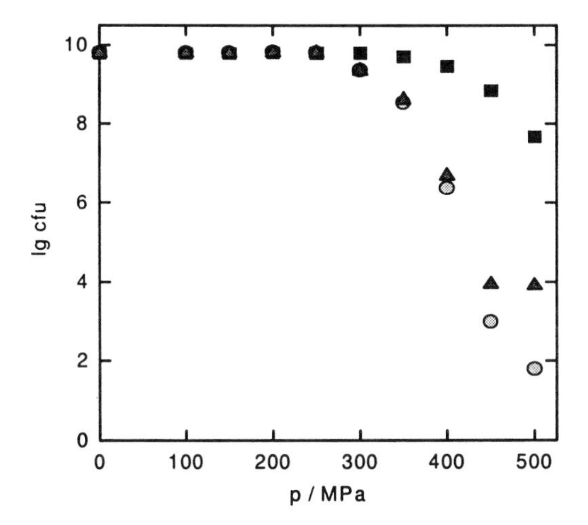

Fig. 12.3. Barotolerance of *P. denitrificans*. Experimental conditions: 2.5 min at 25°C and different pressures. ▲ Data are corrected for pressure and heat effects before the final pressure, indicated on the axis, was reached. ● Raw data. ■ Inactivation effects during pressure set up

aeruginosa [27]. Miyagawa and Suzuki obtained similar results for the inactivation of the enzymes trypsin and chymotrypsin. They called this upper pressure limit the critical pressure and determined values of 800 and 650 MPa for trypsin and chymotrypsin, respectively, independent of pH and temperature [28, 29]. This similarity suggests a major role of proteins during the inactivation of bacteria. Moreover, the critical pressure indicates a complex mechanism of inactivation in both cases.

For bacteria, there is even more complexity involved as can be seen from Figs. 12.1 and 12.2. The inactivation rate of *E. coli* and *C. renale* increases in two distinct stages. For *E. coli* we see a larger step above 200 MPa and a small step at 400 MPa, for *C. renale* it is a small step at 250 and a very large one above 400 MPa. A new inactivation mechanism seems to start at pressures above 400 MPa. *E. coli* show an interesting difference to the other bacteria. Their inactivation begins at 200 MPa but the rate remains relatively low, even at 500 MPa where the faster inactivation reaction is already in effect. It should be mentioned that *E. coli* is a very versatile and adaptable organism: depending on the growth conditions (with more or less oxygen supply), medium, and pH value they present strongly different inactivation kinetics and rates [30, 31]. Similarly, *E. coli* covers a wide range of pressure sensitivities. Patterson et al. have reported on very resistant *E. coli* strains [32]. Figure 12.4 shows how a 90 min treatment with a sublethal pressure of 52.5 MPa enhances the pressure resistance of *E. coli* bacteria of the strain ATCC 11303.

Figure 12.5 summarizes the results of the barotolerance studies in our laboratory for 15 bacterial species at room temperature, together with information on Gram type and shape of the cell. The bars give the span from the starting to the critical pressures, where the maximal rate is reached. In cases where the critical pressure was explicitly proved, it is indicated by a dot on the high-pressure end of the bars. A second bar shows a second step in the barotolerance curve, i.e., the onset of a

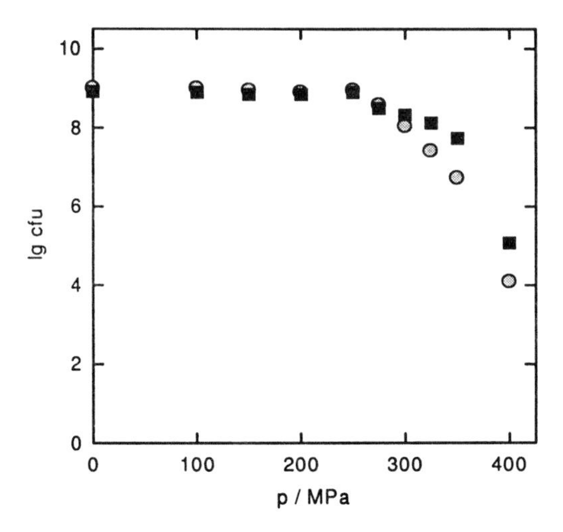

Fig. 12.4. Influence of a pretreatment with sublethal pressure on the barotolerance of *E. coli*. Experimental conditions: 5 min at 37°C and different pressures. ■ 90 min pretreatment at 52.5 MPa and 37°C. ● 90 min pretreatment at 0.1 MPa and 37 °C. The bacteria had been grown with a restricted O_2 supply for 4 h into the exponential phase

new inactivation mechanism with a higher maximal rate than before. Until now this feature was only found for the facultatively anaerobic species *E. coli* and *C. renale*.

The results in Fig. 12.5 show clearly that the most sensitive bacteria are rod-shaped, and the most resistant ones are spheres; between are pleomorphic forms and species consisting of mixed assortments of short rods and cocci. The variability of *E. coli* has already been mentioned above. There is no strong correlation with the Gram type. It is only by coincidence that Gram-positive species predominate on the resistant side: the majority of cocci investigated are Gram-positive. Raster electron microscopy could show that rod-like bacteria are much more mechanically damaged by hydrostatic pressure than are spherical ones. The pressure caused deformations of rods but not of spheres [33]. It seems that pressure induces mechanical stress to the cell wall and spherical bacteria are more stable against it. On the other hand, the reaction of bacteria to pressure resembles that of proteins. Therefore, proteins which are effected by the mechanical stress induced by pressure may be weak point. The best candidates are membrane proteins.

12.3.2 Kinetics of Pressure Inactivation

The time course of inactivation was measured just above the pressure thresholds found in the aforementioned barotolerance curves. Then, the reaction was slow enough to reveal the kinetic details. All data shown are from single experimental runs; their reproducibility was confirmed by at least one second independent run [25].

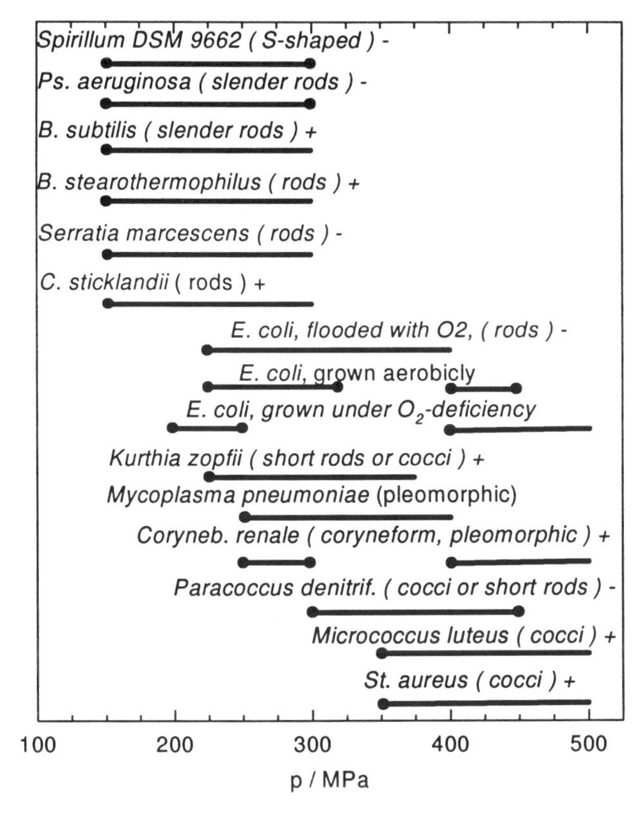

Fig. 12.5. Pressure sensitivity of vegetative bacteria at room temperature (25°C). The bars mark the pressure region from the start of inactivation (*left end*) to the maximal inactivation rate (*right end*). If it is explicitly proven that higher pressure gives no further acceleration of the reaction this is marked (circle at *right end*). The pH values of the pressurized cultures were in the range 7—7.5. In the cases of *Ps. aeruginosa*, *B. subtilis*, *B. stearothermophilus*, and *Serratia marcescens* the pH-value was between 8 and 8.5

Spirillum DSM 9662

Figure 12.6 compares the inactivation of *Spirillum* under pressures of 150 and 200 MPa, both at 25°C. The ordinate gives the number of living bacteria per mL as colony forming units (cfu), the abscissa the time in minutes. The experiment starts with about 10^9 cells/mL. Sterility is reached after 30 min using 200 MPa. It needs 2 h with 150 MPa. From Fig. 12.1 it can be derived that 5 min are enough at 300 MPa. The inactivation proceeds in two phases: a sharp decrease during the first 2 min is followed by a much slower reaction. The population seems to consist of two fractions with different sensitivities to pressure, the more sensitive fraction being increased by pressure. At 150 MPa the inactivation of the more stable fraction is preceded by a lag time, which disappears at higher pressures. Both bacterial fractions die at a first-order rate, i.e., different initial concentrations lead to the same

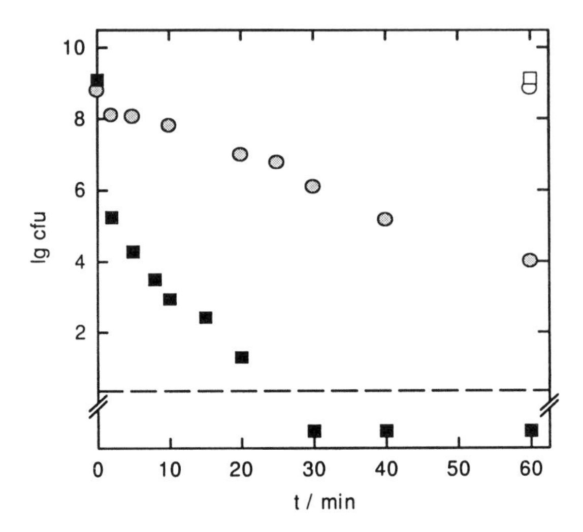

Fig. 12.6. Inactivation of *Spirillum* DSM 9662 at 25°C and at ● 150 MPa and ■ 200 MPa. Controls at ambient pressure (open symbols) and the detection limit (dashed line) are shown

inactivation curves; they are just vertically shifted. The properties described here for *Spirillum* were also found for other bacterial species. They seem to be common features of pressure inactivation [27, 30].

The influence of temperature shown in Fig. 12.7 resembles that of proteins [34] and of the majority of bacteria [27], which are most stable at room temperature. Higher and lower temperatures give faster inactivation.

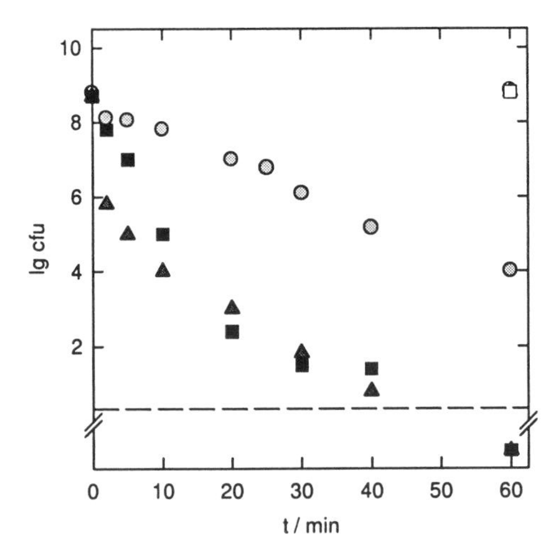

Fig. 12.7. Inactivation of *Spirillum* DSM 9662 at 150 MPa and at ▲ 4°C, ● 25°C, and ■ 40°C. Controls (open symbols) and the detection limit (dashed line) are shown

Kurthia zopfii

For *K. zopfii* (Fig.12.8) the inactivation rates at 4 and 25°C are the same, but after a decrease of the cell number by more than six orders of magnitude they diverge. At 4°C the suspension goes straight on to become a sterile solution while at 25°C biphasic behavior appears, with much slower inactivation of the remaining cells. There is a lag time during the first few minutes which is caused by lumping of the bacteria, with mean aggregation number of 2 [35]. At 40°C the initial rate is very fast, but then it slows down, so that the same inactivation effect is obtained with all temperatures after 40 min.

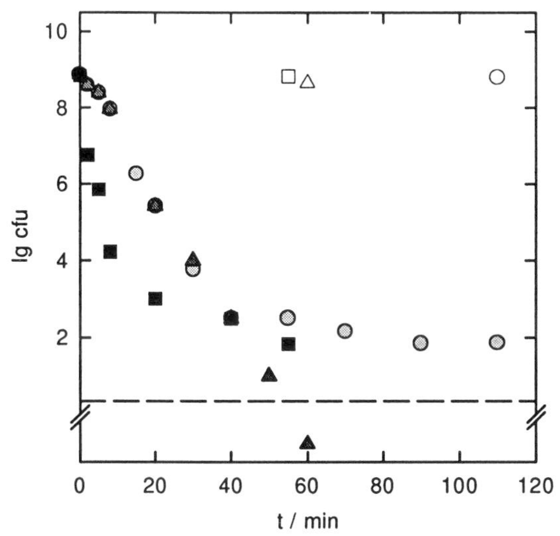

Fig. 12.8. Inactivation of *K. zopfii* at 250 MPa and at ▲ 4°C, ● 25°C, and ■ 40°C. Controls (open symbols) and the detection limit (dashed line) are shown

Corynebacterium renale

The inactivation of *C. renale* (Fig.12.9) shows some similarities to that of *K. zopfii*. Again there is a short lag time at low and room temperature, a simple first-order reaction at 4°C but a biphasic course at 25°C. In contrast to *K. zopfii* 40°C gives a straight line and this changes again at 50°C. Thus, the situation is very complicated. On the way from low to high temperature the inactivation changes two times from simple first-order to biphasic behavior.

Paracoccus denitrificans

Figures 12.10 and 12.11 show the inactivation of *P. denitrificans* at 300 and 350 MPa. Irrespective of the pressure, 4 and 40°C give nearly the same inactivation effect. The curves are biphasic for both temperatures. The initial inactivation is less at room temperature, mainly due to a lag time in the first few minutes. However, at the end 25°C is much more effective than low or high temperatures, the reason being that the first-order kinetics is followed; there seems to be no pressure-resistant fraction at room temperature.

Micrococcus luteus

The inactivation rate of *M. luteus* (Fig.12.12) is nearly the same at all temperatures; there is only a small advantage at 4°C. The shape of the inactivation curves is also very similar. It is biphasic in all cases with a small lag time, which is again most pronounced at room temperature.

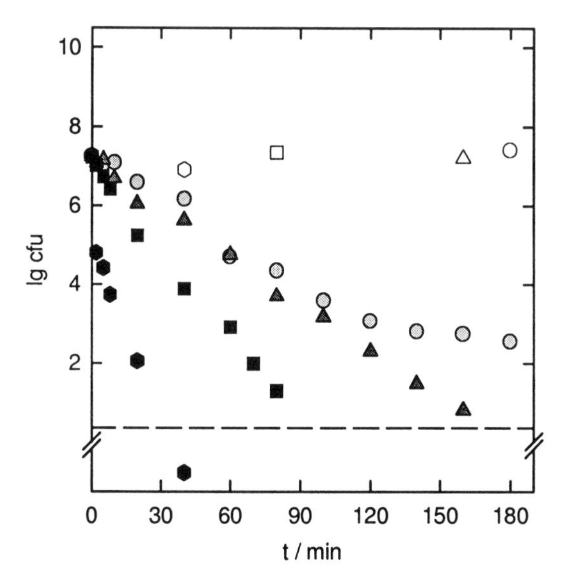

Fig. 12.9. Inactivation of *C. renale* at 250 MPa and at ▲ 4°C, ● 25°C, ■ 40°C and ● 50°C. Controls (open symbols) and the detection limit (dashed line) are shown

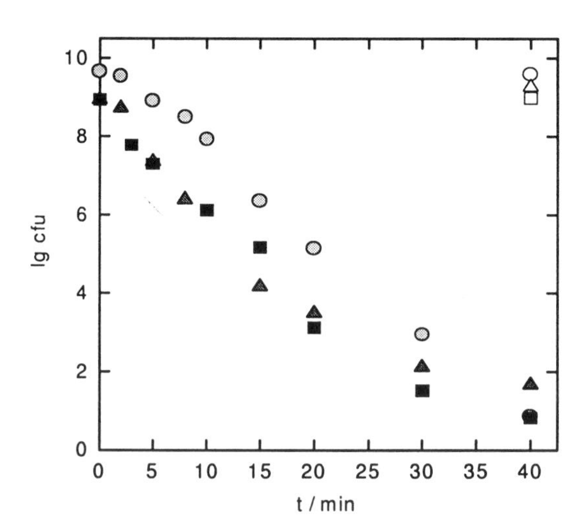

Fig. 12.10. Inactivation of *P. denitrificans* at 300 MPa and at ▲ 4°C, ● 25°C, and ■ 40°C. Controls are shown (open symbols)

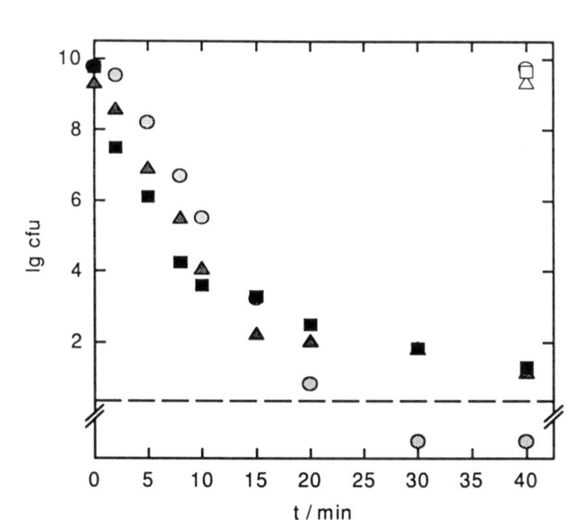

Fig. 12.11. Inactivation of *P. denitrificans* at 350 MPa and at ▲ 4°C, ● 25°C, and ■ 40°C. Controls (open symbols) and the detection limit (dashed line) are shown

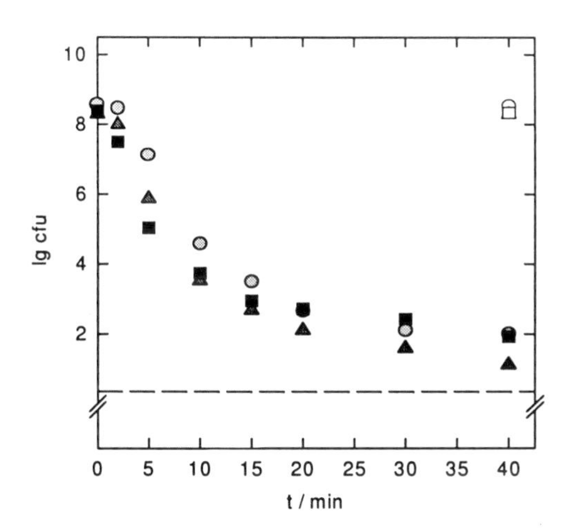

Fig. 12.12. Inactivation of *M. luteus* at 400 MPa and at ▲ 4°C, ● 25°C, and ■ 40°C. Controls (open symbols) and the detection limit (dashed line) are shown

In spite of their large variablility, the kinetic curves of the different species show some common features: (a) If first-order kinetics is found it remains restricted to a part of the whole temperature range only. At other temperatures the curves bend to biphasic behavior. They can be described by the inactivation of a bacterial population consisting of two fractions with different sensitivities to pressure [27]. *K. zopfii* shows simple first-order kinetics at low temperatures, *C. renale* at low temperatures and around 40°C, and *P. denitrificans* only at room temperature. For *Ps. aeruginosa* it was found below 35°C, and for *E. coli* above this temperature [27]. The question arises why the shape of inactivation curves is such a subtle function of temperature, but also of the medium and in some cases of the growth conditions [30]. It may be suggested that the reasons are small membrane transformations caused by changes of the temperature or by other altered environmental conditions. (b) The inactivation is most effective at low (*K. zopfii* and *M. luteus*) or high temperature (*C. renale*). It was shown for many other bacterial species that room temperature usually results in much slower inactivation. This study gives one of the rare exceptions. *P. denitrificans* is sterilized after 25 min at 350 MPa and 25°C but not at 4 or 40°C, in spite of the fact that the initial rate is higher in the latter cases. But the rates at 4 and 40 °C slow down, the curves in Fig. 12.11 are bent and show tailing, and therefore the first-order reaction at 25 °C is first. (c) With the exception of *Spirillum*, lag phases are visible in the first few minutes of the inactivation. These are most pronounced at room temperatures and disappear at higher temperature. The lag phase can be explained by aggregates of bacteria in all the cases given here. It is likely that the connection between bacteria is loosened at higher temperatures. In addition, the lag phase becomes more invisible the higher the rate is, and therefore it is most prominent at room temperature, where the inactivation rate is lowest. *Spirillum* do not aggregate, but their inactivation shows a lag phase for the resistant fraction. The same was found in case of *E. coli* at special growth and medium conditions [30]. There is no explanation for this at the moment.

12.3.3 Stainability of *E. coli* Cells and Electron Microscopy

The cells were stained using ethidium bromide and controlled in the light microscope. Living cells did not fluoresce after staining but cells fixed with ethanol (dead cells) did. Also, all the cells treated at 25°C for 10 min with 300 or 500 MPa absorbed the fluorescent dye [36]. It has been shown that ethidium bromide can slowly penetrate through the membrane into living cells but is thereafter actively pumped out [34, 37]. Cells are stained if the efflux pumps are inhibited. Our observations show that the inhibition by pressure is irreversible. Thus, high pressure seems to damage the transport mechanism.

Staining with propidium iodide yielded more complicated results. These results are quantitatively described by flow cytometry in Figs. 12.13 and 12.14. For each set of results about 50,000 cells were analyzed. The figures show the numbers of counted cells on the ordinate versus the fluorescent intensity on the abscissa. Figure 12.13 shows the controls without pressure. In Fig. 12.13a the bacteria were treated with 70% ethanol but were not stained; in Fig. 12.13b the living bacteria were stained; and in Fig. 12.13c the cells were first treated with ethanol and then stained. The fluorescence intensity of the unstained bacteria remained below 10 (Fig. 12.13a). The staining of living cells looks similar, only a minor part (4.7%) of the cells fluoresced stronger, with intensities between 10 and 100 (Fig. 12.13b). All the cells fixed with ethanol (Fig. 12.13c) were penetrated by the stain and fluoresced with high intensity. Figure 12.14a and b show the results for *E. coli* stained after a pressure treatment with 300 MPa at 25°C for 10 and 20 min, respectively. Compared with the control, the fraction of highly fluorescent cells increased from 4.7 to 14.6% only. This was independent of the time of pressure-treatment. Fixing the pressure treated cells with ethanol (Fig. 12.14c) resulted in the same fluorescence pattern as in Fig. 12.13c; the membrane is now penetrated by propidium iodide. In contrast to ethidium bromide, propidium iodide cannot permeate the membrane of living cells, probably due to the higher charge of the cation (two charges instead of one in the case of ethidium bromide). Therefore, the staining experiments lead to the conclusion that pressure which is lethal to *E. coli* damages the transport system of the bacterial membrane enough to inhibit the active efflux of ethidium bromide but not enough to allow a substantial influx of propidium iodide. In constrast, Smelt et al. found a correlation between stainability with propidium iodide and pressure inactivation of *Lactobacillus plantarum* [38]. Thus, there may be small differences between different bacterial species.

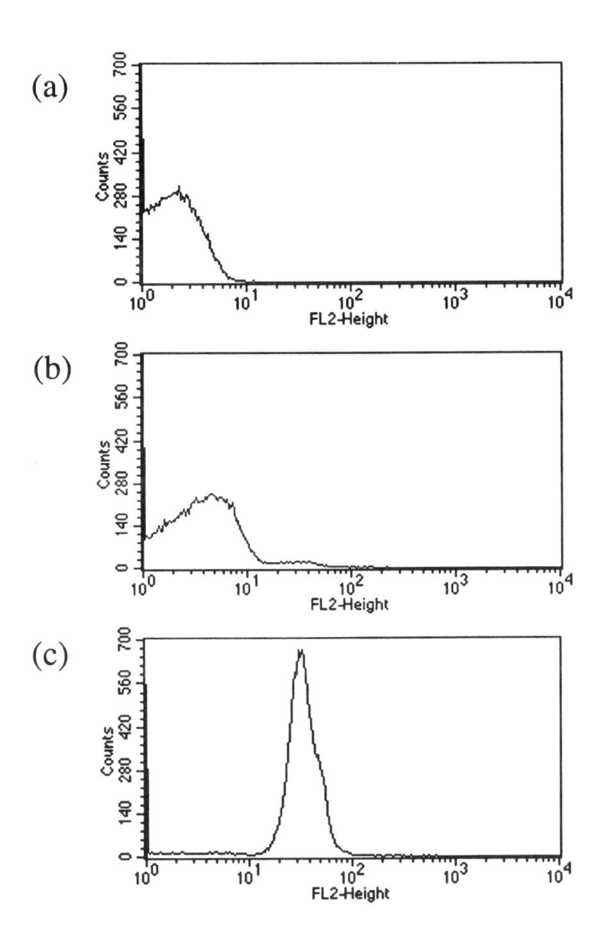

Fig. 12.13. Distribution of the fluorescence intensity of *E. coli* cells measured by flow cytometry (about 50,000 cells were counted). The logarithm of the fluorescence intensity is given on the abscissa, the number of counted cells on the ordinate. (**a**) Cells fixed with ethanol, (**b**) Living cells, not fixed, stained with 40 μg Propidium iodide/mL. (**c**) Cells fixed and stained [first treatment (**a**) then treatment (**b**)]

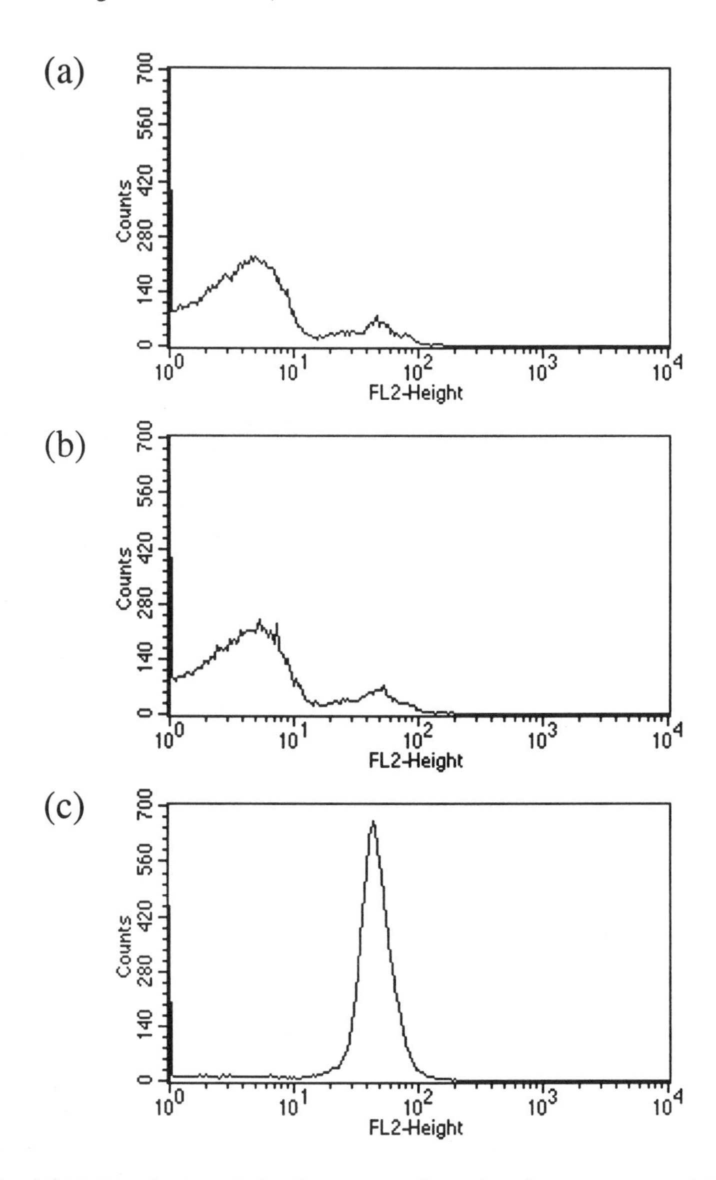

Fig. 12.14. Distribution of the fluorescence intensity of pressure-treated *E. coli* cells measured by flow cytometry. (**a**) Cells treated with 300 MPa at 25°C for 10 min, then stained with 40 µg propidium iodide/mL.(**b**) The same as a), but pressure treated for 20 min. (**c**) Cells treated with 300 MPa at 25°C for 20 min, then fixed with ethanol and after that stained

Figure 12.15 shows thin sections of *E. coli* cells under the electron microscope before and after pressure treatment. Fig. 12.15a is the unpressurized control and Fig. 12.15b a control with a nonlethal pressure of 150 MPa for 15 min. In both cases the interior presents a nearly homogenous distribution of the cell material. Five minutes of 250 MPa, which is lethal for the cell, lead to a segregation inside the cell (Fig. 12.15c). This is more pronounced with 5 min of 450 MPa (Fig. 12.15d). The white spots are DNA, the dark ones are clotted proteins [26]. Similar changes, as shown in Fig. 12.15c and d can be obtained using several methods other than the application of high pressure [39–42]. The common denominator of these methods is the microscopic damage of the cytoplasma membrane, with a consequent loss of homeostasis and a lowering of the pH value inside the cell.

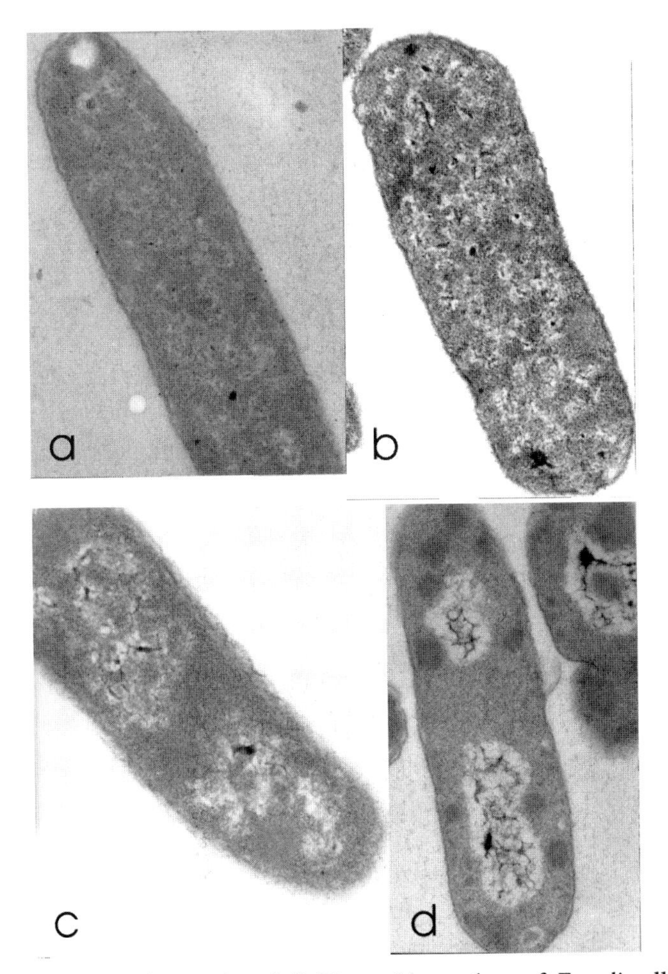

Fig. 12.15. Electron micrographs of 40-90 nm thin sections of *E. coli* cells. (a) Control without pressure (living cell), 38,500×; (b) control with non-lethal pressure of 150 MPa for 15 min, 40,000×; c) lethal pressure treatment with 250 MPa for 5 min, 38,500×; (d) lethal pressure treatment with 450 MPa for 5 min, 25,200×. The temperature during pressurization was 25°C

Thus, the initial points of attack of high pressure seem to be the transport proteins of the cytoplasmic membrane.

12.4 Conclusions

The staining experiments of *E. coli* cells and their inspection under the electron microscope support strongly the hypothesis that transport systems of the membrane are damaged by high hydrostatic pressure. The critical spots of the cell are suggested to be proteins which are involved in the microscopic transport mechanisms. Further confirmation of this view comes from the analysis of the complicated kinetics of inactivation and of the barotolerance curves. The kinetics resembles that of proteins in many aspects. The differences from that of proteins are caused by pecularities of the microorganisms, mainly by the fact that living bacteria can be counted over ten orders of magnitude. As a consequence, even very small fractions of various organisms can be detected by their differing reaction to pressure.

The minimum pressure for inactivation of living bacteria at room temperature extends from 150 to 350 MPa. This pressure limit is usually lower at high and low temperatures. After a fast increase above this limit the rate of inactivation approaches a maximal value at 300 to 500 MPa, dependent on the species. The most sensitive cells are slender rods, the most resistant are cocci. The cell wall seems to be of no importance. *Mycoplasma pneumoniae,* which has no cell wall, has a resistance similar to other pleomorphic, but cell-wall-protected bacteria [43]. However, the cell wall may be a disadvantage in this respect. It is easily deformed by pressure [33] and may pass this deformation on the adjacent membrane, thus introducing stress via the membrane to membrane proteins. The pressure resistance of certain strains of a bacterial species could therefore be explained by more resistant proteins in the critical positions and/or by a better stabilizing membrane and/or different contact of the membrane to the rigid cell wall.

The barotolerances of *E. coli* (Fig. 12.1) and *C. renale* (Fig. 12.2) decrease in two distinct steps. Obviously, the mechanism of inactivation is faster at higher pressures. It seems clear that living cells can be killed by pressure in different ways, depending on the pressure range. Therefore, in cases with a relatively low inactivation rate, another, faster reaction path will take over.

Acknowledgements. We would like to thank Mr. Klaus Hexel, DKFZ Heidelberg, for helping with the flow cytometry. Financial support was provided by the EU, project no. AIR1-CT 92-0296.

References

12.1 A. Certes: Compt. Rend. **98**, 690 (1883)
12.2 A. Certes: Compt. Rend. 99, 385 (1884)
12.3 P. Rogers: Arch. Physiol. Normale Pathol. 7, 12 (1895)
12.4 G.W. Chlopin, G. Tamman, Z. Hyg. Infektionskrankh. 45, 171 (1903)
12.5 B.H. Hite: Bull. W. Virginia Univ. Agric. Expt. Sta. 146, 1 (1914)

12.6 J. Basset, S. Nicolau, M.A. Macheboeuf, Compt. Rend. 200, 1882 (1935)
12.7 J. Basset, A. Gratia, M.A. Macheboeuf, P. Manil, Proc. Soc. Expt. Biol. Med., 38, 248 (1938)
12.8 W.P. Larson, T.B. Hartzell, H.S. Diehl, J. Infect. Dis. 22, 271 (1918)
12.9 G.W. Gould, A.J.H. Sale, in The Effects of Pressure on Organisms, Symposia of the Society for Experimental Biology, ed. by M.A. Sleigh, A.G. Macdonald 26 (Cambridge University Press, Cambridge 1972) pp.147
12.10 B. Sojka, H. Ludwig, Pharm. Ind. 56, 660 (1994)
12.11 B. Sojka, H. Ludwig, Pharm. Ind. 59, 355 (1997)
12.12 B. Sojka, H. Ludwig, Pharm. Ind. 59, 436 (1997)
12.13 A.M. Zimmerman (ed.) High Pressure Effects on Cellular Processes (Academic Press, New York and London 1970)
12.14 M.A. Sleigh, A.G. Macdonald (eds.) The Effects of Pressure on Organisms (Cambridge University Press, Cambridge 1972)
12.15 H.W. Jannasch, R.E. Marquis, A.M. Zimmerman (eds.) Current Perspectives in High Pressure Biolog Biology (Academic Press, London 1987)
12.16 P.B. Bennet, I. Demchenko, R.E. Marquis (eds.) High Pressure Biology and Medicine (University of Rochester Press, Rochester 1998)
12.17 R. Hayashi, in Engineering and Food, ed. by W.E.L. Spiess, H. Schubert (Elsevier Appl. Sc., Amsterdam 1989) pp.815
12.18 C. Balny, R. Hayashi, K. Heremans, P. Masson (eds.) High Pressure and Biotechnology 224, (Coll. Ins. John Libbey Eurotext, Montrouge 1992)
12.19 R. Hayash, C. Balny (eds.) High Pressure Bioscience and Biotechnology, Progress in Biotechnology 13 (Elsevier, Amsterdam 1996)
12.20 K. Heremans (ed.) High Pressure Research in the Biosciences and Biotechnology (Leuven University Press, Leuven 1997)
12.21 N. S. Isaacs (ed.) High Pressure Food Science, Bioscience and Chemistry (The Royal Society of Chemistry, Cambridge, 1998)
12.22 H. Ludwig (ed.) Advances in High Pressure Bioscience and Biotechnology (Springer, Berlin Heidelberg 1999)
12.23 P. Butz, J. Ries, U. Traugott, H. Weber, H. Ludwig, Pharm. Ind. 52, 487 (1990)
12.24 H. B. Steen, M. W. Jernaes, K. Skarstad, E. Boye, Meth. Cell Biol. 42, 477 (1994)
12.25 C. Schreck, Dissertation, University of Heidelberg, 1998
12.26 C. Schreck, W. Herth, H. Ludwig, in preparation
12.27 H. Ludwig, W. Scigalla, and B. Sojka, in High Pressure Effects in Molecular Biophysics and Enzymology, ed. by. J L. Markley, D.B. Northrop, C.A. Royer, (Oxford University Press, Oxford 1996) pp.346
12.28 K. Miyagawa, K. Suzuki, Rev. Phys. Chem. Jpn. 32, 43 (1963)
12.29 K. Miyagawa, K. Suzuki, Rev. Phys. Chem. Jpn. 32, 51 (1963)
12.30 C. Schreck, G.van Almsick, H. Ludwig, in Processing of Foods: Quality Optimisation and Process Assessment, ed by F.A.R. Oliveira, J.C. Oliveira , Chapter 18 (CRC Press, Boca Raton 1999) pp.313
12.31 H. Ludwig, C. Schreck, in High Pressure Research in the Biosciences and Biotechnology, ed. by K. Heremans (Leuven University Press, Leuven 1997) pp. 221
12.32 M.F. Patterson, M. Quinn, R. Simpson, A. Gilmour, in High Pressure Bioscience and Biotechnology, Progress in Biotechnology, ed. by R. Hayashi, C. Balny 13 (Elsevier, Amsterdam 1996) pp.267

12.33 H. Ludwig, P. Butz, H. Weber-Kühn, Deutsche Apotheker Zeitung 51/52, 2774 (1990)
12.34 K. Suzuki, Rev. Phys. Chem. Jpn. 29, 91 (1960)
12.35 G. van Almsick, C. Schreck, H. Ludwig, in Basic and Applied High Pressure Biology IV, ed by J.-C. Rostain, A.G. Macdonald, R.E. Marquis 5 (Medsubhyp Int. 1995) pp.69
12.36 G. van Almsick, Dissertation, University of Heidelberg, 1997.
12.37 H.M. Shapiro, Practical Flow Cytometry, 3rd ed. (Wiley-Liss Inc., New York 1995)
12.38 J. P. P. M. Smelt, A.G.F. Rijke, A. Hayhurst, High Pressure Res. 12, 199 (1994)
12.39 C. N. Cutter, G. R. Siragusa, J. Food Protect. 9, 977 (1995)
12.40 C. A. Cherrington, M. Hinton, G. R. Pearson, I. Chopra, J. Appl. Bact. 70, 161 (1991)
12.41 K. Shimada, K. Shimahara, Agric. Biol. Chem. 12, 3605 (1985)
12.42 Y. Nitzan, M. Gutterman, Z. Malik, B. Ehrenberg, Photochem. Photobiol. 1, 89 (1992)
12.43 C. Schreck, G. Layh-Schmitt, H. Ludwig, Pharm. Ind., 61, 759 (1999); C. Schreck, G. Layh-Schmitt, H. Ludwig, Drugs made in Germany 42, 84 (1999)

13 Dynamics of Cell Structure by Pressure Stress in the Fission Yeast *Schizosaccharomyces pombe*

Masako Osumi[1], Mamiko Sato[2], and Shoji Shimada[3]

[1]Department of Chemical and Biological Sciences, Faculty of Science, Japan Women's University, 2-8-1 Mejirodai, Bunkyo-ku, Tokyo 112-8681, Japan
E-mail: mosumi@sakura.jwu.ac.jp
[2]Laboratory of Electron Microscopy, Japan Women's University, 2-8-1 Mejirodai, Bunkyo-ku, Tokyo 112-8681, Japan
[3]Oriental Yeast Co., Ltd., 3-6-10 Azusawa, Itabashi-ku, Tokyo 174-8505, Japan
E-mail:shimada@oyc.co.jp

Abstract. Study of the effect of hydrostatic pressure on yeast cells revealed the impact of ultrastructural changes including microtubules and actin cytoskeletons. We also found that the fission yeast *Schizosaccharomyces pombe* is more sensitive to pressure stress than the budding yeast *Saccharomyces cerevisiae* using conventional electron microscopy (CEM), immunoelectron microscopy (immuno-EM), and fluorescence microscopy (FM). To investigate the influence of pressure stress on the cell cycle of *S. pombe,* we used the cells of a cold-sensitive mutant, *nda3* KM311, of *S. pombe*, which were arrested highly synchronously at a step similar to mitotic prophase under restrictive temperature at 20°C, for 4 h. We describe here that the morphological changes in actin cytoskeleton were caused by acceleration of pressure stress in *nda3* mutant cells related to induction of diploidization in *S. pombe*. When the *nda3* cells were incubated at the restrictive temperature of 20°C, large cells (diploid cells) appeared on a dye plate after pressure stress of 150 MPa. These cells made up over 40% of the colonies on the plate. *Nda3* cells were first aerobically grown at 30°C in YPD liquid medium to mid-exponential phase, transferred to restrictive temperature at 20°C for 4 h, and then shifted to a permissive temperature at 36°C for 15 min. The cells grown at 20°C had an abnormal ('leaf like') nucleus profile surrounded by normal nuclear membrane. After pressure stress treatment at 100 MPa the nuclear membrane was damaged and the matrix of mitochondria had an electron-dense area. At 150 MPa, other altered features were apparent: the nuclear membrane was broken over a broad area; the vacuoles had fused into large pieces in cells grown at both 20°C and 36°C. The influence of pressure stress on actin cytoskeleton in *nda3* cells was revealed by FM. In the cells grown at 20°C, actin patches were concentrated in the central region and actin rings were seen. Even at 100 MPa specific actin distribution was lost. Long and fine actin cables were seen all over the cells: large actin patches remained in the center of the cell and covered the actin rings, then they changed into thick and short cables at 150 MPa; they finally decomposed but the actin ring was visible even with faint fluorescence. Immuno-EM also showed this phenomenon. These results confirmed the process of degradation in actin cytoskeleton of *nda3* cells by pressure stress.

13.1 Introduction

Hydrostatic pressure has dramatic effects on cytokinetic and mitotic activities of microorganisms, especially yeast cells, which has been recognized as a useful model organism for analyses of the response to stress [1−3]. As with the budding yeast *Saccharomyces cerevisiae*, hydrostatic pressure above 100 MPa kills the fission yeast *Schizosaccharomyces pombe* cells rapidly [4], because it induces a physiological imbalance that causes internal structural damage, which renders them incapable of growing. Our current investigations by conventional electron microscopy (CEM) and fluorescence microscopy (FM) revealed that the membrane systems, especially the nuclear membrane, were most susceptible to pressure stress even at 100 MPa. Under FM, cell-cycle-specific organization of cytoskeletal elements (actin and tubulin cytoskeletons) was also found to be altered when *S. pombe* cells were exposed to pressure stress at 100 to 200 MPa, as seen in *S. cerevisiae* cells [5, 6]. This finding suggests that the structure of cytoskeletal elements, which is related to the nuclear division apparatus, might be severely damaged by pressure stress under the same conditions. The breakdown of nuclear division apparatus by pressure stress thus confirms the induction of diploidization in *S. pombe* as well in *S. cerevisiae*. In fact, we recently found that diploidization of the fission yeast *S. pombe* was induced at a frequency 35% higher than that of *S. cerevisiae* under the same pressure of 200 MPa [7].

In this study we used CEM, immunoelectron microscopy (immuno-EM), FM and dye plate-colony color assay to characterize in detail this stress on *S. pombe* cells. We also used the cold-sensitive *nda3* mutant [8], in which cells were arrested highly synchronously at a step similar to mitotic prophase under restrictive temperature (20°C, 4 h) [9], focusing on the effect of pressure on the distribution of actin cytoskeleton and the appearance of microtubules through the cell cyclerelated to induction of diploidization in *S. pombe*.

13.2 Experimental Methods

13.2.1 Yeast Strain and Cultivation

The JY1 strain of *S. pombe* (*L972h*$^-$, wild-type) and a cold-sensitive mutant *nda*3-KM311 (*h*$^-$ *leu1*) [8] were used. *Nda*3 cells were aerobically grown at 30°C in YPD liquid medium to mid-exponential phase. For temperature shift-up experiments (20→36°C), *nda*3 cells were first grown in YPD medium at 30°C for 16 h, and when the cell concentration reached 5×10^6 cells/ml, suspended cells (5×10^7 cells/ml) were shaken at restrictive temperature at 20°C for 4 h, then the cultures were transferred to permissive temperature at 36°C and harvested at appropriate times [9].

13.2.2 High-Pressure Treatments

Suspended cells (approximately 5×10^7 cells/ml) were put in small polyethylene bottles and placed in a high-pressure apparatus, NKK-ABB. They were treated with hydrostatic pressure of 100~200 MPa for 10 min at room temperature. The decompressed samples were immediately fixed to prepare specimens for FM, immuno-EM and CEM as described below [2, 10].

13.2.3 Colony-Forming Ability

The colony-forming ability of pressurized cells was determined by spreading the cell suspension on YPD agar plates, and the number of colonies appearing on the plates was counted after 3~4 days of incubation at 30°C [10].

13.2.4 Dye Plate-Colony Color Assay

Pressurized cells were spread on a dye plate containing Ponceau 3R and aniline blue. The colony color of pressure-affected cells (diploid cells) became violet, while that of unaffected cells remained pink [7].

13.2.5 Fluorescence Microscopy

Samples were fixed with 3.7% formaldehyde for 2 h. Microtubules were visualized by indirect FM by staining with anti α-tubulin monoclonal antibody and FITC conjugated anti-rat IgG. DNA was stained with propidium iodide or 4,6-diamidino-2-phenylindiole (DAPI) [5, 11]. Actin cytoskeleton was visualized by FM by staining F-actin with rhodamine-conjugated phalloidin.

13.2.6 Conventional Electron Microscopy by Freeze-Substitution Fixation

Samples were fixed with a mixture of 0.5% glutaraldehyde (GA) and 3% paraformaldehyde (PFA) for 1 h at room temperature. Then the cells were frozen by liquid propane for the sandwich method [12]. The frozen samples were transferred to substitution fluid (anhydrous acetone containing 2% OsO_4), maintained at −80°C with solid OsO_4/acetone for 48 h, then moved to −20°C for 2 h, to −4°C for 2 h and finally to room temperature for 1 h. They were washed three times with anhydrous acetone, and embedded in a Quetol 653 mixture. Ultrathin sections were stained with 6% uranyl acetate and lead citrate, and examined with a JEM 1200EXS transmission electron microscope (TEM) at 120 kV.

13.2.7 Immunoelectron Microscopy by Frozen Thin-Sectioning

GA−PFA fixed cells were treated with 1% sodium metaperiodate for 15 min and 50 mM ammonium chloride for 30 min at room temperature, then washed with 50 mM Tris-buffered saline [3]. The cells were mounted with 2% agarose, infused with a 20% polyvinyl pyrrolidone/1.84 M sucrose mixture in 0.1 M PBS, then frozen with propane, and sectioned by Reichert FCS at −100°C. To visualize microtubules, α-tubulin monoclonal antibody was used as a second antibody. To visualize actin cytoskeleton, sections were first incubated with anti-actin monoclonal antibody, and then with 10 nm colloidal gold-conjugated specific antibody [6].

13.3 Results and Discussion

13.3.1 Response of *S. pombe* Cells to Pressure Stress

Table 13.1 summarizes the effects of hydrostatic pressure on cytoskeletal elements, ultrastructure, colony-forming ability, and the maximum induction of diploidization of *S. pombe* cells compared with *S. cerevisiae* cells. First of all, above 100 MPa, the survival curve of *S. pombe* cells as determined by their colony-forming ability displayed a drastic decrease in proliferation, and at a pressure over 200 MPa, the cells were almost completely inactivated (Fig. 13.1,-■-); in *S. cerevisiae*, at pressures over 150 MPa, this ability was sharply reduced with increasing pressure, and only above 250 MPa did the cells lose all ability to proliferate (Fig. 13.1,-●-). Thus, *S. pombe* cells were more sensitive to pressure stress than *S. cerevisiae* cells [6].

Table 13.1. Comparison of effects of pressure between *S. pombe* and *S. cerevisiae*

Method	Characteristic	Remarks	Minimum pressure (MPa)	
			S. pombe	*S. cerevisiae*
FM	Microtubules	Abnormal distribution	100	150
		disappearance	150	250
EM	Ultrastructure	Nuclear membrane altered	100	150
		dense matrix in mitochondrion	150	200
	Microtubules	Disappearance of immuno gold particles	100	150
FM	Actin cytoskeleton	Abnormal distribution disap-	50	150
		pearance	150	150
Plate counting	Colony-forming ability	Complete loss	200	250
Dye plate-colony color	Colony color change	Maximum occurrence of large cells	200	200

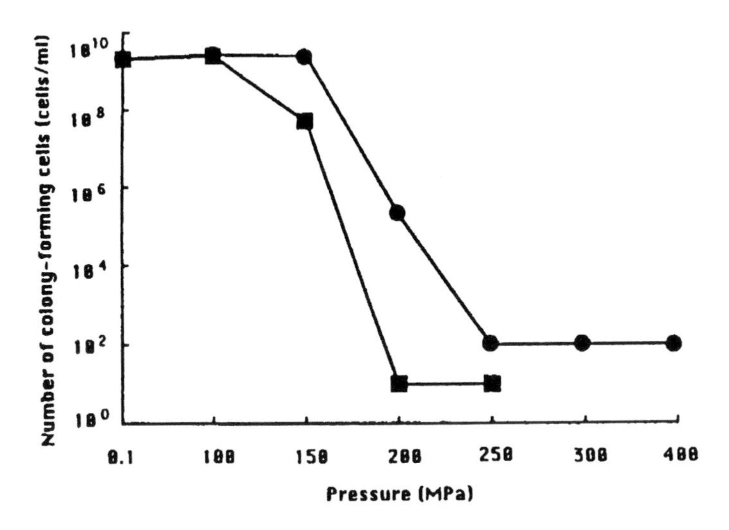

Fig. 13.1. Effect of hydrostatic pressure stress on colony-forming ability of *S. pombe* and *S. cerevisiae*. Results shown here are typical of those from three experiments. -■-, *S. pombe*; -●-, *S. cerevisiae*

13.3.2 Induction of Diploidization in *S. pombe*

We then investigated whether or not a similar phenomenon of direct diploidization by pressure stress occurred in the fission yeast *S. pombe* as in *S. cerevisiae* [7]. In fact, as recently reported, diploid cells derived from the wild strain *JY1* [7] and cold-sensitive mutant *nda3* KM311 (Fig. 13.2) were induced at a frequency of over 35% under high pressure at 150 to 200 MPa [9]. Especially in the latter case, large diploid cells (Fig. 13.3) from the cells incubated for 4 h at a restrictive temperature of 20°C were formed on a dye plate after pressure stress at 150−200 MPa. These cells made up over 40% of the population, although spindle formation and chromosome movement were blocked.

13.3.3 Influence of Pressure Stress on the Cold-Sensitive *nda3* Mutant

The viability of pressure stress on the cell was determined by the colony-forming ability when the cell was plated on agar and incubated for 3 days at 30°C. The influence of pressure stress on cells grown at three different temperatures is shown in Fig. 13.4. Sensitivity to pressure stress differed with the culture conditions. The survival curve sharply decreased with increasing pressure, and around 100 MPa cells cultured at 20°C for 4 h after shift-down displayed a slight decrease in proliferation, while cells cultured both at 30°C for 16 h and at 36°C for 15 min after shift-up were unaffected. Above 200 MPa, however, most cells did not survive after

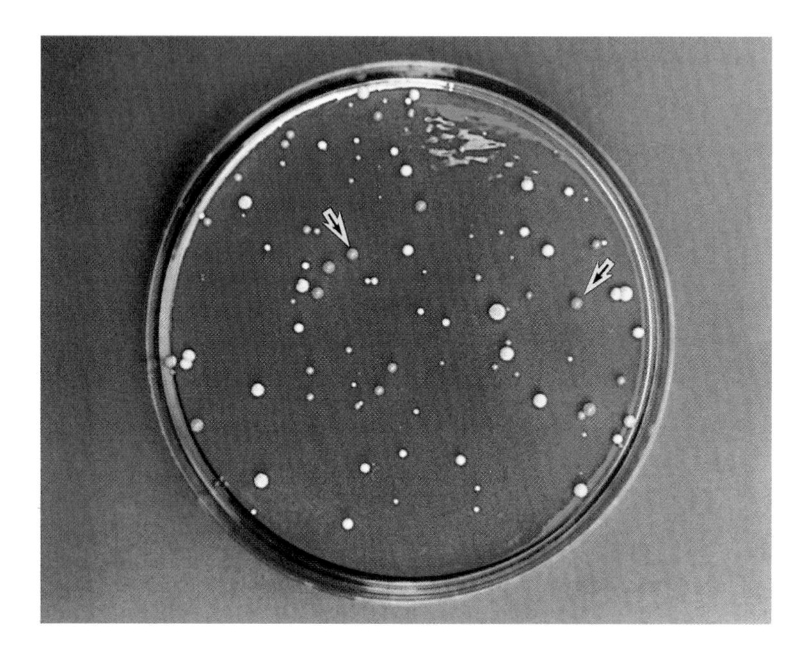

Fig. 13.2. Evidence of induction of diploidization (←) in *nda3*-KM311 cells incubated at 20°C for 4 h detected by dye plate-colony color assay (150 MPa)

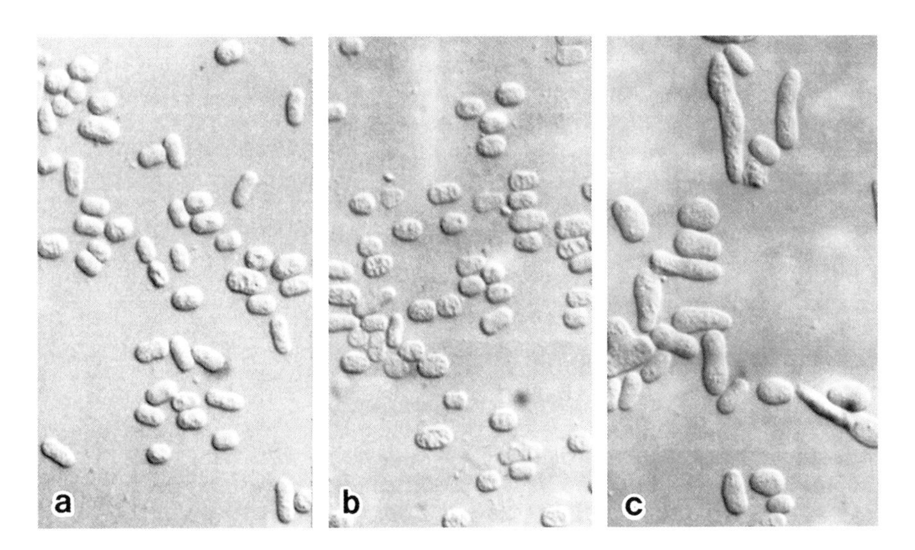

Fig. 13.3. Light microscopic images of *nda3* (150 MPa) derived from colonies formed on a dye plate were taken using Nomarski optics. (**a**) No pressure treatment (all colonies were pink). (**b**) Small cells derived from a pink colony.(**c**) Large cells derived from a violet colony

incubation at 30°C for 16 h (-●-), subsequent culture of 20°C for 4 h after shift-down (-■-), and further incubation at 36°C for 15 min after shift-up (-▲-) [9].

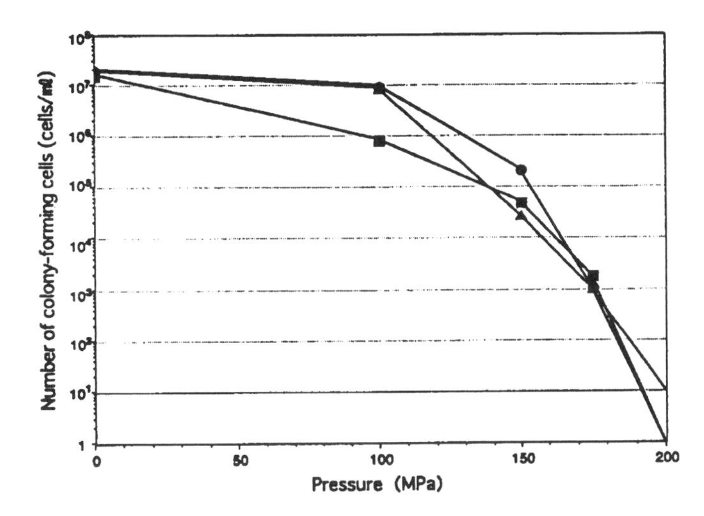

Fig. 13.4. Effect of hydrostatic pressure stress on colony-forming ability of *S. pombe nda3* KM311 cells grown at 30°C for 16 h (-●-), subsequently grown at 20°C for 4 h after shift-down (-■-), and further incubated at 36°C for 15 min after shift-up (-▲-)

13.3.4 Properties of the Cold-Sensitive *nda3* Mutant Cytoskeleton

When *nda3* cells were grown at 30°C for 16 h, shifted-down to a restrictive temperature at 20°C for 4 h and then shifted-up to 36°C without pressure stress (0.1 MPa), cell-cycle-specific organization of microtubule and actin cytoskeleton was as shown in Fig. 13.5. The cell-cycle-specific organization of spindle microtubule (—) and cytoplasmic microtubule (—) at 30°C was the same as reported by Kilmartin and Adams [13] and that of actin cytoskeleton (---) as reported by Marks and Hyams [14]. Cells grown at 20°C for 4 h were arrested highly synchronously (about 70~80%) at a step similar to mitotic prophase; 15 min after the temperature shifted to a permissive one, the spindle microtubule appeared and elongated and there was normal mitotic progression.

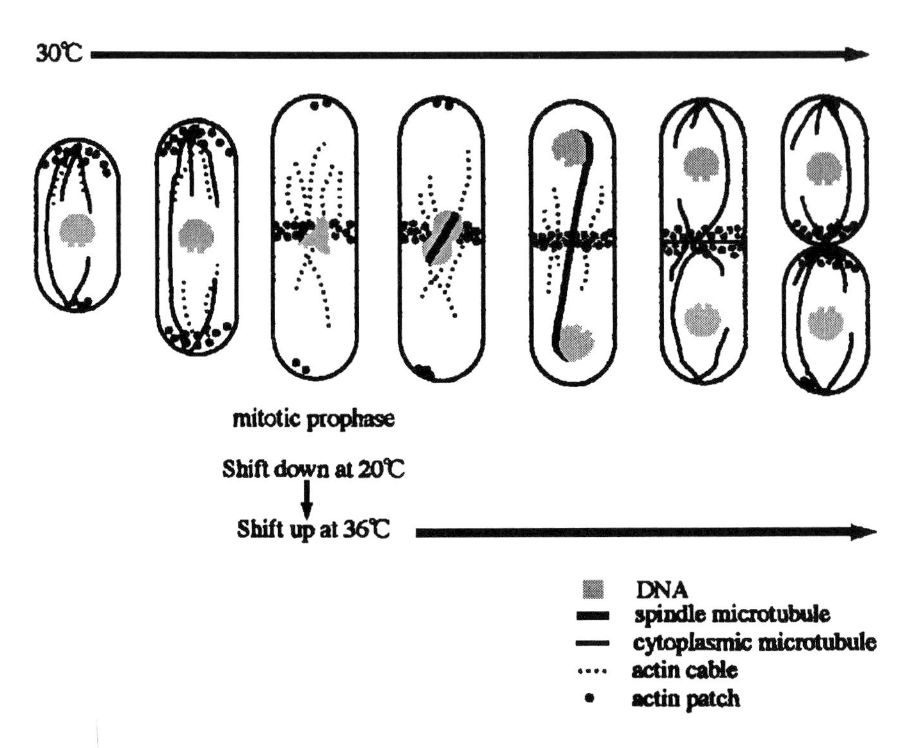

Fig. 13.5. A modified schematic of the changes in nucleus, microtubules and actin cytoskeletons during the cell cycle of *nda3* cells [8] from the data described by Kilmartin and Adams [13] and Marks and Hyams [14]

13.3.5 Dynamics of the Cold-Sensitive *nda3* Mutant Cytoskeleton

These findings in *nda3* of *S. pombe* cells were confirmed by FM and immuno-EM techniques (Figs. 13.6 to 13.8). Figure 13.6 shows FM images of microtubules of *nda3* stained with anti tubulin (a) and rhodamine-conjugated phalloidin at 30°C (b). Three distinct patterns of microtubule organization appear in the mitotic and non-mitotic division cycle: short spindle microtubules (①), elongated ones (②) and cytoplasmic microtubules (③) [5]. At the beginning of the cell cycle described by Marks and Hyams [14] all actin patches are clustered at the end of the cell (①), and in the bidirectional growth stage (②) they are present at both ends, as shown in Fig. 13.6b. These patches are located in the center of the cells at the post-mitotic stage (③) showing an apparent actin ring, and septum formation and cytokinesis follow [5, 6].

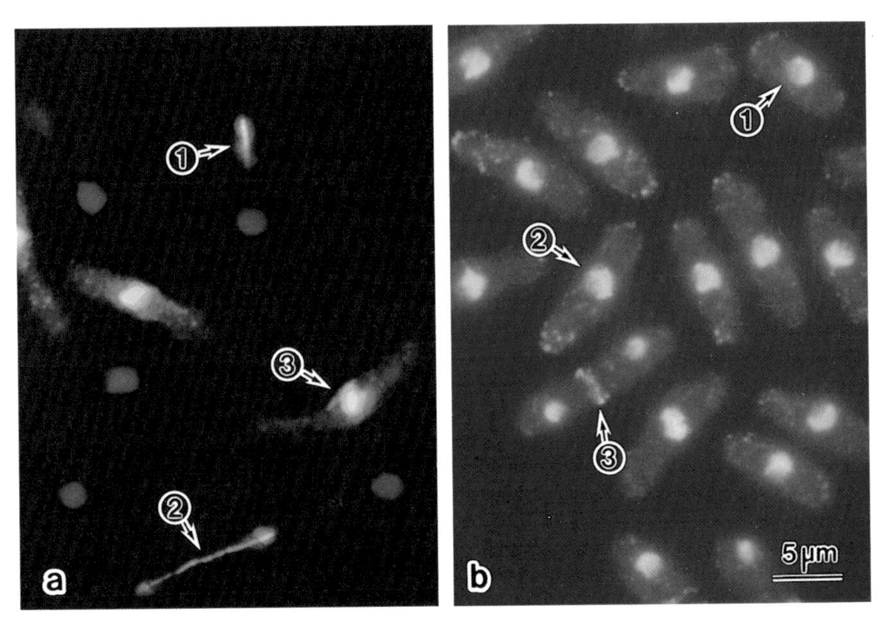

Fig. 13.6. Change in nucleus and cytoskeleton during the cell cycle. FM images of double-staining of DNA (orange) and α-tubulin (green) (**a**), and DNA (blue) and actin cytoskeleton stained with rhodamine-conjugated phalloidin (**b**)

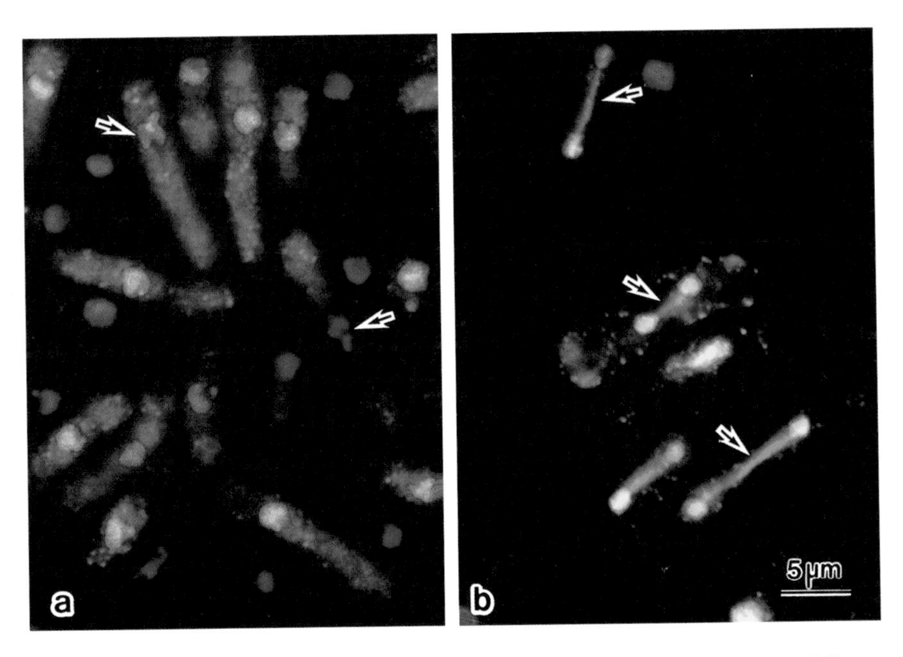

Fig. 13.7. FITC and propidium iodide stained image of *nda3* cells grown at 20°C for 4 h after shift-down (**a**) and further incubated at 36°C for 15 min after shift-up (**b**)

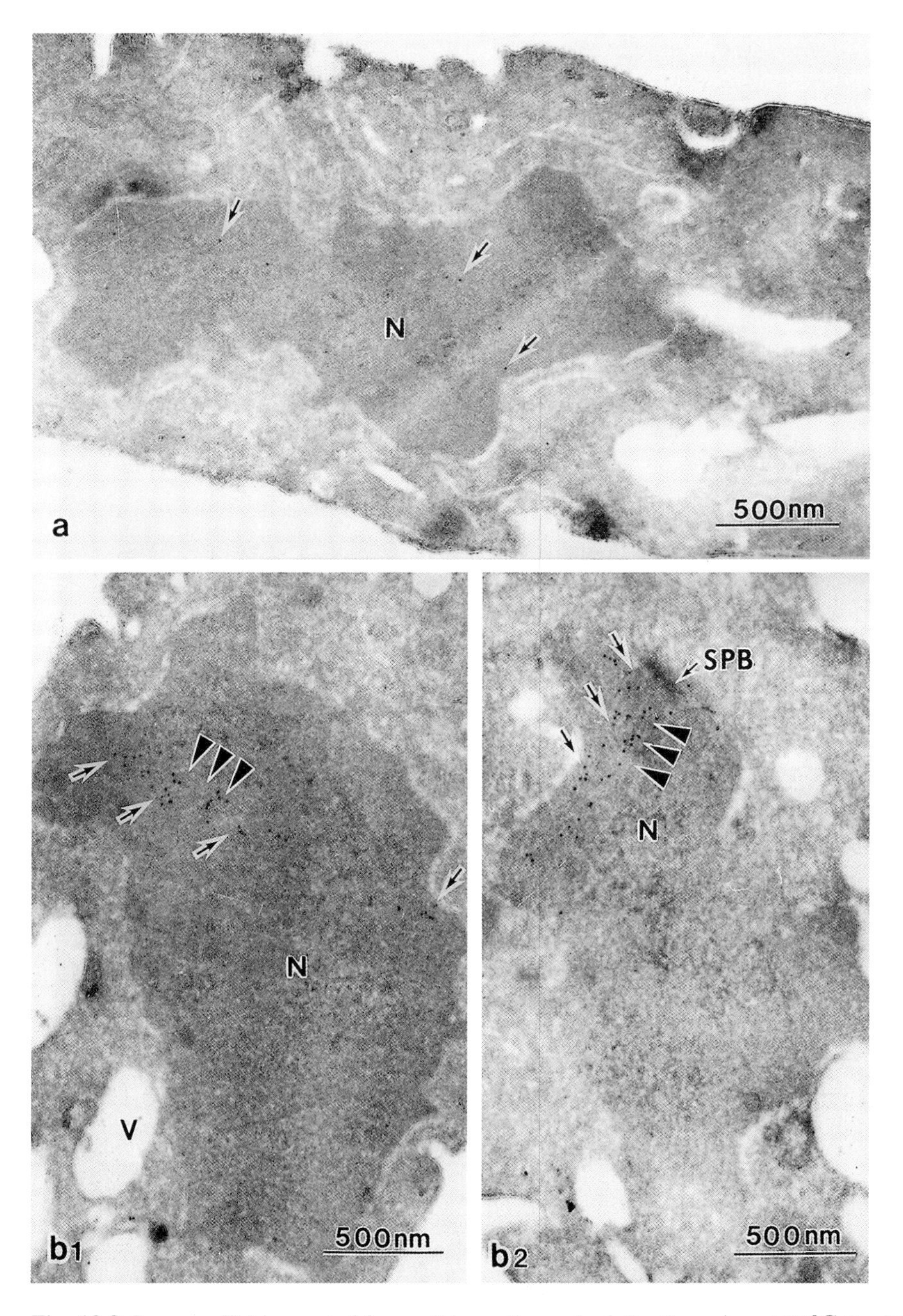

Fig. 13.8. Immuno-EM images of frozen thin-sections of *nda3* cells grown at 20°C for 4 h after shift-down (**a**) and further incubated at 36°C for 15 min after shift-up (**b**). Ten nm colloidal gold particles (←) for anti α-tubulin were used. Arrowheads show microtubules. N, nucleus; SPB, spindle pole body; V, vacuole

When the *nda3* cells are grown at a restrictive temperature of 20°C they are arrested highly synchronously at a step similar to mitotic prophase (Fig. 13.7a). Propidium iodide staining and α-tubulin staining showed three condensed chromosomes (←) but no spindle. These cells are shifted to a permissive temperature at 36°C and after 15 min culture a spindle appears and elongates (Fig. 13.7b, ←). This phenomenon was confirmed by immuno-EM using anti α-tubulin conjugated 10 nm gold particles [16].

Figure 13.8 shows frozen thin sections of *nda3* cells grown at 20°C for 4 h (a) and shifted to permissive temperature of 36°C for 15 min (b_1 and b_2). The gold particles (←) for anti α-tubulin diffused into the nuclear matrix at 20°C. But in cells grown at 36°C for 15 min after shift-up, gold particles appeared along with the microtubules (◀) in the nucleus (Fig. 13.8) [16].

13.3.6 Transmission Electron Microscopic Images of the Ultrastructure of Pressure Stress Cells

The thin-sectioned image of the GA−PFA-fixed, freeze-substituted cells in 30°C showed a normal profile of wild type cells [4]. The nucleus was surrounded by membrane, and many mitochondria and numerous small vacuoles existed in the cytoplasm (Fig. 13.9). There were three types of cells depending on the cell cycle: single nucleus of original cell size (①); elongated nucleus and/or two nuclei (②) of the elongated cell size; and nascent septum formation (③).

When grown at a restricted temperature of 20°C for 4 h, the cells were arrested at the mitotic stage as shown in Fig. 13.7a [9]. The nucleus surrounded by membranes was an abnormal shape, the profile was leaf-like [15] and electron-dense areas in a large part of the nuclear matrix (←), as well as numerous small vacuoles which included electron-dense materials, were visible (Fig. 13.10a). The shift-up cell at 36°C (Fig. 13.10b) is the same as the normal culture in Fig. 13.9. Even these cells grown at restrictive and permissive temperature without pressure stress were not influenced and showed the typical ultrastructure of fission yeast. However, when subjected to pressure stress, several changes of the cell ultrastructure were observed at 100 MPa, the nuclear membrane was damaged and looked like the arm of a squid with an opening at the end (Fig. 13.11a, ←). All of the nuclear matrix was electron-dense [1−3]. Electro dense areas appeared in the matrix of the mitochondria (◀) and vacuoles began to fuse together (Fig. 13.11a and b). Profuse endoplasmic reticulum was visible.

At 150 MPa (Fig. 13.12) the damage had progressed but was more severe: the nuclear membrane was broken over broad areas (←), still existed in the mitochondria electron-dense area (◀) but had shrunk, and the vacuoles had fused into large pieces (⇇) at both restrictive and permissive temperatures. This phenomenon is a strange characteristic of vacuoles in the fission yeast *S. pombe*. The influence of pressure stress was almost the same as that at both cultured temperatures [16] and also as those in the cases of *S. cerevisiae* [2] and *Candida albicans* [1].

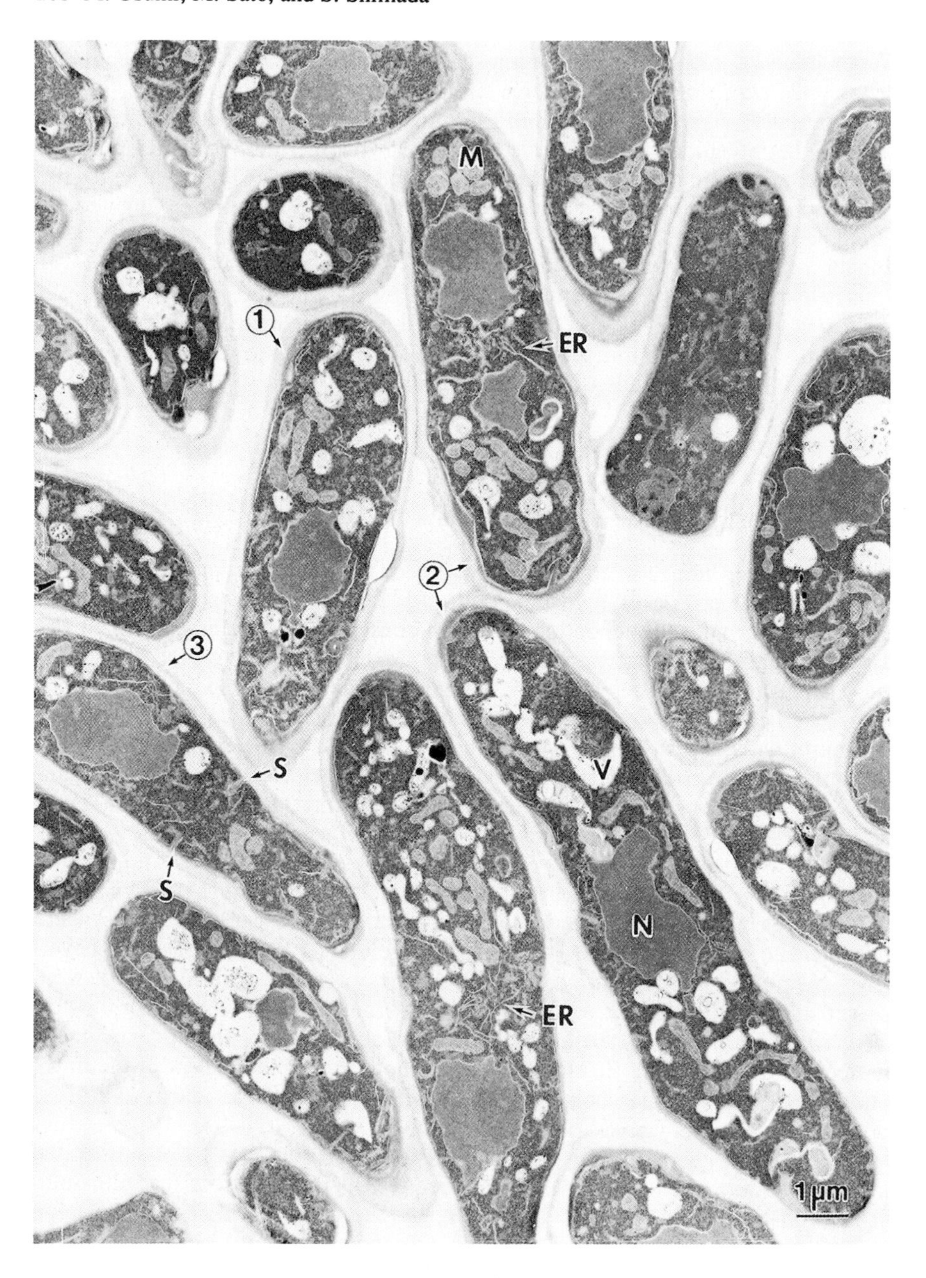

Fig. 13.9. CEM image of *nda3* cells grown at 30℃. There are three types of cells during cell cycle. The original cell has a single nucleus (①); the elongated cell has a elongated nucleus and/or two nuclei (②), and there is nascent septum formation (③, ←). Abbreviations: CW, cell wall; CM, cell membrane; ER, endoplasmic reticulum; M, mitochondrion; N, nucleus; S, septum; V, vacuole

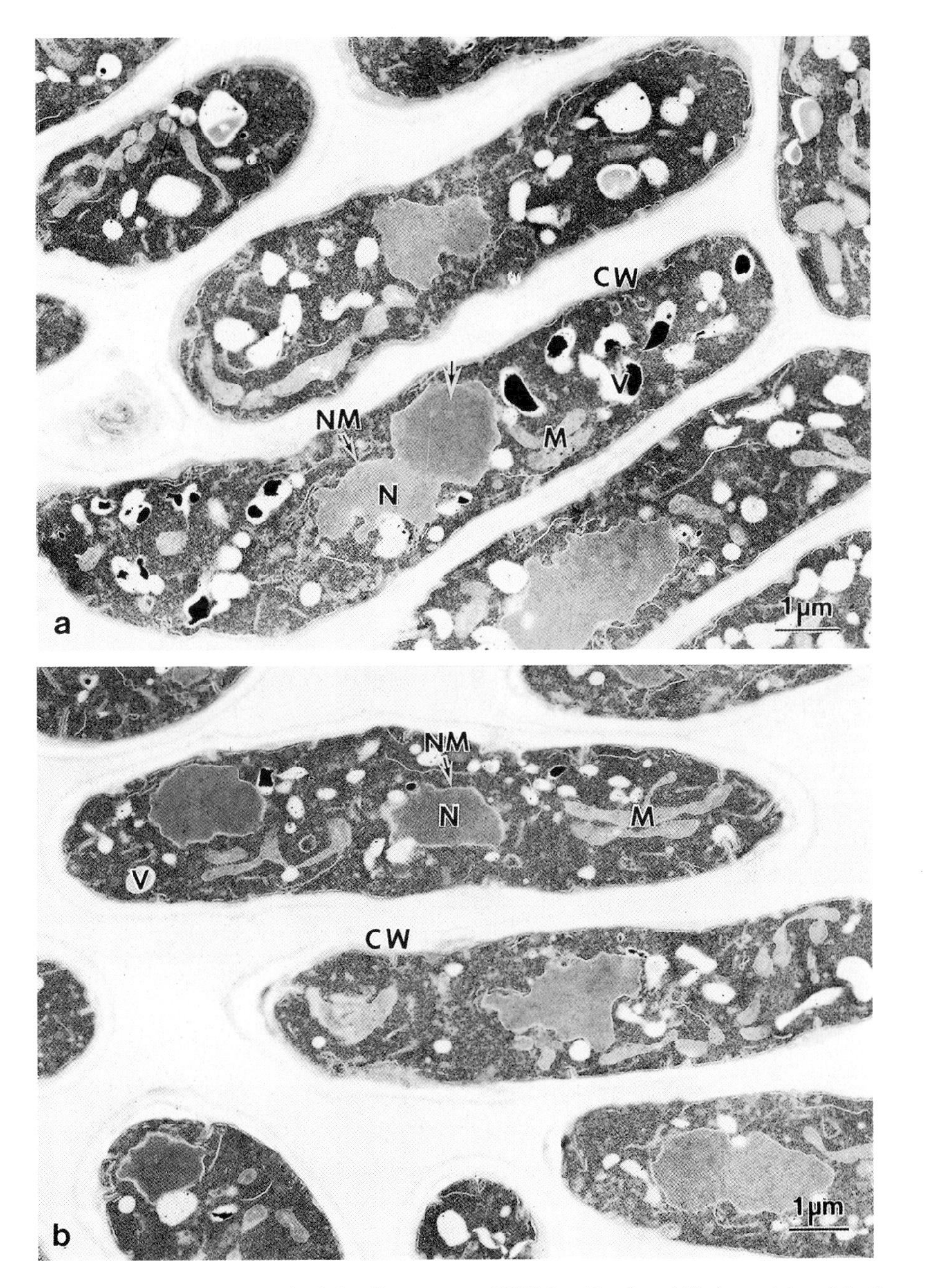

Fig. 13.10. CEM images of *nda3* cells grown at 20°C for 4 h after shift-down (**a**) and further incubated at 36°C for 15 min after shift-up (**b**) without pressure stress. CW, cell wall; ER, endoplasmic reticulum; M, mitochondrion; N, nucleus; NM, nuclear membrane; V, vacuole

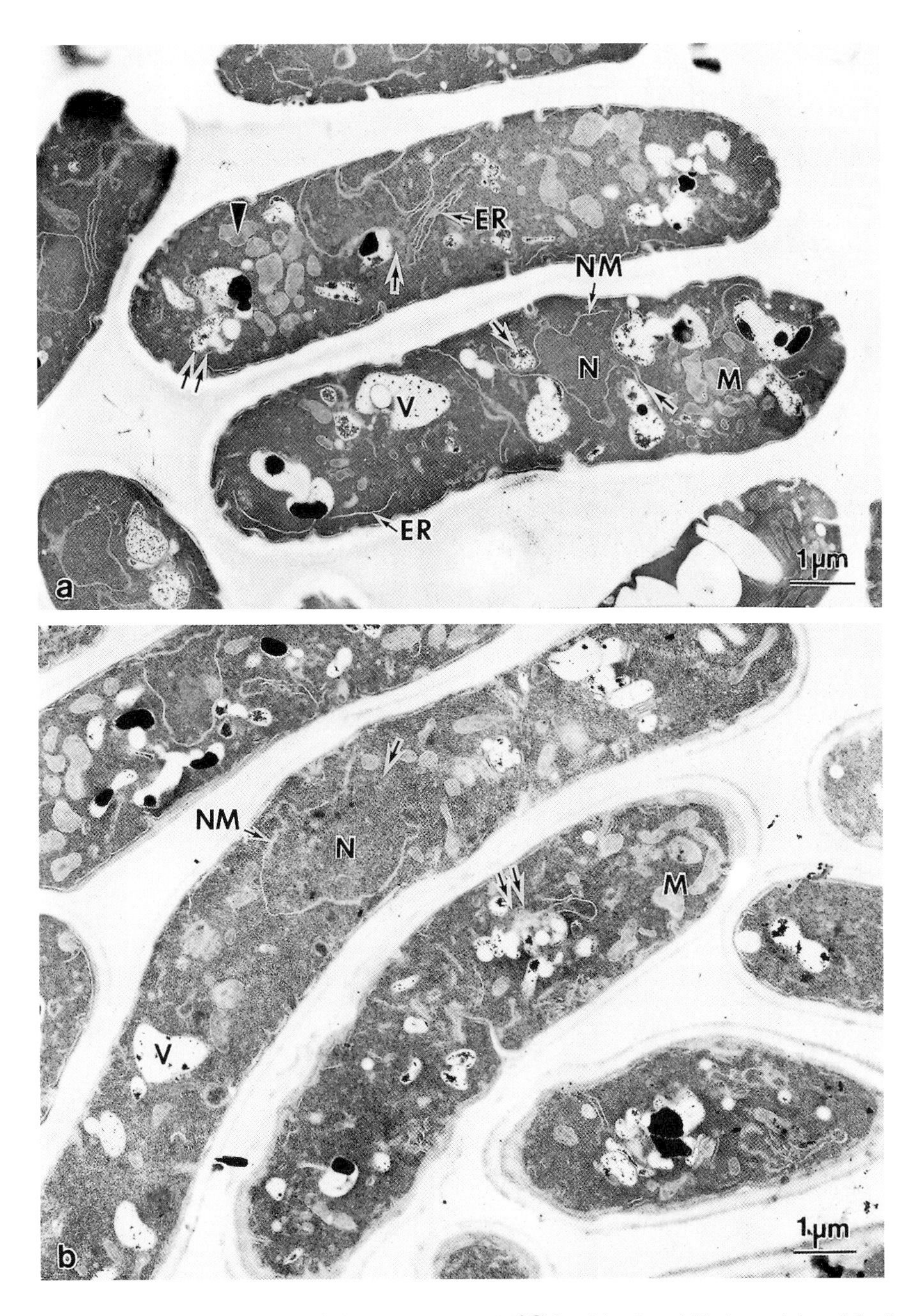

Fig. 13.11. CEM images of *nda3* cells grown at 20°C for 4 h after shift-down (**a**) and further incubated at 36°C for 15 min after shift-up (**b**) treated with a hydrostatic pressure of 100 MPa. Arrows show that nuclear membrane is damaged and opened. Arrowhead shows the matrix of the mitochondrion has an electron-dense area

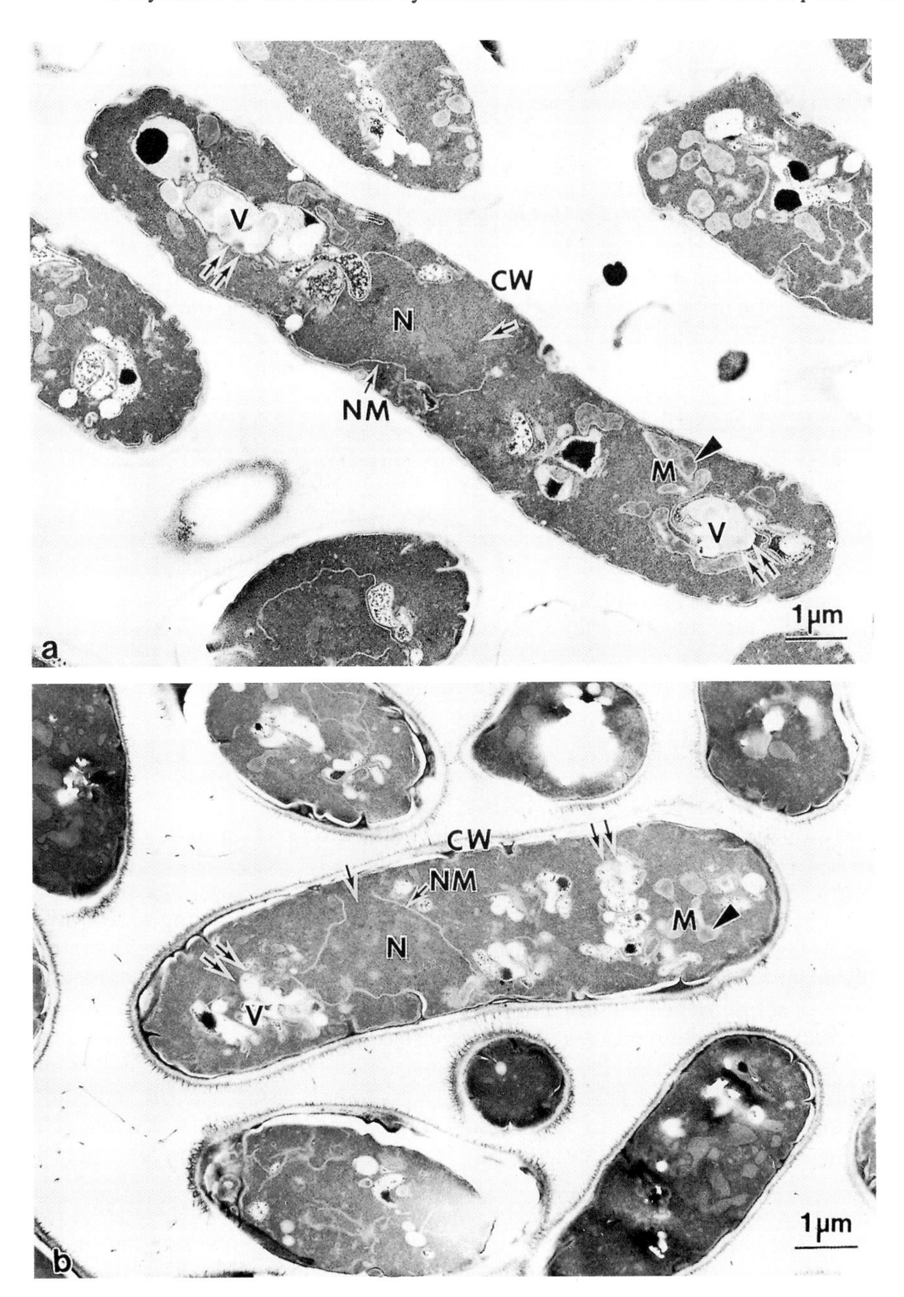

Fig. 13.12. CEM images of *nda3* cells grown at 20°C for 4 h after shift-down (**a**) and further incubated at 36°C for 15 min after shift-up (**b**) treated with a hydrostatic pressure of 150 MPa. Arrows show damaged nuclear membrane and open end. Arrowhead shows an electron-dense area in the mitochondrial matrix

13.3.7 Changes in Actin Cytoskeleton Induced by Pressure Stress

By FM the cells arrested at restrictive temperature at 20°C for 4 h showed actin patches concentrated in the central region; actin rings were also visible (Fig. 13.13a,←), and the actin ring was detected in 70 to 80% of cells.

The immuno-EM image of *nda3* cells grown at 20°C without pressure also shows concentrated colloidal gold particles for anti F-actin in the center of the cell (Fig. 13.13b, ←), for instance, beneath the cell membrane and inside the cytoplasm, as shown in the FM image (Fig. 13.13a, ←).

Drastic changes in actin cytoskeletons occurred, depending on the rate of pressure stress (Fig. 13.14a–c). At 100 MPa, mitotic prophase specific actin distribution was lost. Instead, long and fine actin cables appeared throughout the cytoplasm (←); large actin patches (◀) remained in the center of the cells and subsequently covered the actin ring (Fig. 13.14a); they changed into thick, short cables at 150 MPa (Fig. 13.14b,←); they finally decomposed but the actin ring appeared even with faint fluorescence (Fig. 13.14c, ←) at 200 MPa.

The actin rings were observed in the center of the 20→36°C shift-up cells in FM and immuno-EM images (Fig. 13.15). Similarly, about 70% of shift-up cells with the actin ring was detected after 15 min (Fig. 13.15a, ←). Colloidal gold particles (←) are visible at the area along the nascent septum formation (Fig. 13.15b). The influence of the pressure stress on these cells was the same but more apparent: several profiles of actin cytoskeleton were observed at 100 MPa (Fig. 13.16a)—a fine filamentous form (←), thick and short cable (⇇); a shorter dot actin profile appeared at 150 MPa (Fig. 13.16b, ←); and a ring-shaped actin appeared in the center of the cell at 200 MPa (Fig. 13.16c, ←).

The influence of the pressure stress on the actin cytoskeleton in *nda3* cells cultured at 36°C for 15 min after shift-up was the same as the process of degradation in cells grown at 20°C for 4 h. These results also showed the process of decomposition of actin cytoskeleton in *nda3* cells [6, 16]. The remaining actin ring at 200 MPa suggests that this actin cytoskeleton is less sensitive to pressure stress than actin patches and cable, or that it may be the reason for the different actin-associated proteins in actin patches, cables, and rings [16].

Recently, Mabuchi's group [17, 18] reported the subcellular localization and possible function of actin, and the actin cytoskeleton-related proteins Cdc8 tropomyosin and actin-related protein 3 (Arp3) in *S. pombe*, using specific antibodies and gene disruption. The actin cytoskeleton containing Cdc8 tropomyosin and actin cable has a function in the contractile ring formation, e.g., in a sea urchin egg [18], because myosin II plays a role in the cytokinesis of *S. pombe*. Arp3 was co-localized with F-actin in the patches but not in the actin ring or cables [17]. This suggests that each associated protein of the actin patch, cable and ring may elict a different response under pressure stress. Further elucidation of the influence of the stress is necessary.

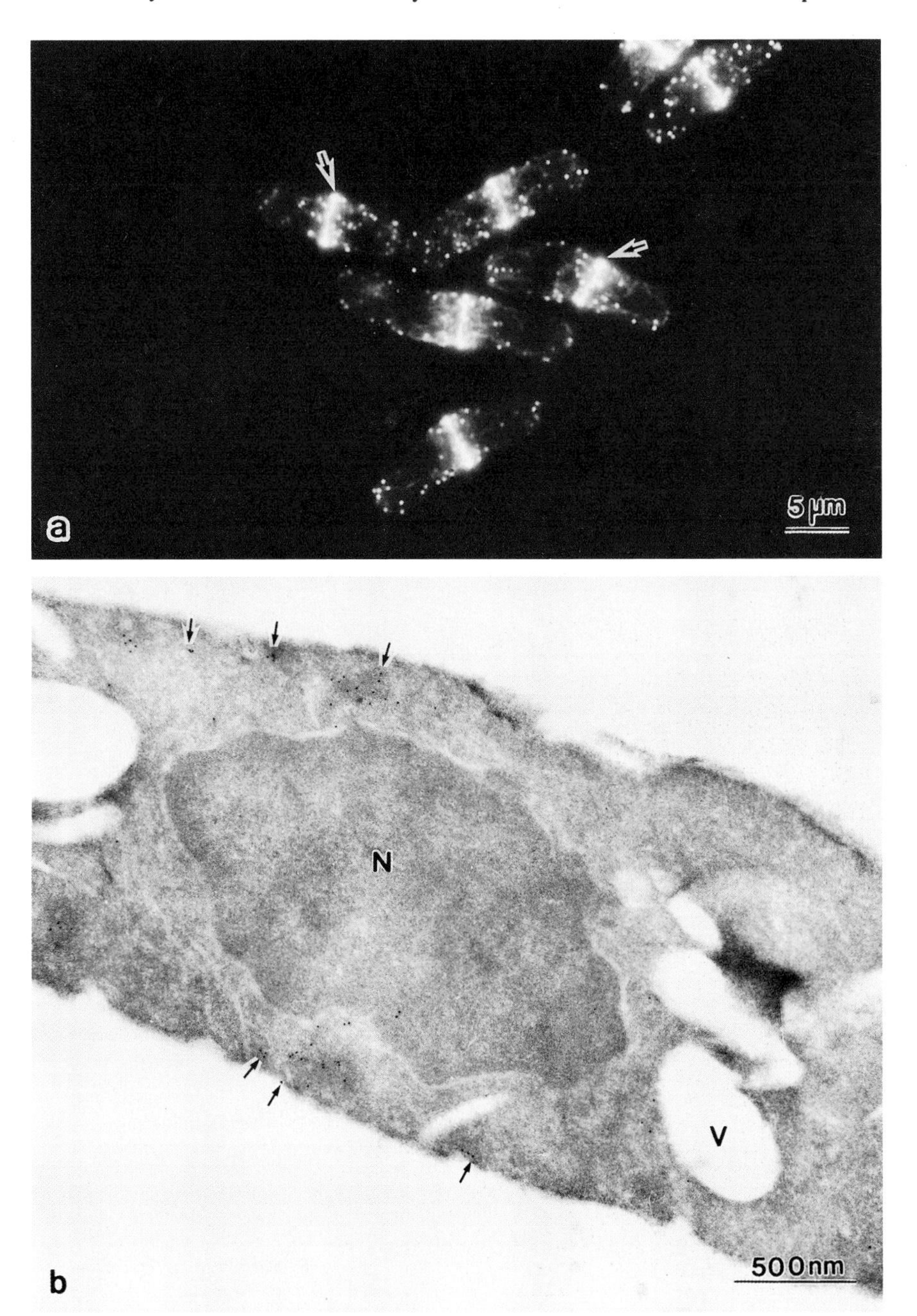

Fig. 13.13. FM (**a**) and immuno-EM (**b**) images of *nda3* cells grown at 20°C without pressure stress. Immuno-EM images used ten nm colloidal gold particles (←) for anti actin. N, nucleus; V, vacuole

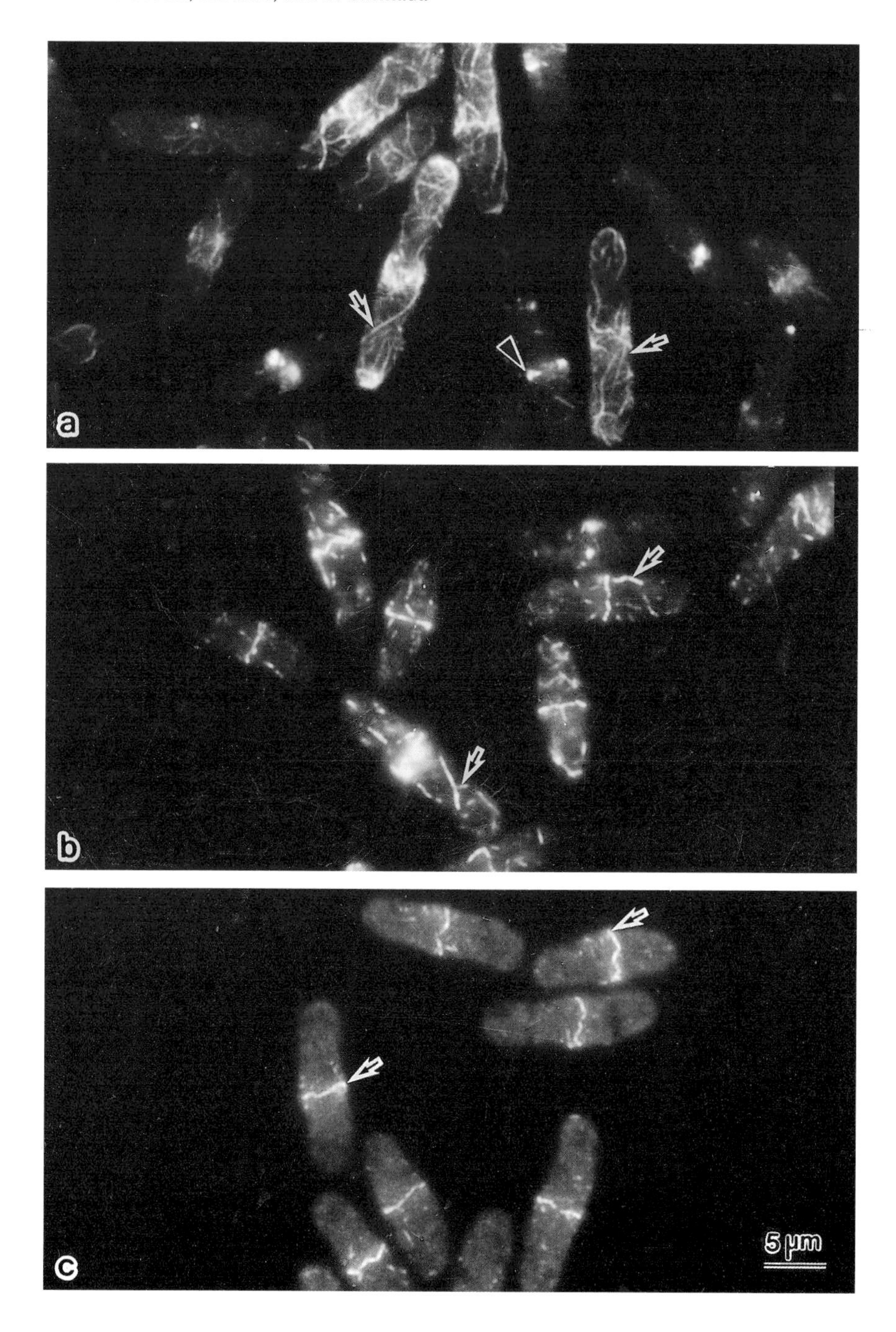

Fig. 13.14. Change in organization of actin cytoskeleton in cells grown at 20℃ for 4 h after-shift-down caused by a hydrostatic pressure of (**a**) 100, (**b**) 150 and (**c**) 200 MPa

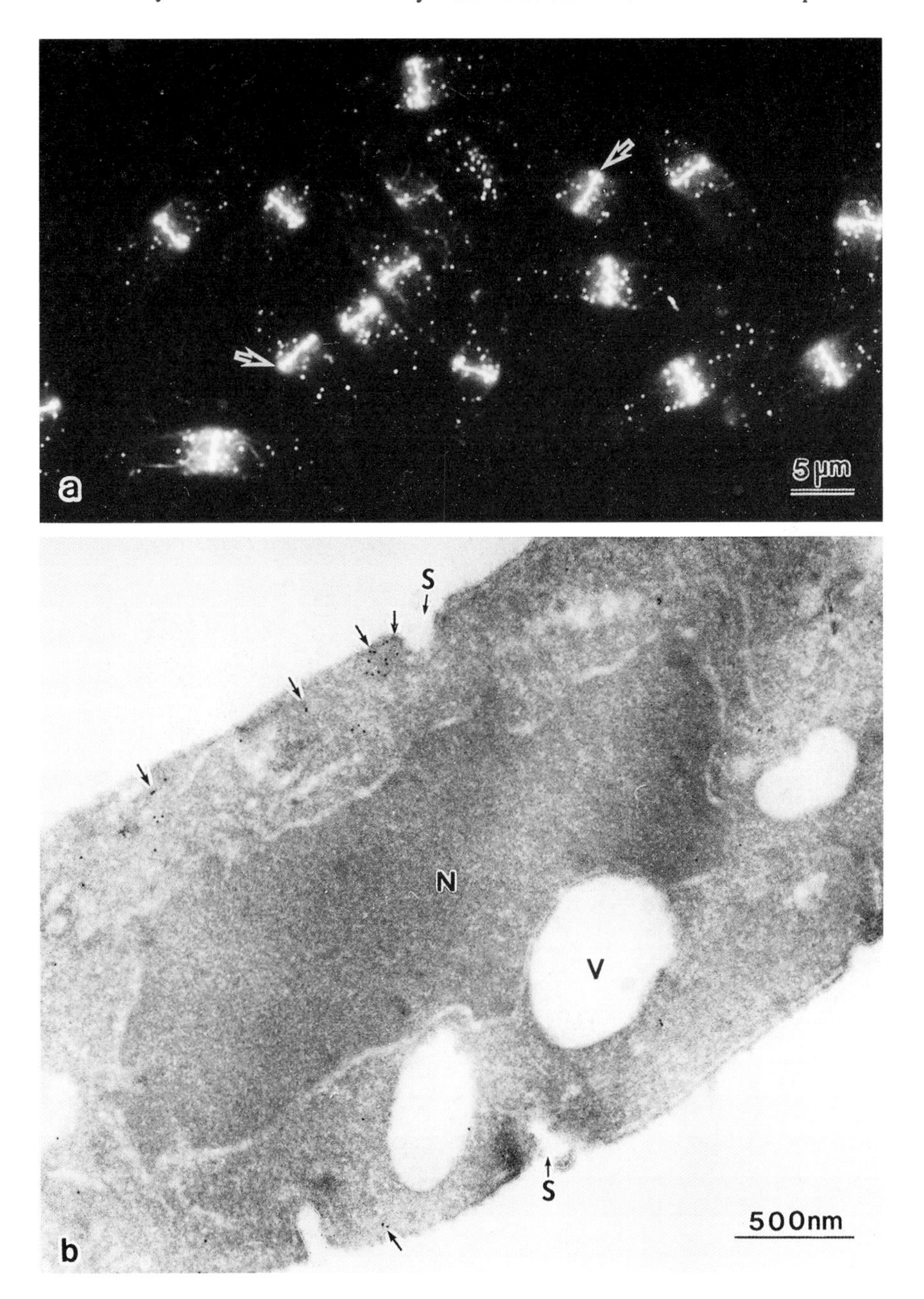

Fig. 13.15. FM (**a**) and immuno-EM (**b**) images of *nda3* cells grown at 20°C for 4 h after shift-down and further incubated at 36°C for 15 min after shift-up without pressure stress. Ten nm colloidal gold particles for anti actin were use d. N, nucleus; S, septum; V, vacuole

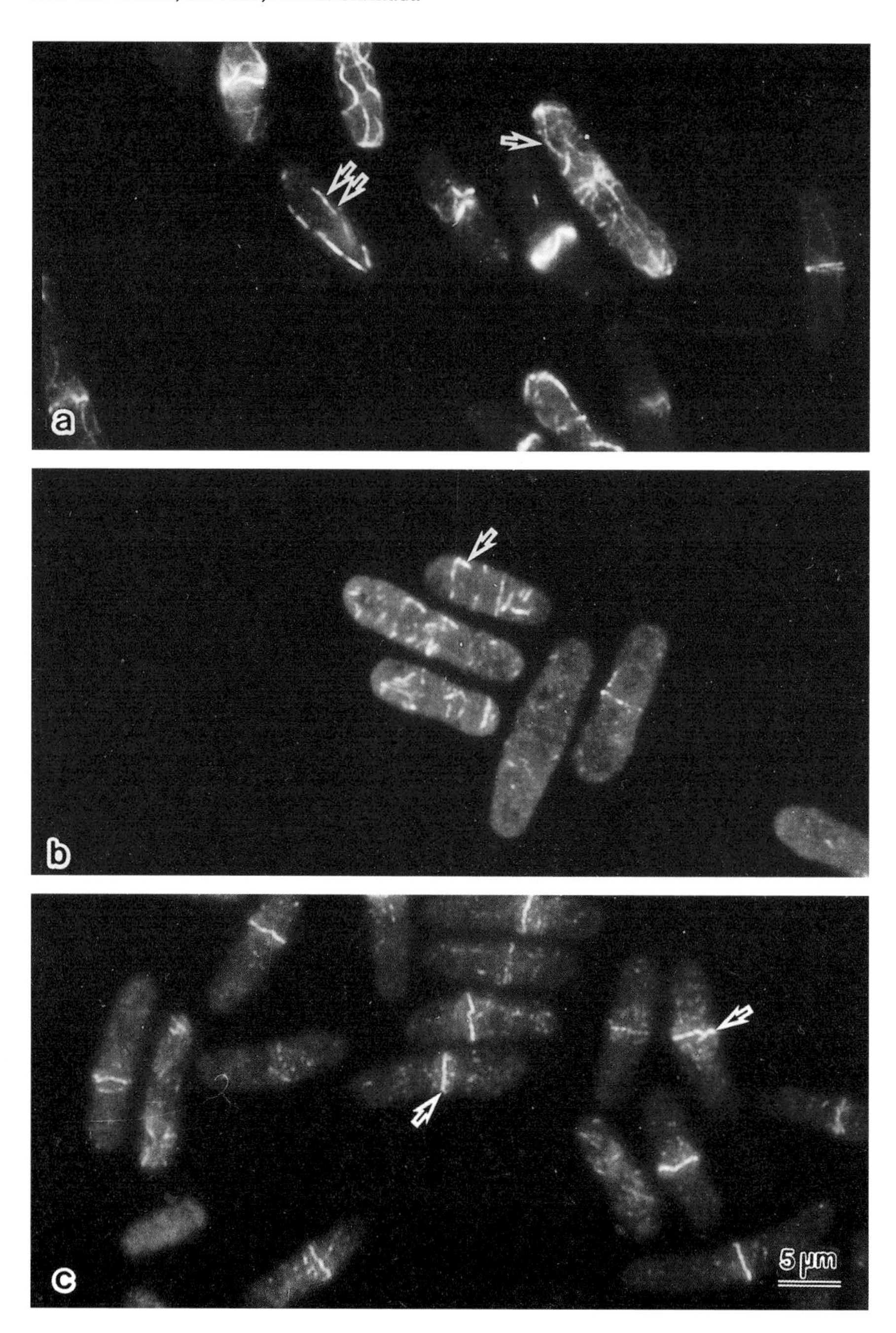

Fig. 13.16. Change in organization of actin cytoskeleton in cells grown at 20°C for 4 h after shift-down and further incubated at 36°C for 15 min after shift-up caused by a hydrostatic pressure of (**a**) 100, (**b**) 150 and (**c**) 200 MPa

13.4 Conclusions

In the non spindle *nda3* cells grown at 20°C, the results of the survival curves showed that the sensitivity increases with pressure stress, that is, non spindle cells were susceptible to this stress. The process of degradation in actin cytoskeletons of these cells by pressure stress was confirmed to be the same as in cells grown at both 20°C and 36°C after shift-up [16], similar to the wild strain of *S. pombe* [6].

Acknowledgements. We would like to thank Professor Dr. M. Yanagida, Kyoto University, for providing us the cold-sensitive *nda3* KM311. We also thank Ms. A. Aizawa, Ms. S. Amino, Ms. J. Morimoto and Ms. M. Konomi for their technical assistance.

References

13.1 M. Osumi, N. Yamada, M. Sato, H. Kobori, S. Shimada, R. Hayashi: in *High Pressure and Biotechnology*, ed. by C. Balny, R. Hayashi, K. Heremans, P. Masson, Vol. 224 (John Libbey Eurotext, Montrouge 1992) pp.9

13.2 S. Shimada, M. Andou, N. Naito, N. Yamada, M. Osumi, R. Hayashi, Appl. Microbiol. Biotechnol. **40**, 123 (1993)

13.3 H. Kobori, M. Sato, A. Tameike, K. Hamada, S. Shimada, M. Osumi, FEMS Microbiol. Lett. **132**, 253 (1995)

13.4 M. Osumi, M. Sato, H. Kobori, Zha Hai Feng, S. A. Ishijima, K. Hamada, S. Shimada: in *High Pressure Bioscience and Biotechnology,* ed. by R. Hayashi, C. Balny, (Elsevier, Amsterdam 1996) pp.37

13.5 H. Kobori, M. Sato, A. Tameike, K. Hamada, S. Shimada, M. Osumi: in *High Pressure Bioscience and Biotechnology,* ed. by R. Hayashi, C. Balny (Elsevier, Amsterdam 1996) pp. 83

13.6 M. Sato, H. Kobori, S. A. Ishijima, Zha Hai Feng, K. Hamada, S. Shimada, M. Osumi, Cell Struct. Funct. **21**, 167 (1996)

13.7 K. Hamada, Y. Nakatomi, M. Osumi, S. Shimada, FEMS Microbiol. Lett. **136**, 257 (1996).

13.8 Y. Hiraoka, T. Toda, M. Yanagida, Cell **39**, 349 (1984)

13.9 M. Sato, H. Kobori, S. A. Ishijima, A. Tameike, M. Morimoto, M. Yanagida, S.Shimada, M. Osumi: in *High Pressure Research in the Biosciences. and Bio technology,* ed. by K. Heremans (Leuven University Press, Leuven 1997) pp.245

13.10 K. Hamada, Y. Nakatomi, S. Shimada, Curr. Genet. **22**, 371 (1992)

13.11 H. Kobori, N. Yamada, A. Taki, M. Osumi, J. Cell Sci. **94**, 635 (1989)

13.12 M. Baba, M. Osumi, J. Electron Microsc. Tech. 5, 246, (1987)

13.13 J. V. Kilmartin, A. E. M. Adams, J. Cell Biol. **98**, 922 (1984)

13.14 J. Marks, J. S. Hyams, Eur. J. Cell Biol. **39**, 27 (1985)

13.15 T. Kanbe, Y. Hirooka, K. Tanaka, M. Yanagida, J. Cell Sci. **96**, 275 (1990)

13.16 M. Sato, A. Aizawa, S. Amino, S. A. Ishihima, S. Shimada, M. Osumi: in *Electron Microscopy* 1998, ed. by H. A. Calderón Benavides, M. José Yacamán, Vol. IV (Institute of Physics Publishing, Bristol and Philadelphia 1998) pp. 267

13.17 R. Arai, K. Nakano, I. Mabuchi, Cell Struct. Funct. **23**, 152 (1998)

13.18 I. Mabuchi, R. Arai, T. Noguchi, F. Motegi, K. Nakano, Cell Struct. Funct. **23**, 70 (1998)

Subject Index

Printing: Saladruck, Berlin
Binding: H. Stürtz AG, Würzburg